Smart Device Recognition

Hui Liu · Chengming Yu · Haiping Wu

Smart Device Recognition

Ubiquitous Electric Internet of Things

Science Press
Beijing

Springer

Hui Liu
Institute of Artificial Intelligence
and Robotics
School of Traffic and Transportation
Engineering
Central South University
Changsha, China

Chengming Yu
Institute of Artificial Intelligence
and Robotics
School of Traffic and Transportation
Engineering
Central South University
Changsha, China

Haiping Wu
Institute of Artificial Intelligence
and Robotics
School of Traffic and Transportation
Engineering
Central South University
Changsha, China

ISBN 978-981-33-4924-7 ISBN 978-981-33-4925-4 (eBook)
https://doi.org/10.1007/978-981-33-4925-4

Jointly published with Science Press
The print edition is not for sale in China (Mainland). Customers from China (Mainland) please order the print book from: Science Press.

This Springer imprint is published by the registered company Springer Nature Singapore Pte Ltd.
The registered company address is: 152 Beach Road, #21-01/04 Gateway East, Singapore 189721, Singapore

Preface

The development of modern society poses new and high challenges to the regulation capability, intelligence, and digitization of the electric grid. Building the Ubiquitous Electric Internet of Things (UEIOT) is an effective way to solve technical problems of the electric grid and break through the bottleneck. It makes the electric users, grid companies, electric generation companies, suppliers, and their equipment all connected. Then, the shared data is generated for service users, electric grids, electric generation, suppliers, government, and society. It creates greater opportunities for the development of more market players.

Due to the rapid development of the UEIOT, the non-intrusive device recognition methods which obtain various device information by analyzing indoor level gathered signal have become the focus in the electric energy saving. Besides, smart device identification is the core of the technology and equipment. Using data science to realize non-intrusive smart device identification is of great significance for energy conservation and the development of mechanical and electrical control technology. It is significant to build the application framework of various data identification technologies in non-intrusive device recognition.

The book introduces a series of state-of-the-art device identification methods, which can provide ideas for doctoral students and researchers and encourage further research. In the book, various methods of intelligent device identification are introduced in detail, including machine learning, deep learning, intelligent clustering, optimization model, integrated learning, single-label and multi-label identification models, etc. Besides, a large number of experimental simulations are carried out. In addition, the book also illustrates some traditional device recognition solutions in Chap. 2 for comparison, which are based on physical methods or template matching method. Not limited to the application in the field of energy conservation, the book also analyzes the potential application of intelligent device identification in mechanical and electrical system optimizations, environmental pollution detection, and so on. In general, the book provides an important reference for the development of data science and technology in non-intrusive device recognition and would promote the application of intelligent device identification methods in industry.

The studies in the book are supported by National Natural Science Foundation of China, National Key R&D Program of China, the Innovation Drive of Central South University of China. The publication of the book is funded by the graduate textbook project of Central South University of China. In the process of writing the book, Rui Yang, Chengqing Yu, Shuqin Dong, Chao Chen, Jiakang Wang, Zijie Cao, Yucheng Yin, Zeyu Liu, and other team members have done a lot of experimental verification and other work; the authors would like to express heartfelt appreciations.

Changsha, China
October 2020

Prof. Dr. -Ing. habil. Hui Liu

Contents

Nomenclature

AAM	Advanced Asset Management
AC	Air Conditioner
AdaBoost	Adaptive Boosting
ADO	Advanced Distribution Operation
AE	Auto Encoder
AMI	Advanced Metering Infrastructure
AMR	Automatic Meter Reading
ANN	Artificial Neural Network
AR	Auto Regressive
ATO	Advanced Transmission Operation
AUC	Area Under Curve
BBKH	Biogeography Based Krill Herd
BOA	Butterfly Optimization Algorithm
BPMLL	Backpropagation Neural Networks Multi-Label Learning
CART	Classification And Regression Tree
CDMs	Committee Decision Mechanisms
CLS	Concept Learning System
CNN	Convolutional Neural Network
DAQ	Data AcQuisition module
DBSCAN	Density-Based Spatial Clustering of Applications with Noise
DE	Differential Evolution
DER	Distributed Energy Resource
DT	Decision Tree
ELM	Extreme Learning Machine
EMI	Electro Magnetic Interference
EMS	Energy Management System
FA	Firefly Algorithm
FFT	Fast Flourier Transform
FGKM	Fast Global K-Means
FHMM	Factorial Hidden Markov Model

FN	False Negative
FNN	Feedforward Neuron Network
FP	False Positive
FSM	Finite State Machine
GA	Genetic Algorithm
GLR	Generalized Likelihood Ratio
GMM	Gaussian Mixture Model
GOF	Goodness Of Fit
GRU	Gated Recurrent Unit network
GWO	Grey Wolf Optimization
HEMS	Home Energy Management System
HMM	Hidden Markov Model
IoT	Internet of Things
ISO	International Standardization Organization
ITU	Internet Telecommunication Union
KNN	K-Nearest Neighbor
LPBoost	Linear Programming Boosting
LPWA	Low Power Wide Area
LSTM	Long Short-Term Memory
MDMS	Measurement Data Management System
MLKNN	Multi-Label K-Nearest Neighbors
MLP	Multi-Layer Perceptron
MOGWO	Multi-Object Grey Wolf Optimization
MOPSO	Multi-Object Particle Swarm Optimization
NBM	Naive Bayesian Model
NEI	National Energy Internet
NILM	Non-Intrusive Load Monitoring
NSGA-II	Non-dominated Sorting Genetic Algorithm-II
NTF	Nonnegative Tensor Factorization
PCA	Principal Component Analysis
PELs	Plugged-in Electric Loads
PI	Power Index
PLAID	Plug Load Appliance Identification Dataset
PSO	Particle Swarm Optimization
RAkEL	RAndom k-labELsets
Ranking SVM	Ranking Support Vector Machine
RBM	Restricted Boltzmann Machines
RG	Recursive Graph
RNN	Recurrent Neural Network
ROC	Receiver Operating Characteristic
RPA	Recurrence Plot Analysis
SAE	Sparse Auto-Encoder
SEM	Sequential Expectation-Maximization
SOM	Self-Organizing Map
SVD	Singular Value Decomposition

SVM	Support Vector Machine
TN	True Negative
TP	True Positive
UEIOT	Ubiquitous Electric Internet Of Things
URLLC	Ultra-Reliable and Low Latency Communications
WM	Washing Machine
WMRA	Wavelet Multi-Resolution Analysis
WPD	Wavelet Packet Decomposition
WRG	Weighted Recursive Graph

Chapter 1
Introduction

1.1 Overview of Ubiquitous Electric Internet of Things (UEIOT)

With the continuous expansion of the electric grid and the continuous increase in the capacity of electric generation equipment, a large number of distributed new energy continue to emerge. The number of connected entities of the electric grid is increasing and it has become the central link of the energy revolution, the hub of energy transmission and conversion. The development of the times poses new and higher challenges to the flexible regulation capability, the intelligent and digital development of the electric grid. Building the Ubiquitous Electric Internet of Things (UEIOT) is an effective way to solve electric grid technical problems and break through the bottleneck of electric grid development. It is also a comprehensive innovation and profound transformation of electric grid companies. The UEIOT will drive the collaborative innovation and development of the upstream and downstream industrial chains of electric grid companies. Then it leads the high-quality construction of the National Energy Internet (NEI). To explain UEIOT clearly, it needs to start with the Internet of Things (IoT). The term of the IoT was first proposed by Kevin Ashton (Ashton 2009). In the current situation, the IoT is defined as the following: The Internet of Things is an infrastructure that can connect things, people, systems, and information resources. By integrating with intelligent services, it can process physical information and virtual information and react accordingly. The Internet Telecommunication Union (ITU) also defines the IoT accordingly: let each target object connect to the network through the sensing system to realize the communication and connection between people anytime and anywhere. Then it extends to the on-demand information acquisition, transmission, storage, cognition, decision-making, and use of services between people and things, and things and things (Wortmann and Flüchter 2015). The IoT organically integrates people, things, and systems to meet the needs of timely communication and information processing. Later, with the development of IoT technology, all people and things can realize information interaction without

H. Liu et al., *Smart Device Recognition*,
https://doi.org/10.1007/978-981-33-4925-4_1

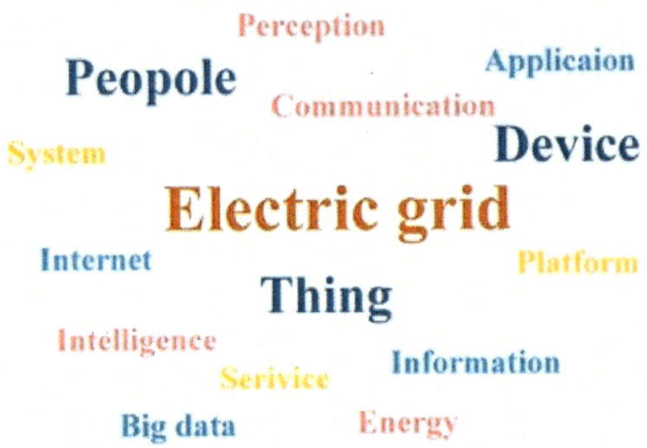

Fig. 1.1 Closely related links of the UEIOT

distinguishing time and location, that is, the network exists at all times. The emergence of the UEIOT pushes the connection between the electric grid and people, things, and equipment to a new level. The UEIOT can be regarded as the product of the development and integration of the electric grid + Internet of Things (Yang et al. 2019c). Ubiquitous electric Internet of Things relies on some technologies, including mobile interconnection, Artificial Intelligence (AI), and advanced communications to realize the interconnection of everything and human-computer interaction in the electric system (Yang et al. 2019b). In the end, a system using comprehensive perception technology, efficient processing of information, flexible and convenient application capabilities was built. The links involved in the construction of the UEIOT intelligent system can be briefly summarized as shown in Fig. 1.1.

The UEIOT makes the Internet of Things technology run through it, enabling information to be shared in a true sense through various communication technologies and technological means. The UEIOT transmits valuable electric data to various industries to achieve sustainable development in the electric domain. This is not only technological innovation. It also provides new ideas for corporate management. The UEIOT has the ability of self-perception and intelligent processing. It can perceive the defect situation of each device in the system, and transmit the information to the maintenance control system at a high speed, and finally realize automatic repair of some faults. The UEIOT enables interconnection, perception, and feedback between industries, which makes the electric market more flexible. This approach provides a more stable, efficient, flexible, and intelligent energy system for the production and life of people. The UEIOT realizes the interconnection of everything and human-computer interaction for various businesses of electric operation through advanced technologies such as big data, intelligent computing, cloud platform services, edge computing, IoT, mobile interconnection, block chain, and so on. It connects electric users, electric grid companies, electric generation companies, suppliers, equipment, and people and things. Then generate shared data to serve users, electric grids, electric generation, suppliers, and the government and society, and it create greater opportunities for people.

1.1.1 Features of Ubiquitous Electric Internet of Things

The UEIOT has the characteristics of holographic perception, ubiquitous connection, open sharing, integration, and innovation. The holographic perception refers to the comprehensive perception of the status of equipment and users in all links of the electric grid and the full penetration of services. Ubiquitous connection refers to the realization of the full-time and spatial ubiquitous connection of grid equipment, users, and data. Open sharing refers to the sharing and integration of grid data and creating a shared platform for the progressing of the new energy industry and market users. Convergent innovation refers to the deep fusion of ubiquitous electric Internet of Things and a strong smart grid. Producers and consumers jointly participate in the innovation of electric grid business to support electric grid companies to create higher levels of value. These characteristics enable it to be clearly distinguished from other things. The specific details of globalization, ubiquity, openness, and innovation can be explained as follows (Yang et al. 2019b):

- Globalization
  Continuing the advantages of the Internet, the UEIOT can coordinate all resources that can be communicated. The UEIOT integrates and connects these information resources to build a real-time and highly interconnected system.
- Ubiquity
  Relying on the most cutting-edge sensor technology and network transmission technology, the UEIOT system can realize the perception Internet of Things in multiple scenarios, so as to flexibly respond to different levels of data communication services. This advantage not only allows users to freely choose the level of deployment and realizes the exchange of information between subjects, but also provides data-level support to optimize dispatch services.
- Openness
  The realization of interconnection allows desensitization data to be shared on the entire unified business platform. Finally, a data-sharing network that can serve different industries and different platforms is constructed. This reflects the value and significance of data in the Internet era fully.
- Innovation
  With the help of the rapidly developing Internet, people can fully utilize the coordination and hub characteristics of the smart grid to achieve the perfect integration of data from different sources. On this basis, other innovative products with data as the core competitiveness are designed. This fully taps the value of the system and achieves mutual benefit and win-win results.

1.1.2 Composition of Ubiquitous Electric Internet of Things

In the intelligent age, the existence of the UEIOT has played a vital role in the development of the electric industry and other related industries. The scale expansion of the electric industry and the scale increase of electric transmission and distribution of the electric grid need to rely on the support of the UEIOT technology. The technological development and popularization of the UEIOT can bring many better changes to human life, including improving the economics of grid operation, promoting the development of clean energy, enhancing the ability of grid asset management, and so on. This will revolutionize energy (Zheng et al. 2012).

According to the system and architecture of the UEIOT, different levels can be used to classify it. It mainly includes 4 different levels: holographic sensing layer, ubiquitous network layer, shared platform layer, and multiple application layer. Advanced sensors, controls, and software applications connect millions of devices and systems on energy production, transmission, and consumer ends to form a perception layer of the UEIOT. Basic network facilities such as the Internet and satellite communication network complete the access and transmission of data from the perception layer. This forms the network layer of the UEIOT. The information resources in the network are integrated into an interconnected information network platform through calculation, so as to solve the problems of data storage, retrieval, utilization, mining, security, and privacy protection, and finally form the platform layer. Integrate the production data, operation data, and management data of electric companies to discover the rules of operation and tap the potential value. Finally, the application layer is formed. The perception layer and transmission layer of the UEIOT extend the company's perception ability, information acquisition ability, operation monitoring, and analysis ability, which makes the company's management decision-making clairvoyant and smooth. The four different levels of diagrams are provided as Fig. 1.2.

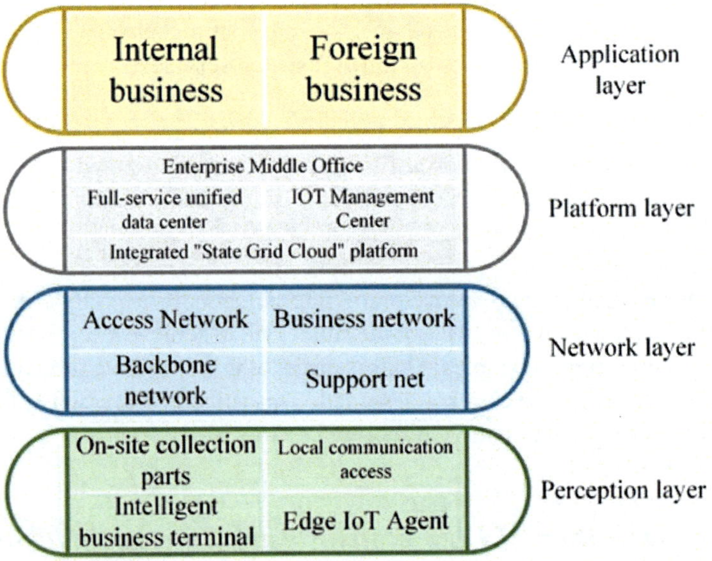

Fig. 1.2 Hierarchical division of UEIOT system architecture

- Holographic sensing layer
  The main object of the holographic sensing layer is not the UEIOT, but the intelligent terminal unit. Its existence provides various possibilities for the development of the IoT, and users can use the operability of the terminal unit to vertically extend the service scope of the IoT.
- Ubiquitous network layer
  The ubiquitous network layer is the foundation of the entire system. It can provide the most reliable basic guarantee for the secure transmission of information. At the same time, it can realize data intercommunication and resource scheduling, which provides more possibilities for the system to meet complex business requirements.
- Shared platform layer
  The shared platform layer system can not only perform real-time monitoring of data from different sources, rational use of resources, intelligent maintenance, and other operations, but also use deep learning technology to learn user behavior and habits to achieve resource scheduling automation.
- Multi-application layer
  The multi-application layer can enhance the diversity and innovation of the system business. With the help of the platform layer's advantages in data, the application layer can design and realize the integration of different businesses such as energy, production, and services. Finally, it achieves the purpose of improving the business coordination capabilities of the entire system, and then realizing data business.

1.1.3 Application Prospect and Value of Ubiquitous Electric Internet of Things

The UEIOT is an important foundation for the construction of the energy Internet, and it is also an important hub bridge for grid development and social applications. In short, the application prospects of the UEIOT can be reflected in ensuring the safe operation of electromechanical networks, effective development of renewable energy, and promotion of integrated energy services (Borgia 2014).

(1) *Ensure the safe operation of the electric grid*
UEIOT can solve the problems of unbalanced energy distribution, irrational grid structure, and poor regulation ability in the national electric. On the one hand, the use of advanced technologies such as data sharing, intelligent decision-making systems, and big data analysis can promote grid-level data and intelligence operation and maintenance, so as to realize real-time monitoring and effective management of equipment in the electric grid. On the other hand, the real-time monitoring of tidal energy, solar energy, load, and the establishment of a unified control system for electric energy can improve the flexibility and regulation capability of the electric grid. In addition, deepening the application of the UEIOT physical ID can realize the connection of upstream and downstream information in all links of electric grid device planning and design, procurement, construction, maintenance, and operation.

(2) *Promote clean energy consumption*
Currently, many renewable energy sources have not yet been highly developed. Due to the instability of wind and photovoltaic electric generation, some areas lack frequency and phase modulation units, and the lack of flexibility of the electric grid may easily lead to the abandonment of wind and electricity. A major role of UEIOT is to build a virtual electric plant intelligent management and cloud control platform. It can calculate, analyze, and store distributed new energy data and electricity load information, so as to optimize the interconnection of distributed new energy, electric dispatching systems, and electric trading platforms. Finally, the goal of regional coordinated control, load increase, and decrease as needed, and reduction of the impact of distributed new energy grid connection is achieved.

(3) *Promote comprehensive energy services*
UEIOT creates a "user-centric" comprehensive energy service platform. The comprehensive energy service platform will complete multiple tasks around three aspects: energy management, electric demand response, and electric trading. These tasks include: (a) Provide energy Internet user services; (b) Expand new energy consumption models; (c) Meet users' diversified energy needs; (d) Guide users to participate in the experience of integrated energy services; (e) Promote the popularization of new comprehensive energy models, systems, and business formats.
UEIOT is another way for grid companies to achieve smart, diversified, and ecological. The UEIOT establishes the connection bridge between the equipment and electric users, electric grid enterprises, electric generation enterprises, and suppliers. Relying on the advantages of electric grid hubs and shared platforms, the UEIOT provides more value services for the progressing of the entire industry and the market, while also creating greater opportunities for the times. The combination of the smart grid and the UEIOT can form an electric energy Internet platform with complementary advantages and complement each other. Together, they form a "three-in-one" energy Internet of new energy flow, business flow, and data flow. The application framework of the UEIOT is provided in Fig. 1.3.

UEIOT can achieve multiple tasks, including:

- Improving the friendly grid-connection level of distributed new energy through virtual electric plants and multi-energy complementation;
- Realize extensive interconnection and data sharing of electric transmission and distribution equipment, and improve grid operation efficiency;
- Focus on customers and optimize customer service;
- Create data sharing services to serve society.

UEIOT is a significant means to effectively solve the challenges faced by the electric system. In summary, relying on the advantages of the UEIOT, the development of smart device identification technology will be unprecedentedly strengthened.

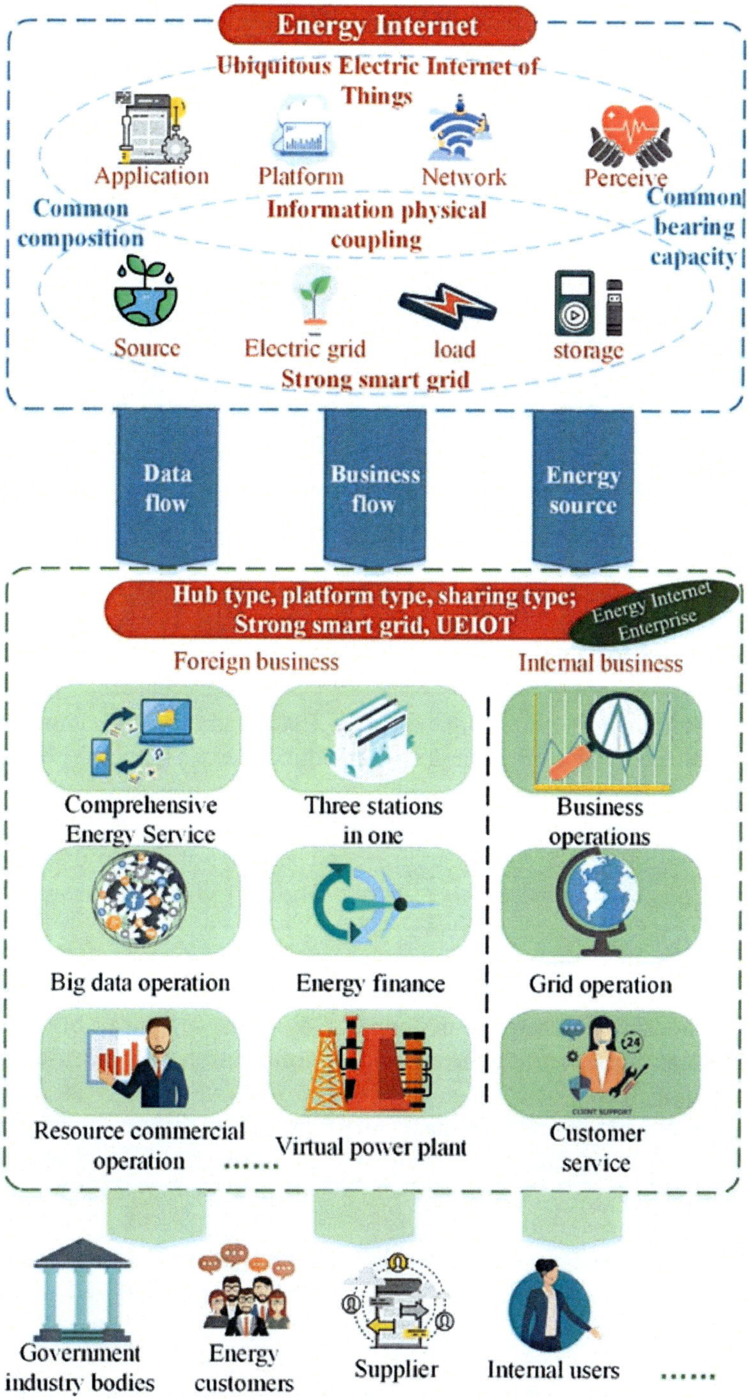

Fig. 1.3 Application framework of UEIOT

This book also focuses on the recognition of smart devices, and demonstrates the methodology and effects of device recognition in detail through actual algorithm cases.

1.2 Key Techniques of UEIOT

The foundation and core of the UEIOT is the electric system. To realize the efficient and accurate operation of the system, it needs to involve multiple cross-domains, including communication, recognition, computing, sensing, cloud data, information security, etc. This chapter mainly introduces several key technical links involved in the UEIOT: Sect. 1.2.1 *Smart electric device recognition,* Sect. 1.2.2 *Internet of things,* Sect. 1.2.3 *Big data analysis,* Sect. 1.2.4 *Cloud platforms,* Sect. 1.2.5 *Computational Intelligence,* Sect. 1.2.6 *Smart model embedding, and* Sect. 1.2.7 *Others.*

1.2.1 Smart Electric Device Recognition

In the era of highly developed big data, the UEIOT and other technologies promote the network access of a large number of terminal electric devices. Network traffic has become huge and complex. If the huge number of network electric equipment involved can be effectively identified, the efficiency of network operation and management will be greatly improved. The crisis of device data flow can be effectively alleviated, and the construction of smart cities will go further (Chen 2019).

Intelligent equipment recognition technology completes the acquisition of electric load energy consumption information by collecting and analyzing load data such as voltage and current entering urban buildings. With the support of big data, if users can understand the electric consumption information of each electrical appliance in real-time, the energy-saving work can be carried out to the greatest extent. The rational use of electric resources has a major impact on the economic development of the entire society. Effectively increasing the utilization rate of electric energy and rationally planning the distribution of electric energy resources can promote the development of a resource-saving society.

Currently, there are two main technology methods for realizing electric load monitoring, namely, invasive load monitoring method and non-invasive load monitoring method. For the intrusive load monitoring method, it completes load monitoring and analysis by installing a monitoring device for each electric load, and then performing a series of real-time monitoring, data storage, and data transmission. The advantage of this method lies in the high degree of restoration of the collected data and accurate data analysis. However, due to a large number of electrical loads in buildings, this method requires a large quantity of monitoring devices. The high installation and maintenance costs restrict the development of this method. Corresponding to the invasive monitoring method is the non-invasive monitoring method. The non-intrusive load monitoring method only needs to install a monitoring device at the electric supply entrance of each user to complete the real-time collection of electric information such as total voltage, current, active electric, reactive electric, and apparent electric. Finally, the artificial intelligence recognition algorithm is used to identify the type of load. The advantages of this method are simple equipment installation and maintenance, low cost, and high application type. As a hot research direction in the energy field based on non-intrusive smart device identification methods, this technology has an important enlightening significance for electric users and electric companies.

For electric users, smart electric device identification technology helps users have a more detailed understanding on the electric usage of different electric loads at various times. The user can reasonably plan the usage time of the electric load according to the electric energy usage information of the electric load. After comparing and analyzing the energy consumption information between different loads, users can purchase more energy-efficient loads in a targeted manner. In addition, users can verify the energy-saving performance of the energy-saving plan formulated and the energy-saving load purchased. In general, under the premise that their normal production is not affected, users can use smart device identification technology to reduce electric consumption.

For electric companies, smart electric device identification technology is beneficial for electric companies to have a more realistic understanding of the composition of electric system loads. Through the construction of a more accurate electric load model, the flexible load with demand response capability is selected from the total electric load. Analyze the degree of load demand response, so as to guide the demand-side response. At the same time, users are guided to actively participate in electric supply and demand adjustment measures such as "tiered tariffs" and "peak-cutting and filling valleys" advocated by electric companies to improve electric utilization efficiency. In general, smart electric device identification technology helps optimize the planning and operation of electric companies, and it plays a vital role in the development of smart electric companies.

1.2.2 Internet of Things

The concept of the IoT was first proposed in UK and originated from the Troy coffee pot incident. In the following days, Bill Gates again mentioned the IoT in "The Road to the Future" and had a certain idea of the IoT. Some things in his vision have become real existence in today's society, such as digital products, electronic products, and smart products. In the following years, the IoT gradually had a clearer concept. For the first time, engineers in the UK mentioned connecting the IoT with the Internet, thus realizing the connection of the world on the IoT. In the end, the concept of the IoT was accurately proposed by the International Telecommunication Union. The so-called IoT is to connect each target object through the network, which can realize the communication between people at any time. Moreover, the IoT is not limited to people, but also expands the acquisition of information between people and things, things, and things (Sun et al. 2019).

As the world is actively mastering IoT technology, the IoT technology has developed rapidly. The IoT technology has penetrated into human life. As the most complex and largest system at present, the electric system is affecting the lives of people all over the world. Meanwhile, people's demand for electric systems is also increasing. Therefore, to ensure the intelligence and safety of the electric system, the combination of the electric system with the IoT technology has become an inevitable trend.

The IoT in the context of UEIOT is a system that can realize the basic electric grid, personnel, and the perception of the surrounding environment, and control the electric grid. The interaction and assistance between entities make the objects have a close connection, a high degree of collaboration, perception, and control capabilities. The IoT promotes the deep integration of the physical world and the digital world, and accelerates industrial development. To achieve the establishment of a data integration platform for the UEIOT, it is necessary to integrate "big data + cloud computing + artificial intelligence" technology. The IoT platform can provide some value-added functions such as data mining and analysis, edge computing, etc. to promote the development of various scenarios and industry developers that are highly related to the IoT. The State Grid Corporation will integrate emerging communication technologies such as "big data, cloud computing, IoT, mobile internet, and artificial intelligence" with the new electric system to connect all equipment, machines, materials, and people for electric production and consumption together. This ensures the smooth operation of the electric network and can provide effective services for the company's daily operation control. After years of technological and theoretical changes, the UEIOT system has become more and more mature.

1.2.3 Big Data Analysis

Big data technology originated from Google (Chen et al. 2014). In response to the ever-expanding mass data storage problems faced by search engines, Google proposed a set of big data platforms based on distributed parallel clusters. It mainly includes a distributed file system, massively parallel computing programming model, and distributed database. Benefiting from big data technology innovation, some data that seemed difficult to collect and utilize are beginning to be easily integrated and utilized. The advent of the big data era has three reasons: first, the ability of human beings to store data has been continuously enhanced; second, the ability of humans to produce data has been continuously enhanced; third, the ability of humans to use data has been continuously enhanced.

With the construction of the UEIOT, the amount of data will further increase. The data stores have typical characteristics of big data. Realizing the construction goals and various tasks of the ubiquitous power Internet of Things, making enterprise management leaner, power grid operation safer, better customer service, and building a mutually beneficial and win-win energy ecosystem must ultimately be accomplished through data. Therefore, UEIOT is a network that carries big data, and big data is its core content. The application of big data technology to play the role and value of massive data, and use data to drive production, operation, management, and innovation is the foundation of UEIOT's construction. Starting from the construction of UEIOT, the promotion of big data applications should focus on improving the following five aspects:

- Improve data management and control capabilities.
  Formulate and improve enterprise-level data models and data standards. Clarify the corporate data management responsibility interface. Clarify data types through data inventory. Improve data quality through data governance. Promote the open sharing of data by formulating negative lists and sharing lists.
- Improve data collection capabilities.
  Make full use of modern information technology, advanced communication technology, and sensor technology. Formulate structured and unstructured data collection specifications. Finally, the status of each power system's link is fully perceived, the data collection and recording capabilities are comprehensively improved, and the continuous enrichment of data resources is realized.
- Improve data aggregation capabilities.
  Build a full-service data center and build a structured and unstructured data storage platform that meets the requirements of big data processing. The platform has storage capabilities for relational data and various types of massive data such as text, voice, image, and video. It can realize the orderly aggregation of all enterprise data.
- Improve data computing capabilities.
  Construction of enterprise data center. Build a one-stop computing platform for data acquisition, cleaning, analysis, presentation, and result storage. The platform has a parallel computing cluster composed of multiple computers and provides diverse data analysis tools. The platform has the data loading capabilities and data processing capabilities required for enterprise-level big data computing.
- Improve data application capabilities.
  Create an atmosphere where data is used to speak, manage, make decisions and innovate. Cultivate a multi-level data analysis team, strengthen data application training, continuously explore data application requirements, and enrich data analysis models. Effectively promote the use of data to drive production, operation, management, and innovation.

The construction effectiveness of the UEIOT in supporting big data can be measured by the "three online", that is, whether data online, computing power online, and algorithm online are achieved. Data online means that people who use data can use tools to understand what data is stored in a full-service data center, and easily obtain the data they need through a one-stop platform. Computing online means that data analysts can obtain computing and storage capabilities through a one-stop platform, whether they are doing light data analysis or massive data analysis, to meet the needs of data mining and analysis. The algorithm online refers to a one-stop platform that integrates commonly used statistical analysis tools and professional analysis tools in the industry. It can provide analysts with convenient and quick calls to mine the value of analysis data.

The development of UEIOT should focus on the continuous promotion of big data applications at multiple levels:

- Effectively promote the application of big data by business personnel within the enterprise.
  The implementation entities of this type of big data application are various business personnel at all levels of the State Grid Corporation. Its implementation object is structured data. Its implementation technology uses traditional screening, filtering, association, matching, etc. Its implementation task is to give full play to business personnel through a one-stop computing platform to analyze and respond to existing structured data. To promote the application of big data at this level is to consolidate the mass base, which can stimulate the creativity of grassroots to solve problems through data, promote and test the construction effect of the UEIOT.
- Effectively promote the application of big data within the enterprise's professional teams.
  The implementation subject of this kind of big data application is the professional data analysis team such as the Big Data Center of State Grid Corporation. The focus of its implementation is structured data. Its implementation technology requires special tools such as data mining software and statistical software. Its implementation task is to dig and analyze the laws hidden under the data that are difficult to find. Promoting the application of big data at this level is the focus of data value creation, which can fully reflect the improvement of data application capabilities and lead the direction of UEIOT in the construction of big data support.
- Effectively promote internal and external integration of prospective big data applications.
  The implementation subject of this kind of big data application is the specialized cooperation team inside and outside the enterprise. Its implementation object focuses on massive structured data and unstructured data. Its implementation technology requires the use of specialized technologies such as the Hadoop platform, Spark memory computing, and deep learning. Its implementation task is to drive the intelligent development of enterprises in the fields of production, operation, and management through massive structured and unstructured data deep learning. Most of the big data applications at this level are forward-looking and exploratory. Once it achieves results, it will bring about a huge improvement in the quality and efficiency of the company in one aspect, and bring new business formats.

The construction of UEIOT by grid companies is a process of applying big data technology and also a process of promoting big data applications. The construction of UEIOT will fully support the application of big data, and the application of big data will fully reflect the effectiveness of the construction of the UEIOT. The goals and tasks of UEIOT construction have clear phase characteristics. But playing the role and value of big data is an enterprise-level strategy that requires long-term adherence.

1.2.4 Cloud Platforms

With the increase of the proportion of electric energy in the terminal energy consumption, the requirements of economic and social development for the quality of electric energy and the intellectualization of the power grid are constantly improved (Chen et al. 2019). Speeding up the construction of the Energy Internet requires the development of cloud platforms. By mining the value of big data, the work objectives, and work content requirements of various tasks can be completed in time. Cloud platforms need to be resilient to access pressures and quickly respond to remote resource failures. Electric grid companies build cloud data centers and cloud platforms. They use the cloud platform's flexible scalability, fault self-healing, and cross-data center resource integration capabilities to support the transformation of multidimensional lean management systems, so as to achieve better-operating results, increase security in various dimensions, and help the construction of the energy Internet.

The concept of the cloud platform can be divided into narrow and broad senses. In a narrow sense, the cloud platform is a network that can provide corresponding resources. In a broad sense, the cloud platform is a service that integrates resources, realizes automatic management through software, and quickly provides targeted resources on demand. Cloud computing has high-efficiency computing power. It can process tens of thousands of data in a very short time and provide extremely fast network services. The cloud platform system includes cloud data, cloud platform center, and client, which are shown in Fig. 1.4.

The cloud platform is an information-sharing platform with data such as power system business, regulation business, grid flow distribution, and information flow distribution. It has the characteristics of virtualization, dynamic scalability, on-demand configuration, high flexibility, high reliability, and high-cost performance. The cloud platform can provide data storage and computing services for a wide range of industries to quickly meet the needs of users.

(1) *The role of cloud platform*

- Integration of resources
  Cloud platform technology automatically collects, integrates, and accurately classifies electric grid operating data according to user needs. The cloud platform uses virtualization technology to replace the programs on the client computer with a virtual server. When the customer sends demand information to the remote terminal, the remote computer sends appropriately targeted information to the user according to the demand.

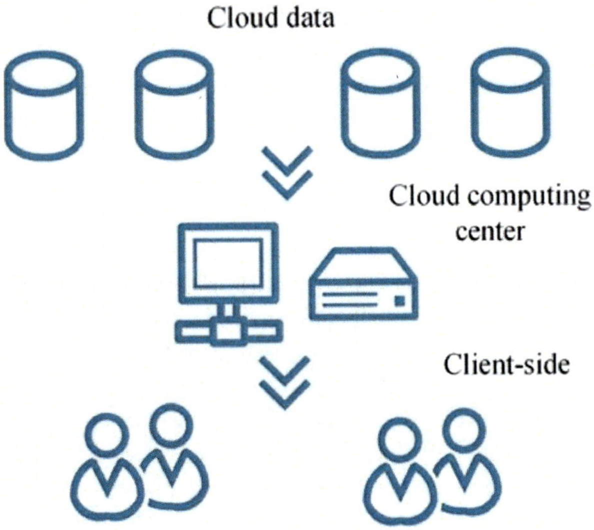

Fig. 1.4 Cloud platform system

- Improve efficiency and economy
  The cloud platform searches data accurately and quickly, which improves work efficiency. It has no requirements on time and location, as long as the device and the network are connected, the required data can be accessed, and the efficiency is high. Cloud platform technology can simplify maintenance and release, and reduce user costs. It can integrate independent computing platforms to establish regional and even national systems. In addition, the cloud platform can integrate idle storage and computing resources, reduce the access to new information equipment, save capital investment, and be economical.

(2) *Application of cloud platform in the electric grid*
The biggest role of cloud platforms in the operation of power grids is to share data, analyze suggestions, and assist decision-making. It provides relevant dynamic, easy-to-expand, and virtualized big data for the electric grid operation mode, assisting dispatchers to make the best adjustments to the electric grid operation mode in time, and realize risk warning. The cloud platform can also use technical means to make the best adjustment strategy for electric grid failures. After the dispatcher confirms it, it automatically adjusts the electric grid operation mode, improves the electric grid control ability, and maximizes the use of electric energy. The application design of the cloud platform in the electric grid operation mode is shown in Fig. 1.5, including basic cloud, application cloud, coordination cloud, and service cloud.

Among them, the basic cloud integrates the temperature, light, and heat signal data of the electric grid operation and the database of each system, unified management, and forms a data service. The application cloud access to the electric system

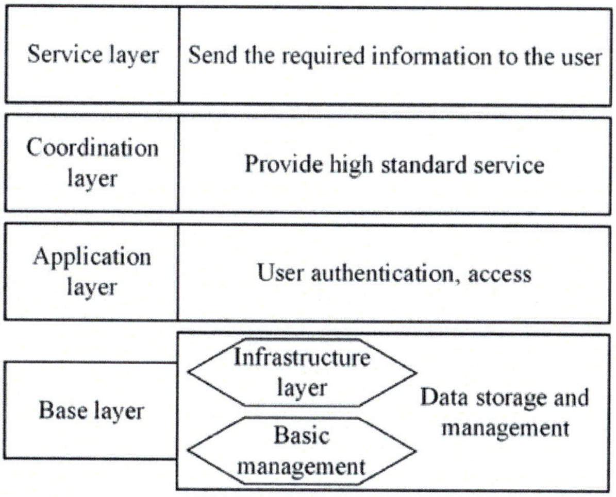

Fig. 1.5 Cloud platform structure diagram in the electric system

provides structured and unstructured data covering the dispatching field. The coordination cloud provides electric flow calculation, transient stability, and short-circuit calculation services. It can stabilize, adjust, and design electric grid operating data, so that the electric grid can run to the best state. The coordination cloud also provides parameter preparation, data preparation, status notification, comprehensive maintenance status, fault status, load status, and other complete scheduling plans. In addition, the coordination cloud can also integrate various data, assist dispatchers to adjust the operation mode of the electric grid, and promptly discover the shortcomings of the current electric grid operation, and provide optimization strategies and solutions. Finally, the coordination cloud judges the rationality of the regional power grid architecture and gives a reasonable rectification plan. The service cloud serves as the entrance of the cloud platform in the grid operation mode, that is, the interface entrance. The detailed diagram of the cloud platform application is shown in Fig. 1.6.

With the support of big data and cloud platform, the construction of UEIOT with the electric system as the core is unstoppable. Reasonable use of limited hardware resources to meet UEIOT's analysis, calculation, and storage is an urgent requirement for electric systems. The introduction of the cloud platform can realize the need of supporting the scheduling and operation of storing big data and completing application analysis. So as to complete the effective use of the storage resources of the entire system, give full play to the analysis and calculation level of the existing system, improve the online analysis capability of electric system dispatch, enhance the level of real-time control, and provide strong technical support for enhancing the reliability of smart electric grids (Xu et al. 2016; Tarutani et al. 2016).

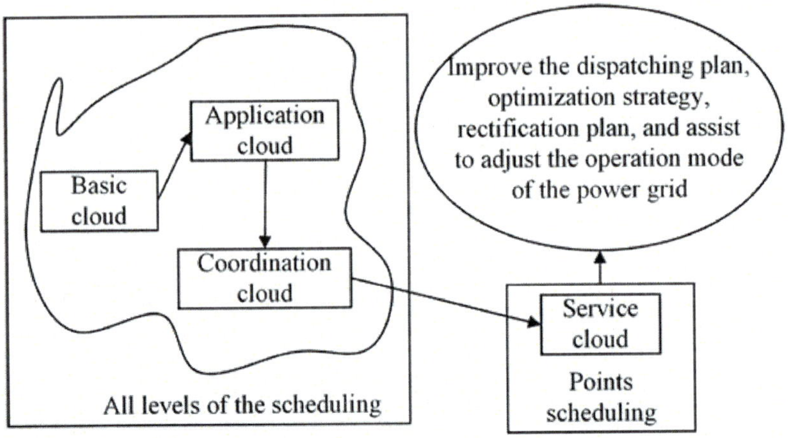

Fig. 1.6 Application of cloud platform

1.2.5 Computational Intelligence

Currently, there is no clear definition of computational intelligence. Computational intelligence is the definition given by American scientists from the perspective of computational intelligence systems. If a system has the following characteristics, then it belongs to the scope of computational intelligence. These features are summarized as the system's processing of numerical data is only at the bottom level, and the pattern recognition components contained in it do not use the relevant knowledge of the meaning of artificial intelligence, and it has a certain degree of computational adaptability, error tolerance rate, and close to human calculation speed and error rate, etc. In addition, from the perspective of disciplines, computational intelligence is a unified discipline concept formed on the basis of relatively mature development in the three fields of neural networks, evolutionary computing, and fuzzy systems. Among them, the neural network is a structural simulation method of human intelligence. It interconnects artificial neurons to construct an intelligent mechanism that mimics the biological nervous system. This mechanism is mainly based on the combination of Artificial Neural Network (ANN) systems.

Computational intelligence refers to the establishment of functional connections based on data and calculations (Konar 2006). So as to solve the problem to realize the simulation and understanding of intelligence. Computational intelligence also refers to the use of computational science and technology to simulate human intelligence structures and behaviors. Computational intelligence emphasizes the realization of the inherent intelligent behavior of organisms through computational methods. Computational intelligence can be recognized as a low-level intelligent cognition. The difference between computational intelligence and artificial intelligence is that the cognitive level drops from the middle level to the low level. The middle-level system contains knowledge (exquisite), while the low-level system does not. When a system only involves numerical (low-level) data that does not apply knowledge in

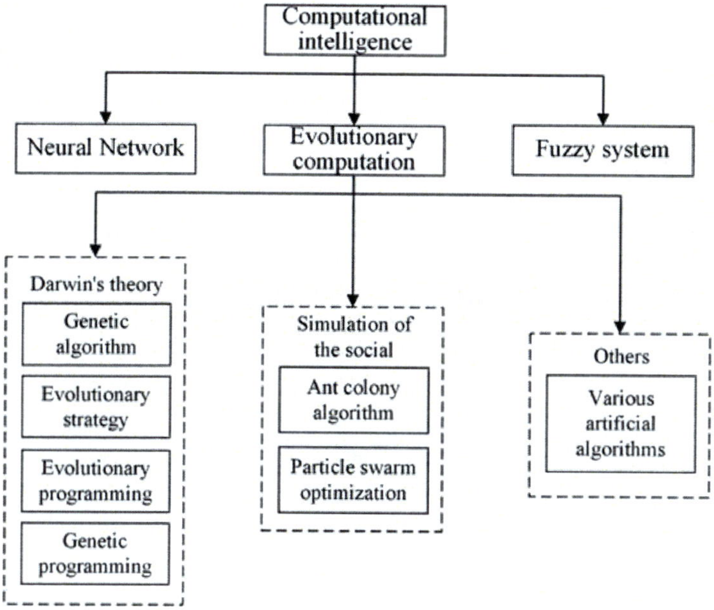

Fig. 1.7 Components of computational intelligence

the sense of AI, and it can exhibit the following features: (a) Computational adaptability; (b) Computational fault tolerance; (c) Approaching human speed; (d) Error rate similar to humans. If an intelligent computing system adds knowledge (exquisite) value in a non-numerical way, it becomes an AI system. Bezdek proposed the concept of computational intelligence in 1992 (Bezdek 1998). In his view, computational intelligence does not depend on knowledge but on digital data provided by manufacturers. Besides, AI application knowledge boutique. The ANN should be called a computational neural network. The Components of computational intelligence are presented in Fig. 1.7.

Computational intelligence has the following three distinct characteristics:

- Intelligent characteristic
  The intelligent features including the self-adaptability and self-organization of the algorithm, which makes it highly versatile.
- Parallel characteristic
  The algorithm is basically an optimized solution to the problem in the way of group cooperation, which makes it easy to deal with large-scale parallel processing tasks.
- Robust characteristic
  The algorithm is not sensitive to initial parameter conditions. Excellent robustness helps it find the optimal solution under different conditions.
  The principle of computational intelligence is to effectively extract computational models and optimize the design of intelligent algorithms based on relevant problem characteristics and targets. At the same time, it can use related modules

for collaborative operations. Such as knowledge and method improvement, fixed-point information data exchange, effective accumulation of information perception, and task scheduling implementation. In short, the application of computational intelligence technology can effectively realize the intelligent processing of UEIOT information, and its application advantages are very significant. The specific analysis of computational intelligence applications in UEIOT is given as follows (Long et al. 2020):

- Information and status collection function
  This function is mainly used to display and process relevant data information collected by sensors. This module effectively displays the collected information and various equipment operating status information, randomly assigns human-machine interface agents, and timely feedbacks information data to users.
- Information data extraction function
  This function collects environmental information data in sensor nodes, sends them to the genetic neural network, and provides relevant data information to the state collector. At the same time, it also has the function of managing data and information perception agents.
- Service extraction
  This function includes two sub-functions, namely command generation, and node management functions. Among them, the command generation function can generate commands to directly control sensor actions. The command generation function is responsible for generating commands in the sub-module. This function can be directly used for the behavior control of various devices in the Internet of Things. The node management function is mainly based on information nodes and sensor nodes to manage all child nodes. The node management function can enable the core module to manage all the sub-node modules, thereby functioning based on data information nodes and sensor nodes. For example, when electrical equipment stops running, this function can be used to detect problems based on sensors in time and feedback information to users in time.
- Data mining
  Based on the data mining function, the purpose of communication between the intelligent computing module and the database can be achieved. The IoT system is based on genetic fuzzy neural network calculation rules and capabilities, and stores data in a database.
- Command classification
  Users can use terminal devices to input corresponding commands to complete tasks such as standardized processing, matching database information, and calculating matching results through genetic fuzzy neural networks. This function can provide interface implementation based on user and smart terminal access. In the application of the genetic algorithm, to effectively simplify the description problem, it is necessary to use code strings of different lengths. For example, in practice, genetic algorithms are used to optimize the design of the fuzzy controller rule base. Usually, the number of rules is not known in advance, but the change can be described based on the length of the individual chromosomes of the rule. Using the genetic algorithm can optimize the structure of the artificial neural network.

In this process, if the tree items of each layer are unknown, the same number of chromosome lengths can also describe changes. The UEIOT is a new network that combines modern AI with the application of computing intelligence technology. AI endows computers with management ability and emotion similar to humans.

1.2.6 Smart Model Embedding

The design of the perception layer of the traditional Internet of Things mostly uses a single-function terminal, that is, an intelligent Internet terminal is installed inside the electrical equipment, and then a specific protocol is used to transmit data. This method is increasingly unable to meet the requirements of more equipment access. Traditional equipment must be modified to get its data (Bauerle et al. 2014). This has great limitations on the development of social informatization. Universal access to various IoT terminals is now a widely used device access method (Jia et al. 2019). When the design architecture of the UEIOT system access module and the terminal adaptation method are accessed under this architecture, multiple types of terminals can be accessed conveniently and quickly.

This adaptive access method requires a lot of manual labor to confirm the device type (Zhang et al. 2020). That is, the information of the device needs to be manually initialized while the terminal is connected, so it cannot automatically identify the type of the connected device. In addition, due to inevitable human error and other reasons, it will cause equipment information errors and delays, and eventually generate a large amount of invalid data (Tu et al. 2019). This is not conducive to the development of the IoT with rapid information changes.

The embedding of the artificial intelligence model excavates the potential characteristics of the electric load information from a deeper level, and efficiently completes the identification of smart devices. Currently, many scholars have made analysis and research on the identification of equipment types. They mainly use data-driven algorithms to mine the potential characteristics of electrical equipment parameter data (Rashid and Louis 2019; Huang et al. 2019b), and identify equipment by obtaining relevant legal and mode information of user equipment (Mets et al. 2015; Li et al. 2016). The application methods involved include Markov chains, decision trees, etc. (Arif et al. 2017; Rashid et al. 2019; Rosdi et al. 2014; Hou et al. 2019; Chicco and Ilie 2009; Chen et al. 2018; Liu et al. 2017). These methods avoid the waste of traditional methods for manual use, and can effectively make data analysis and equipment identification for specific equipment. However, most of the current models analyze specific electrical equipment. These models are highly targeted but have weak robustness and complex implementation, so they cannot be flexibly applied to a large number of electrical equipment. In addition, the device identification method based on historical data is poor in timeliness. Although it can achieve the purpose of device identification, it is too slow in the face of the rapid information change of the Internet of Things. Therefore, a simple and easy-to-understand universal device identification model is constructed, and the model is applied to the Internet of Things system to identify devices in real-time, which greatly improves the efficiency of device identification. It can ensure the timeliness and validity of data collected by the UEIOT (Wang et al. 2019).

1.2.7 Others

- Smart chip
  After the UEIOT technology entered the public's field of vision, types of equipment including metering, protection, electric transformation, control, monitoring, electric consumption, etc. poured into the electric system (Wong et al. 2010). The operation of electrical equipment continuously generates data. Gradually, traditional industrial collection equipment cannot meet the needs of modern industry. These data collection devices with poor reliability and low accuracy are difficult to match with highly intelligent terminal devices. To solve this problem, miniature smart sensors and smart terminals based on smart chips have been developed. The new equipment has the characteristics of high precision, low power consumption, miniaturization, and intelligent computing. On the one hand, it excellently completes the collection, extraction and transmission of equipment information. On the other hand, terminal intelligence is realized through local edge computing to complete local self-control. It is not like traditional facilities that only have telemetry and not the remote control. In recent years, the Cambrian chip developed by the Chinese Academy of Sciences and the Edge TPU chip developed by Google both have big data analysis and edge computing functions. The advent of these smart chips effectively improves the level of terminal intelligence.
- 5G and Low Power Wide Area (LPWA)
  A comprehensive communication network can meet the transmission needs of large amounts of electronic data (Gohil et al. 2013). The electric system is widely distributed, so it often involves complex topographic conditions. As a result, the communication system cannot cover all areas, and on-site information cannot be disseminated. IoT has two main types of technologies in the current stage. One is a wired information technology based on power carrier and Ethernet, and the other is a wireless information technology based on low-energy wide area networks and 5G. Although the traditional information and communication technology has a wide distribution, its collection frequency is small, the connection is difficult, and it is not easy to supply electric energy in some areas, so it cannot be applied to many scenarios. It can be seen that communication methods based on wireless networks are the main means to realize ubiquitous communication in the IoT. 5G technology is regarded as the cornerstone of the IoT. In addition, the 5G sheet network communication technology can be adaptively adjusted according to different scenarios. This is the key to solving the holographic perception and ubiquitous connection of the electric system. For control services such as intelligent distributed distribution automation and millisecond precision load control, it can choose Ultra-Reliable and Low-Latency Communication (URLLC) chips

with ultra-low latency. Besides, for large-scale power information collection and smart car charging stations, enhanced mobile bandwidth slices can be used. 5G chip network technology can choose different chips for different applications. The realization of data transmission on the 5G network channel and access to the same 5G data platform will surely become the main means to realize UEIOT communication.

1.3 Smart Device Recognition in UEIOT

With the wide application of UEIOT in smart manufacturing, smart medical, smart city, and other fields, a large number of UEIOT devices have flooded into cyberspace, including monitoring equipment, industrial control equipment, network equipment, and office equipment. There are nearly a million IoT devices added to cyberspace every day (Xia et al. 2012). The Internet of Things devices have built a bridge between people and things, and things and things, and have become important assets in cyberspace.

As the core technology of the UEIOT support side, intelligent load identification and monitoring occupy an important research position in residential electricity management. Smart device identification can obtain the relevant information of each electricity load in the range by analyzing the total electricity meter data in a specific area. Such as the number of loads, the type of each load, the working status, and the corresponding energy consumption usage. The information obtained by the smart device identification method has great practical value for all power users. For ordinary users, if the device details obtained through load monitoring can be fed back in time, it will help guide users to use electricity reasonably, thereby achieving energy saving and consumption reduction, and saving electricity expenses. For electric companies, non-intrusive load monitoring can achieve a fine-grained perception of the load components without significantly increasing the input. Thereby improving the accuracy of electric system load forecasting, and improving the safety and economy of the electric grid. It is also conducive to more accurate modeling of user behavior to achieve differentiated and precise services for users. For electrical equipment manufacturers, the equipment status and corresponding energy consumption information provided by non-intrusive load monitoring can provide a basis for equipment fault diagnosis or further realize predictive maintenance of equipment. Non-intrusive load monitoring belongs to the interdisciplinary application of computer, communication, electronics, and electrical engineering, and can be widely used in fields such as building energy-saving, smart cities, and smart grids. It aims to fully perceive the residential electricity load, accurately predict the user's electricity consumption behavior, and realize the independent identification of the operating state of electrical appliances and the management of energy consumption. So as to formulate personalized intervention strategies to guide users to use electricity more reasonably. The application and popularization of smart meters provide the possibility for the analysis and identification of residents' load characteristics. Smart meters can effectively

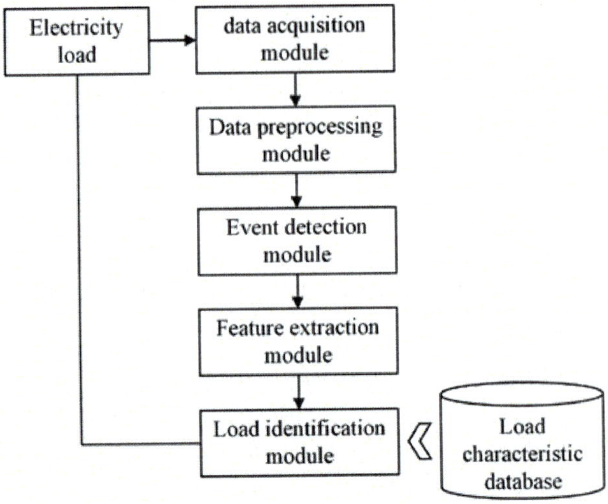

Fig. 1.8 Non-invasive load identification technology

collect data and information. With the technical help of big data resource sharing and analysis and mining platforms, it will guide users to diversify and rationally use electricity.

Recently, many international scholars have devoted themselves to the research of non-intrusive load identification algorithms. To accurately and effectively classify the power load information of the intricate load, it is necessary to use smart device recognition technology. Years of technological development has summarized several key steps for smart device identification, which are event detection module, feature extraction module, and device identification module. After the total load information is collected, the processed data is used to detect whether there is a change in the operating state of the electrical equipment. Then extract a series of load characteristics for load identification before and after the state change, and finally identify the corresponding electrical equipment. The overall technical process is presented as Fig. 1.8.

1.3.1 Data Acquisition Module

The essence of NILM is load decomposition, that is, the user's total load electric information is obtained through entrance detection equipment. The total load information is decomposed into each electric device, so the user can know the energy consumption of each electric device, the frequency of use, the start and stop time, and other electricity consumption information. This electricity information has a very high application value. It is the backbone of the development of smart grids and will bring many benefits to society.

The non-intrusive load monitoring and decomposition system complete the visit to the user by installing monitoring equipment and obtaining the total power consumption, and then the task of identifying the load type and working status of the equipment is completed. Upload this information to the system to understand the user's consumption behavior, load working status, and corresponding power consumption.

Data collection is the first step in non-intrusive load monitoring and decomposition. Its purpose is to obtain temporary and steady-state signals of the user's total load. Measurement errors are inevitable, but they should be minimized. In general, the measurement error comes from two aspects. On the one hand, the measurement devices are not uniform, that is, the use of different measurement devices for the same electrical appliance to collect information results in non-uniform result standards. On the other hand, the data collected by the sensor is missing due to compression and improper operation of the transmission process.

The electrical physical quantities such as the current and voltage of the residential load can be measured by the Non-Intrusive Load Monitoring and Decomposition (NILMD) device. These data can be regarded as signals which carrying power information. They contain various types of load component information with different characteristics. By performing corresponding calculations and feature extraction and classification of these data, the NILMD system can realize the function of load decomposition.

1.3.2 Event Detection Module

The essence of the event detection algorithm is mutation point detection. That is an algorithm for identifying the point where the jump change exceeds the threshold in the adjacent points of the electrical signal sequence. The mutation point detection can usually be performed based on low-frequency electrical signals and high-frequency electrical signals. The low-frequency electrical signal mainly captures steady-state events, and judges the occurrence of steady-state events through the difference in working state characteristics over a long period of time. For example, calculating the difference in the average value of the active power before and after the time period can be used as the basis for the start and stop of the equipment, and the start and stop of the equipment can be preliminarily determined based on the change range. In addition, reactive power, apparent power, resistance, reactance, etc. can all be used as characteristics for judging steady-state events (Zoha et al. 2012). Compared with low-frequency electrical signals, high-frequency electrical signals can judge and capture transient events through the difference of electrical signals in a shorter time. Detecting high-frequency signals require a higher sampling frequency and require higher hardware configuration, so it is expensive. However, high-frequency signals can provide critical information that low-frequency signals cannot obtain. For example, two appliances that consume the same power may have very different instantaneous on-currents. The high-frequency signal can obtain the key information provided by the analysis of the transient characteristics, and then determine the actual operation of the equipment in the load. Currently, there are three methods for event detection, including expert heuristics, probability models, and template matching (Parson et al. 2014; Esa et al. 2016).

The expert heuristic event detection method uses the load information to be tested and makes rules based on prior conditions to detect mutation points. Hart proposed a method of load judgment using characteristic variation (Hart 1992). They use active power and reactive power as the characteristics to distinguish household appliances. When the change of the current and later characteristic values is close to the prior value, it is determined that the moment is the start and stop of the electrical appliance corresponding to the prior value. Alcalá et al. proposed a method based on envelope extraction (Alcalá et al. 2017). This method uses root mean square calculation to get the current envelope. By deriving the change of the envelope curve, the change is transformed into some peaks. After that, the peak value whose amplitude is greater than the threshold is regarded as an event. The experimental comparison shows that the technology method can achieve a similar effect to the probability model without the training process.

Event detection in load identification is defined as detecting changes in the operating state of electrical equipment (i.e. opening and closing of the equipment) (Chang et al. 2010). It usually uses changes in total power consumption as the main basis for state changes. The most direct detection method is to compare the electric variation deviation of adjacent samples. If the electric change deviates from a predetermined threshold, it is determined that the equipment start-stop event occurs. However, in practical applications, the power time series fluctuates randomly due to many factors. With a threshold, it is difficult to distinguish events from noise. In this case, a probability model is proposed.

Generalized Likelihood Ratio (GLR) is a probability model, which was first proposed to solve the problem of sudden change (Luo et al. 2002). The GLR defines the average window before the event, the average window after the event, and the event detection algorithm. Two Gaussian distributions are defined using the average window before the event and the average window after the event. The probability density function of two Gaussian distributions is used to determine the possibility of the existence of the event in the detection window. With similar effects to GLR-based models, event detection models based on Goodness Of Fit (GOF) methods have also been widely studied (Yang et al. 2014). The distribution difference before and after an electrical event can be used for event detection. It can be seen from the experimental results, the early warning performance of the GOF model is better than the expert heuristic method.

The event detection method based on template matching is achieved by comparing the detected total electric signal with the template power signal (Leeb et al. 1995). The basic steps of the event detection method based on template matching are given as follows:

(a) Calculating and analyzing the spectral envelope of the row time series, and convert the high-frequency row U and I time-series into characteristic time series with obvious transient characteristics;

(b) Establishing a template library, and extracting and standardizing typical examples of starting and stopping electrical equipment during the training process;
(c) Utilizing the average change mechanism to trigger the detection event process based on template matching, so as to complete the task of calculating the distance between the detected total electric signal and the template electric signal.

Distance calculation is a key step in the event detection method based on template matching. In non-intrusive load detection, the Euclidean distance measurement method is utilized for template matching operation (Shaw et al. 1998). Nevertheless, the Euclidean distance is difficult to adapt to the sensitive situation of the data signal. That is, if there is a phase difference between two similar signals, a larger Euclidean distance may be obtained. In this case, the time series similarity measurement method is recommended to use the dynamic time warping method (Basu et al. 2015) and another variant method (Maus et al. 2016) for template matching.

Fortunately, template matching methods can play a role in multiple aspects, including event detection and electrical equipment identification. If the typical instances in the library have tags corresponding to the device type and working status, the device type, and device status can be detected through template matching.

1.3.3 Feature Extraction Module

Non-intrusive load monitoring is the process of identifying the existence of a specific device from the total measurement data based on load characteristics, and finally determining its working status and stripping it out one by one. Extracting suitable load characteristics from the collected data is essential for monitoring performance. The load characteristics used for non-intrusive load monitoring can be divided into three categories, namely steady-state characteristics, transient characteristics, and non-traditional characteristics. The specific classifications are shown in Fig. 1.9.

The steady-state characteristics are the characteristics of the equipment in various stable working conditions. Steady-state characteristics include changes in power (including active, reactive, and apparent power), steady-state voltage and current waveforms, V-I trajectory (Voltage-current trajectory), steady-state current and voltage harmonics, etc. Among them, power change is the most commonly used time-domain steady-state feature. Hart et al. utilized the change of normalized average power for event detection and clustered processing to analyze the change of equipment state (Hart 1992). Bhattacharjee et al. used a decision tree to classify devices by setting thresholds for active power (Bhattacharjee et al. 2014). To reduce the cost of monitoring, Chou et al. used the Gaussian Mixture Model (GMM) combined with the Sequential Expectation-Maximization (SEM) algorithm to construct the equipment state classifier based on the power signal collected at low speed (Chou and Chang 2013). The load decomposition method is proposed by Egarter et al., and they

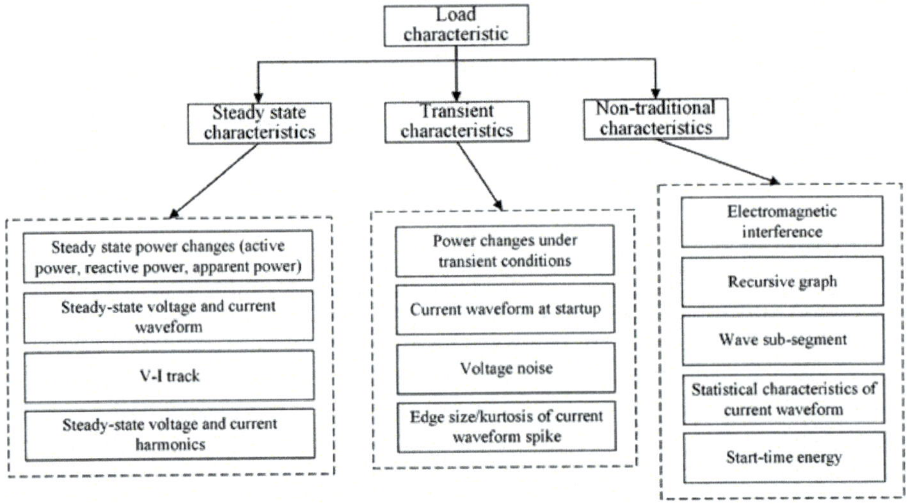

Fig. 1.9 Taxonomy of load features

take the collected power data as input and use a particle filter algorithm to approximate the probability density function of the device state (Egarter et al. 2014). There are also many documents that perform certain conversion processing on the power signal, and use the new features obtained for load classification and decomposition. For example, Figueiredo et al. regard load decomposition as a single-channel signal source separation problem, and use nonnegative tensor factorization (NTF) to obtain the most relevant features for signal separation (Figueiredo et al. 2013). Gabaldón et al. performed a Hilbert transform on the collected total power signal to obtain suitable load decomposition characteristics (Gabaldón et al. 2014).

Using a single feature for load decomposition may overlap the features of different devices. Therefore, multiple features are usually combined in a certain way. For example, the literature treats the load decomposition task as a multi-variable time series problem for processing, and obtains decomposition performance better than a single variable (Wang et al. 2013; Kramer et al. 2015; Wang and Zheng 2011). Basu et al. extract specific statistics from the collected data, such as the duty cycle of the equipment operating state, the variance of power fluctuations in a specific state, etc., for load identification and decomposition (Lu et al. 2012; Basu et al. 2012). Tina et al. obtained the load curve of each device in the family through time-domain transformation, and comprehensively used multiple features for load decomposition (Tina et al. 2014). The V-I trace is a plane figure composed of current and voltage waveforms. Du et al. designed a load decomposition algorithm using the shape of the V-I trajectory as a load feature (Du et al. 2015; Wang et al. 2018; Baptista et al. 2018). Hassan et al. evaluated a scheme that combines V-I trajectory-based load identification with identification algorithms based on other features. In the end, it comes to dynamic, noise, and very similar load equipment (Hassan et al. 2013). The combination scheme can provide better overall prediction accuracy and reliability. Before using the principal component analysis and k-nearest neighbor algorithm, Yang et al. cropped the graphic template of the VI trajectory and reduced the image pyramid, designed and implemented a set of multi-stage load classification algorithms, and verified the effectiveness of the algorithm on the actual data set (Yang et al. 2019a).

In addition to time-domain features, many researchers also use fast Fourier transform to convert time-domain data such as current into frequency-domain features, and combine these frequency-domain features with other time-domain features for load decomposition (Wild et al. 2015; Nguyen et al. 2017). Although the fast Fourier transform can well characterize the frequency characteristics of the signal, it loses part of the time domain information and cannot see the change of frequency with time. To solve this problem, time-frequency analysis methods can be used, such as using discrete wavelet transform or short-time Fourier transformation to obtain load characteristics (Chang et al. 2013a; Su et al. 2011). Among them, Chang et al. used Parseval's theorem to calculate the power spectrum of wavelet transform coefficients of different scales, which efficiently characterized the on/off transient characteristics of the load (Chang et al. 2013a). Chen et al. used convolution and Wavelet Multi-Resolution Analysis (WMRA) to extract new power features from the original power waveform, and used Parseval's theorem to obtain the Power Index (PI) from it. Finally, they complete the task of load identification through the power index (Chen et al. 2013). Su et al. analyzed the feature extraction performance of wavelet transform and short-time Fourier transform through comparative experiments, and concluded that the transient features obtained by wavelet transform have better performance for load identification (Su et al. 2011). In addition, there are also studies using S transform for load identification (Jimenez et al. 2014). Borin et al. performed Stockwell transformation on the collected current data, and then used the converted time-frequency domain data as load characteristics. They use the vector projection method in the field of image pattern recognition for load identification (Borin et al. 2015).

The transient feature is the feature information collected from the process of device state switching. For example, the power change in the transient process, the starting current waveform, the voltage noise, the edge size or kurtosis of the current waveform spike, etc. can all be used as load characteristics (Cole and Albicki 1998). Most transient features are obtained indirectly through transformation techniques such as Fourier transform (Patel et al. 2007). Compared with steady-state characteristics, transient characteristics are more relevant to the characteristics of the device itself, so it can better identify the device (Norford and Leeb 1996). Further, the steady-state and transient characteristics of the load can be combined for load identification and decomposition, so that the advantages of both can be fully utilized (Meehan et al. 2014). Meehan et al. use Committee decision mechanisms (CDMs) to integrate multiple steady-state and transient characteristics and simultaneously use them for load decomposition, which improves the performance of load decomposition (Meehan et al. 2014).

In addition to the above steady-state and transient characteristics, some non-traditional load characteristics can also be used for load decomposition to achieve good decomposition results. Gulati et al. studied the reliability and feasibility of using high-frequency Electro Magnetic Interference (EMI) for load identification through simulation and experiment (Gulati et al. 2014). Popescu et al. use Recurrence plot analysis (RPA) as an analysis tool for nonlinear load current waveforms to extract load characteristics required for load identification (Popescu et al. 2014). Patri et al. use an algorithm based on Shapelets to distinguish the state changes of different loads, so as to realize the load decomposition (Patri et al. 2014). Du et al. converted the load waveform into a quantified state sequence to construct a current waveform Finite State Machine (FSM), and realized the load identification by analyzing the statistical characteristics of the state sequence (root mean square value and residence time) (Du et al. 2014). Since the device's turn-on transient process contains a lot of characteristics related to the characteristics of the device itself, Chang et al. used the turn-on energy as the load characteristic to design the load identification algorithm (Chang et al. 2011). To extract the most significant features for load decomposition from the collected data, Ahmadi et al. obtained a combination of features for load decomposition by performing the Principal Component Analysis (PCA) on current and voltage waveforms (Ahmadi and Martı 2015).

1.3.4 Load Identification Module

After collecting equipment data and determining the characteristics used for load decomposition, the decomposition model can be embedded. Then select the optimal model by designing specific learning strategies and algorithms. Finally, the extracted features are input into the model for reasoning to realize the identification of the load type, the working state, and the estimation of the corresponding power. After years of development of artificial intelligence technology, a large number of intelligent models have been derived that can be used for embedded recognition of power equipment. At present, the models and algorithms of load decomposition can be divided into two categories: methods based on combinatorial optimization and methods based on pattern recognition. The taxonomy of the load disaggregation models and algorithms are shown in Fig. 1.10.

The method based on combination optimization tries to match the observed power measurement value with the possible combination of the device power signal in the device feature library to reduce the matching error as an optimization strategy (Du et al. 2010). The load identification method based on template filter constructs the feature filter by obtaining the feature of each load prior. On this basis, they generate a template filter through 0-1 processing, and use the template filter to filter the collected mixed signals respectively. Then judge the load type and operating status according to the filtering result (Wu et al. 2017). The principle of this kind of method is intuitive, and it has a better decomposition effect for the situation where the number of equipment is small and the equipment characteristics are known, but there are some problems. For example, the essence of the model is NP-complete, and it is difficult to improve the optimization efficiency. This method assumes that the characteristics of each device are required to meet the superimposition, but not

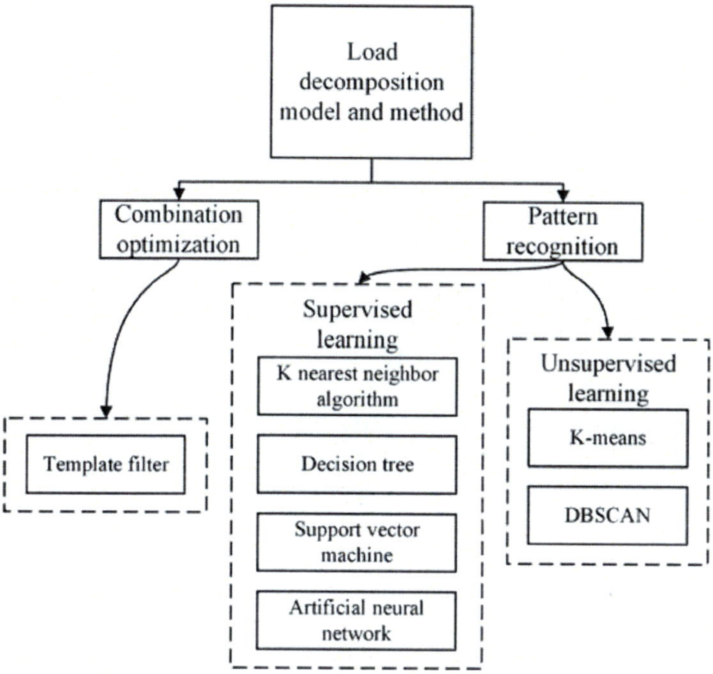

Fig. 1.10 Taxonomy of the load disaggregation models and algorithms

all load characteristics meet this condition. To achieve accurate identification, this method requires a complete device feature library, which is often difficult to meet in actual application scenarios (Cheng et al. 2016).

The method based on pattern recognition directly learns the characteristic pattern of the device from the data, so as to realize the identification and decomposition of the load. According to whether the label information is used in the learning process, two types of supervised learning algorithms and unsupervised learning algorithms can be further divided.

- Supervised learning
  In addition to the total load data, the supervised learning algorithm also needs labeled data such as the status information of each device during the training phase. These tag data can be collected with the assistance of the user or can be obtained by installing independent monitoring devices for each device. K-Nearest Neighbor (KNN), Decision Tree (DT), Support Vector Machine (SVM), Artificial Neural Network (ANN), etc. are all commonly used models in load decomposition (Figueiredo et al. 2012; Nguyen et al. 2015; Su et al. 2016; Chang et al. 2015). Nguyen et al. input the collected active power, reactive power, and apparent power into the decision tree to classify equipment (Nguyen et al. 2015). Through experiments, they found that using the variation of each power component as the input of the decision tree can obtain higher classification accuracy. Lin et al. designed a load decomposition method integrated with ANN and Particle

Swarm Optimization (PSO) (Lin and Hu 2018). Among them, the ANN is used for load identification, the PSO is used to search for the optimal ANN parameters, and the test is performed on actual residential data, and finally an accuracy of more than 90% is obtained.

- Unsupervised learning
  Unsupervised algorithms do not need labels for model training, but directly mine the similarity of features from the data. For example, clustering methods such as K-means and Density-Based Spatial Clustering of Applications with Noise (DBSCAN) are used to extract load characteristics and identify equipment. Yang et al. use data processing tools such as Hilbert transform to pair device switching events (Yang et al. 2015). Some researchers also regard load decomposition as a blind signal separation problem (Gabaldón et al. 2014). They embed domain knowledge into the decomposition process in a certain way to achieve an effective improvement in decomposition performance (Huang et al. 2019a). Unsupervised algorithms meet the needs of many application scenarios where label data is not easy to obtain in field environments, but their accuracy is usually lower than supervised algorithms. Using an ensemble learning method to combine multiple different classifiers or utilizing a multi-label classification algorithm to comprehensively use multiple load features for classification can further improve the accuracy and generalization performance of the load decomposition algorithm (Wu et al. 2019). For example, Liu et al. designed a random forest ensemble classifier for the imbalanced problem of load classification, and verified the performance of this method on the Plug Load Appliance Identification Dataset (PLAID) (Liu et al. 2019). Considering that multiple independent loads may have state changes at the same time, Li et al. proposed three graph-based semi-supervised multi-label load monitoring algorithms, and evaluated the performance of the proposed algorithms on five data sets (Li and Dick 2018).

With the support of smart device recognition technology, UEIOT has shown unique advantages in perception interaction, data sharing, communication, automation, etc. After intelligent load identification and prediction, the massive devices flooding into cyberspace can be accurately sensed. Effective data mining can realize the independent identification of electrical appliance operating status and use energy consumption management. So as to formulate personalized intervention strategies and provide scientific guidance to guide users to use electricity more reasonably.

1.4 Different Strategies for Smart Device Recognition

After years of technological innovation and development, smart device identification technology has derived multiple branches. We use the Web of Science database to detect references about non-intrusive load monitoring and smart device identification. Taking into account several commonly used expressions for device recognition, the topic detection is defined as "TS = (Non-intrusive load disaggregation OR

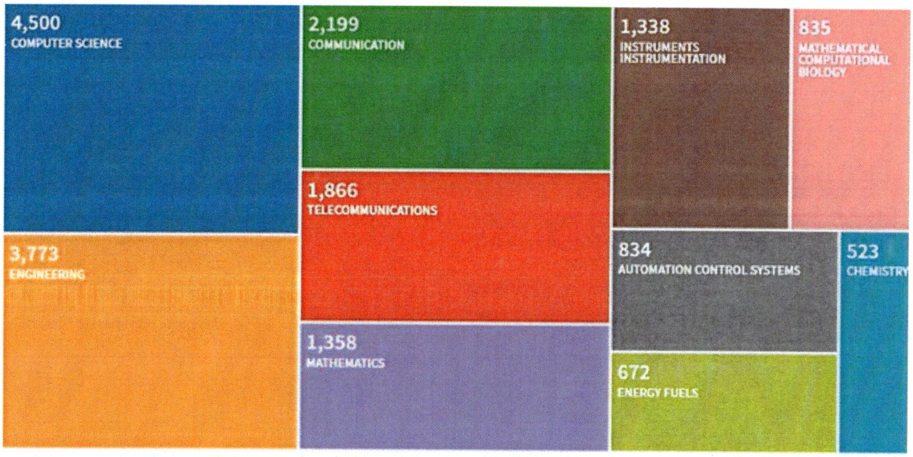

Fig. 1.11 Research direction map of documents based on the subject retrieval of "TS = (Non-intrusive load disaggregation OR Non-intrusive load monitoring OR Non-intrusive Appliance Load Monitoring OR Smart device recognition)"

Non-intrusive load monitoring OR Non-intrusive Appliance Load Monitoring OR Smart device recognition)". As of October 12, 2020, a total of 5,952 highly relevant scientific articles can be found, including 35 highly cited documents and 1 hot paper. In Fig. 1.11, among the 5952 documents retrieved, the top 10 research directions are computer science, engineering, communication, telecommunications, mathematics, instruments instrumentation, mathematical computational biology, automation control systems, energy fuels, and chemistry. It can be seen that smart device recognition is an interdisciplinary subject in many fields such as computer science, engineering, mathematics, and energy. The combination of traditional engineering technology and new generation artificial intelligence technology provides guidance for the development of smart device technology from both theoretical and practical aspects.

It is worth mentioning that there are many different recognition strategies in smart device recognition technology. The representative smart device recognition strategies are clustering strategies, optimizing strategies, ensemble strategies, and deep learning strategies.

1.4.1 Clustering Strategies for Device Recognition

When Professor Hart first proposed the concept of NILM, the clustering method was used to identify electrical points on the P-Q plane (Hart 1992). The research work of load clustering mainly includes the determination of the number of clusters, the analysis, and evaluation of clustering performance, the selection of the initial value of the centroid, and the effective convergence of the algorithm. Some scholars have

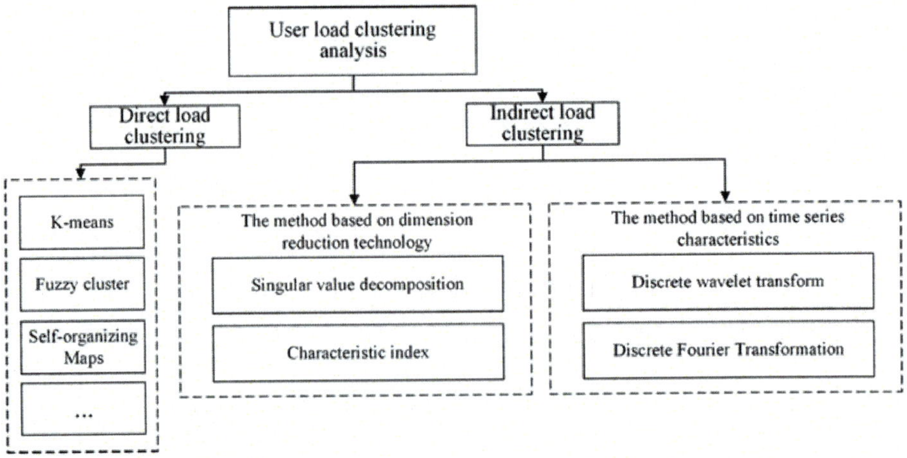

Fig. 1.12 Methods of load clustering

studied the factors that affect load clustering, such as distance calculation methods, and the impact of data dimensionality reduction. Generally, there are two clustering methods, one is direct and the other is indirect. The specific classification results of load clustering are shown in Fig. 1.12.

The direct method performs a clustering algorithm based on the load curve of the electric user. However, due to the large dimension and quantity of the original load data, the requirements for calculation and storage are relatively high during the clustering process. The indirect method is to first process the original load data to reduce its complexity. There are usually two processing methods, one of which is to reduce the complexity of the original load data through dimensionality reduction technology (Varga et al. 2015). Another method is to process the load in the time or frequency domain (Teeraratkul et al. 2018).

The clustering method is a typical unsupervised algorithm. Its basic function is to divide a given data set into homogeneous groups based on the inherent similarity of data points. Objects in the same cluster are more similar to each other, while objects in different clusters have a very little similarity. Clustering methods can be roughly divided into different methods, including partition, hierarchical, density-based, grid-based, model-based, and so on. Several common clustering algorithms are shown in Table 1.1.

1.4.2 *Optimizing Strategies for Device Recognition*

With the progress of smart grids, there are more and more types of electric loads for residential users. As an important part of electric load, the energy conservation method has attracted extensive attention from society. After years of development of

Table 1.1 Common clustering algorithms

Segmentation methods	Hierarchical methods	Density-based methods	Other methods
K-means	Balanced iterative reducing and clustering using hierarchies	Density-Based spatial clustering of applications with noise	Spectral clustering
K-medoids	ROCK clustering	Optics clustering	Gaussian mixture model

intelligent algorithms, some researchers use optimized algorithms for load identification research. Chang et al. utilized the PSO algorithm to optimize the training parameters of the artificial neural network, thereby improving the accuracy and computational efficiency of the load identification model (Chang et al. 2013b). Researchers in the NILM field have also explored many other optimization models, such as genetic algorithm (Hock et al. 2018), sparse optimization (Piga et al. 2016), ant colony optimization (Dorigo et al. 2006), and grey wolf optimization (LV et al. 2019). These single-objective optimization algorithms can achieve certain results, but they cannot obtain the best results in many aspects. Later, after years of technological innovation, many multi-objective optimization algorithms are used for load identification of smart devices. These multi-objective optimization algorithms include the Non-Dominated Sorting Genetic Algorithm II (NSGA-II), Multi-Objective Particle Swarm Optimization (MOPSO), and Multi-Objective Grey Wolf Optimization (MOGWO) algorithm (Liu et al. 2020). The multi-objective optimization algorithm can take into account multiple load objectives that need to be optimized, so as to improve the accuracy of equipment identification in all directions.

1.4.3 Ensemble Strategies for Device Recognition

In the early development stage of smart device identification technology, many scholars proposed effective load identification methods. With the change of technology, research has found that load identification methods based on integrated strategies can further improve the accuracy of equipment identification and the universality of models (Liu et al. 2019). As an important branch of machine learning in AI, ensemble learning has been attracting attention. Ensemble learning refers to the reasonable combination of several weak classifiers to form a much stronger classifier. In the process of ensemble learning, the necessity is to guarantee the diversity of individual parts. According to the sources of diversity, ensemble learning can be categorized as data diversity and learner diversity. Among them, data diversity methods are represented by Bagging (Breiman 1996) and Boosting (Hastie et al. 2009). They obtain different datasets by re-sampling the data and other methods to train classification models. Learner diversity methods mainly adopt heterogeneous

Table 1.2 Intelligent equipment identification method based on ensemble strategy

Ensemble strategy	
Optimization weighting strategy	Boost strategy
Multi-objective evolutionary algorithms (MOEAs) (Li et al. 2019)	Bagging, boosting, and random committee method (Shahriar and Rahman 2013)
Grey Wolf Optimizer (GWO) (LV et al. 2019)	Adaptive Boosting (AdaBoost) (Ren et al. 2019)

classifiers to generate diversity. Different classifiers have diverse performances when tackling different datasets. Nevertheless, the application of ensemble learning in load identification is relatively rare. For instance, (Sun et al. 2020) developed an ensemble clustering-based framework for household load profiling and driven factors identifications. (Himeur et al. 2020) proposed a non-intrusive appliance recognition system based on multiscale wavelet packet tree and ensemble bagging tree classifier, which showed average accuracies of 97.01% and 96.36% for the GREEND and REDD real datasets.

At present, common ensemble strategies in device recognition technology include ensemble based on optimized weights and ensemble methods based on boost strategies. The specific methods are provided as shown in Table 1.2.

1.4.4 Deep Learning Strategies for Device Recognition

As the core technology that directly promotes the progress of machine learning, the deep learning has become the focus of research from all walks of life. This method has an obvious effect on the field of device identification. Deep learning is derived from and developed from traditional machine learning, and its essence is a neural network with multiple hidden layers. It is applied to the hidden layer nodes of the multi-layer network through nonlinear transformation, and integrates low-level features to form high-level abstract features, thereby realizing the expression of data features. In the fields of image recognition and natural language processing, deep learning has been relatively mature applications. Its application frameworks include Keras, TensorFlow, Caffe, Theano, etc. (Jia et al. 2014; Chollet 2018; Jakhar and Hooda 2018; Bergstra et al. 2010; Abadi 2016). The network structure of deep learning is different due to the number of network layers, weight sharing, and edge characteristics.

The proposal and progress of deep learning solve the problem that traditional neural networks tend to converge to local minimums (Hinton et al. 2006). The main features of deep learning can be summarized as follows: massive sample data, network structure with multiple hidden layers, and excellent target data feature extraction ability. The emergence of deep learning is mainly affected by two aspects. On the one hand, due to the large-scale accumulation of data caused by the development of communication and measurement technology, it is difficult for humans to intuitively find the characteristic rules. On the other hand, due to the increase in computing power brought about by the rise of parallel and heterogeneous computing such as large-scale computing clusters and graphics processing units (GPU), the machine's ability to process data has made a great leap. In the research of power system equipment identification methods, the widely used deep learning models mainly include stacked auto-encoder Stacked Auto-Encoder (SAE) (Li et al. 2016b), Deep Belief Network (DBN) (Hua et al. 2015), Recurrent Neural Network (Zheng et al. 2019), Convolutional Neural Network (CNN) (LeCun et al. 1998), Dilated Residual Network (DRN), etc. (Xia et al. 2019). The characteristics and applications of each model are shown in Table 1.3.

From the concept of a strong smart grid to UEIOT, the focus of electricity has been shifted to data value mining at the sensor level. The UEIOT is different from traditional electric grids in that it acquires system information based on massive advanced sensors and collectors to promote flexible interactions in all aspects of the system. The UEIOT includes a four-layer structure of the perception layer, network layer, platform layer, and application layer. The multi-type sensors and intelligent IoT terminals at the perception layer realize the maximum perception of the system and the dissemination and collection of information. The network layer transmits and shares information and data collected by the perception layer. So as to fully improve the intrinsic value of mining data and increase the utilization rate of data. The platform layer realizes the storage, screening, cleaning, and mining of the information transmitted by the network layer, and provides a solid data foundation for the application layer. The application layer uses the data information of the platform layer to realize the application development of UEIOT, improve the energy structure, and enhance the comprehensive utilization efficiency of social energy. Deep learning is a key technology that solves UEIOT's large-scale data processing, data knowledge mining, and promotes the transparent perception and optimization of the entire system. Focus on the specific technical embodiment of deep learning in the

Table 1.3 Features and applications of several typical deep learning models in UEIOT

Model	Advantage	Disadvantage	Application occasion
SAE	Strong perception	Limited by the feedback mechanism	Power equipment fault diagnosis/Image speech recognition
DBN	Transfer learning features	Limited learning depth	Power system transient stability assessment
RNN	Realize complex nonlinear mapping	Prone to gradient dispersion	Power big data fusion and detection
CNN	Strong generalization ability	Higher requirements for GPU and samples	Image recognition of power equipment
DRN	Strong knowledge expression ability	Model verification is complex	Short-term power load forecasting and optimization classification

UEIOT application, which mainly lies in data processing, edge computing, and situational awareness technology. It realizes the interconnection among power users, power companies, and equipment terminals, and generates shared data. At the same time, it also strives to make use of the platform function and sharing function of the grid to create higher service value. The data processing technology is mainly applied to the network layer and the platform layer. The edge computing is mainly applied to the perception layer. The situational awareness technology is mainly applied to the platform layer and application layer.

1.5 Scope of the Book

The smart device identification technology in the context of UEIOT provides a reliable guarantee for energy consumption information management. This book focuses on smart non-intrusive device identification technology in the field of smart grids. It comprehensively introduces a variety of smart algorithms and gives future potential application directions. The book consists of 9 chapters, and the organization structure of each chapter is provided as follows:

Chapter 1: Introduction

This chapter first gives an overview of the UEIOT and discusses its key technologies. Then the application of smart device recognition technology in UEIOT is deeply analyzed. Finally, the key strategies in the smart device recognition technology are summarized.

Chapter 2: Smart non-intrusive device recognition based on physical methods

This chapter first introduces the application of physical characteristics in device identification. Then, the performance of the decision tree algorithm, template matching algorithm, and the current decomposition algorithm is studied in this field. Finally, the performance improvement of the classification algorithm based on the optimization algorithm is studied.

Chapter 3: Smart non-intrusive device recognition based on intelligent single-label classification methods

This chapter first introduces the smart non-intrusive device identification technology in the context of smart single-label. Then a number of comparative experiments are conducted to verify the performance differences of Support Vector Machine (SVM), Extreme Learning Machine (ELM), and Artificial Neural Network (ANN) in smart device recognition.

Chapter 4: Smart non-intrusive device recognition based on intelligent multi-label classification methods

This chapter mainly introduces the application procedure of these multi-label models in non-intrusive device recognition. The most commonly used models, including the Ranking Support Vector Machine (Ranking SVM), Multilabel K-Nearest Neighbors (MLKNN), and Networks Multilabel Learning (BPMLL), are evaluated on a popular public dataset named REDD dataset. Moreover, some comparison experiments are conducted and the comparative analysis is also given in the experiment part.

Chapter 5: Smart Non-Intrusive Device Recognition Based on Intelligent Clustering Methods

This chapter first gives an introduction of two intelligent clustering methods as well as their application in non-intrusive device recognition. Experiments are conducted to evaluate and compare the clustering validity of different clustering algorithms.

Chapter 6: Smart non-intrusive device recognition based on intelligent optimization methods

This chapter first introduces some current researches of optimization-based methods, and then several experiments based on NSGA-II, MOPSO, and MOGWO are carried out to evaluate the model performance of the intelligent optimization methods.

Chapter 7: Smart non-intrusive device recognition based on ensemble methods

This chapter analyzes two common ensemble methods and introduces their methodologies. For an optimized weighting strategy, several single-objective and multi-objective algorithms are tested. For boosting strategies, Adaptive Boosting (AdaBoost) and Linear Programming Boosting (LPBoost) are investigated.

Chapter 8: Smart non-intrusive device recognition based on deep learning methods

This chapter discusses the device recognition methods based on deep learning models and experiments are designed to verify the effectiveness of the algorithms. The load sequence-based and the graph processing-based device recognition algorithms are evaluated separately.

Chapter 9: Potential Applications of Smart Device Recognition in Industry

This chapter first gives a brief overview of the importance of non-invasive smart device recognition technology, then describes in detail the application of this technology in power and energy, complex electromechanical systems, and environmental pollution monitoring. Finally, the application of smart devices recognition in other aspects is expanded.

References

Abadi, M. (2016). TensorFlow: learning functions at scale. In *Proceedings of the 21st ACM SIGPLAN International Conference on Functional Programming* (pp. 1–1).

Ahmadi, H., & Martı, J. R. (2015). Load decomposition at smart meters level using eigenloads approach. *IEEE Transactions on Power Systems, 30*(6), 3425–3436.

Alcalá, J., Ureña, J., Hernández, Á., & Gualda, D. (2017). Event-based energy disaggregation algorithm for activity monitoring from a single-point sensor. *IEEE Transactions on Instrumentation and Measurement, 66*(10), 2615–2626. https://doi.org/10.1109/TIM.2017.2700987.

Arif, A., Wang, Z., Wang, J., Mather, B., Bashualdo, H., & Zhao, D. (2017). Load modeling—A review. *IEEE Transactions on Smart Grid, 9*(6), 5986–5999.

Ashton, K. (2009). That 'internet of things' thing. *RFID Journal, 22*(7), 97–114.

Baptista, D., Mostafa, S. S., Pereira, L., Sousa, L., & Morgado-Dias, F. (2018). Implementation strategy of convolution neural networks on field programmable gate arrays for appliance classification using the voltage and current (VI) trajectory. *Energies, 11*(9), 2460.

Basu, K., Debusschere, V., & Bacha, S. (2012). Load identification from power recordings at meter panel in residential households. In *2012 th International Conference on Electrical Machines* (pp 2098–2104). IEEE.

Basu, K., Debusschere, V., Douzal-Chouakria, A., & Bacha, S. (2015). Time series distance-based methods for non-intrusive load monitoring in residential buildings. *Energy and Buildings, 96,* 109–117. https://doi.org/10.1016/j.enbuild.2015.03.021.

Bauerle, F., Miller, G., Nassar, N., Nassar, T., & Penney, I. (2014). Context sensitive smart device command recognition and negotiation. In *International internet of things summit* (pp. 314–330). Springer.

Bergstra, J., Breuleux, O., Bastien, F., Lamblin, P., Pascanu, R., Desjardins, G., ..., Bengio, Y. (2010). Theano: a CPU and GPU math expression compiler. In *Proceedings of the Python for Scientific Computing Conference (SciPy)* (vol. 3, pp. 1–7). Austin, TX.

Bezdek, J. C. (1998). Computational intelligence defined-By everyone! In *Computational intelligence: Soft computing and fuzzy-neuro integration with applications* (pp. 10–37). Springer.

Bhattacharjee, S., Kumar, A., & RoyChowdhury, J. (2014). Appliance classification using energy disaggregation in smart homes. In *2014 International Conference on Computation of Power, Energy, Information and Communication (ICCPEIC)* (pp. 1–6). IEEE.

Borgia, E. (2014). The internet of things vision: Key features, applications and open issues. *Computer Communications, 54,* 1–31.

Borin, V., Barriquello, C., & Campos, A. (2015). Approach for home appliance recognition using vector projection length and Stockwell transform. *Electronics Letters, 51*(24), 2035–2037.

Breiman, L. (1996). Bagging predictors. *Machine Learning, 24*(2), 123–140.

Chang, H.-H., Chen, K.-L., Tsai, Y.-P., & Lee, W.-J. (2011). A new measurement method for power signatures of nonintrusive demand monitoring and load identification. *IEEE Transactions on Industry Applications, 48*(2), 764–771.

Chang, H.-H., Lee, M.-C., Chen, N., Chien, C.-L., & Lee, W.-J. (2015). Feature extraction based hellinger distance algorithm for non-intrusive aging load identification in residential buildings. In *2015 IEEE Industry Applications Society Annual Meeting* (pp. 1–8). IEEE.

Chang, H.-H., Lian, K.-L., Su, Y.-C., & Lee, W.-J. (2013a). Power-spectrum-based wavelet transform for nonintrusive demand monitoring and load identification. *IEEE Transactions on Industry Applications, 50*(3), 2081–2089.

Chang, H. -H., Lin, C. -L., & Lee, J. -K. (2010). Load identification in nonintrusive load monitoring using steady-state and turn-on transient energy algorithms. In *The 2010 14th International Conference on Computer Supported Cooperative Work in Design* (pp. 27–32). IEEE.

Chang, H.-H., Lin, L.-S., Chen, N., & Lee, W.-J. (2013b). Particle-swarm-optimization-based nonintrusive demand monitoring and load identification in smart meters. *IEEE Transactions on Industry Applications, 49*(5), 2229–2236.

Chen, G., Chen, J., Zi, Y., Pan, J., & Han, W. (2018). An unsupervised feature extraction method for nonlinear deterioration process of complex equipment under multi dimensional no-label signals. *Sensors and Actuators, A: Physical, 269,* 464–473.

Chen, H., Wang, X., Li, Z., Chen, W., & Cai, Y. (2019). Distributed sensing and cooperative estimation/detection of ubiquitous power internet of things. *Protection and Control of Modern Power Systems, 4*(1), 13.

Chen, K. -L., Chang H.-H., & Chen, N. (2013). A new transient feature extraction method of power signatures for nonintrusive load monitoring systems. In *2013 IEEE International Workshop on Applied Measurements for Power Systems (AMPS)* (pp. 79–84). IEEE.

Chen, M., Mao, S., & Liu, Y. (2014). Big data: A survey. *Mobile networks and applications, 19*(2), 171–209.

Chen, Q. (2019). Research on implementation strategy of ubiquitous power internet of things. *Power generation technology, 40*(2), 99–106.

Cheng, X., Li, L., Wu, H., Ding, Y., Song, Y., & Sun, W. (2016). A survey of the research on non-intrusive load monitoring and disaggregation. *Power System Technology, 40*(10), 3108–3117.

Chicco, G., & Ilie, I.-S. (2009). Support vector clustering of electrical load pattern data. *IEEE Transactions on Power Systems, 24*(3), 1619–1628.

Chollet, F. (2018). *Deep Learning mit Python und Keras: Das Praxis-Handbuch vom Entwickler der Keras-Bibliothek*. KG: MITP-Verlags GmbH & Co.

Chou, P.-A., & Chang, R.-I. (2013). Unsupervised adaptive non-intrusive load monitoring system. In *2013 IEEE International Conference on Systems, Man, and Cybernetics* (pp. 3180–3185). IEEE.

Cole, A. I., & Albicki, A. (1998). Data extraction for effective non-intrusive identification of residential power loads. In *IMTC/98 Conference Proceedings. IEEE Instrumentation and Measurement Technology Conference. Where Instrumentation is Going* (Cat. No. 98CH36222) (pp. 812–815). IEEE.

Dorigo, M., Birattari, M., & Stutzle, T. (2006). Ant colony optimization. *IEEE Computational Intelligence Magazine, 1*(4), 28–39.

Du, L., He, D., Harley, R. G., & Habetler, T. G. (2015). Electric load classification by binary voltage–current trajectory mapping. *IEEE Transactions on Smart Grid, 7*(1), 358–365.

Du, L., Yang, Y., He, D., Harley, R. G., & Habetler, T. G. (2014). Feature extraction for load identification using long-term operating waveforms. *IEEE Transactions on Smart Grid, 6*(2), 819–826.

Du, Y., Du, L., Lu, B., Harley, R., & Habetler, T. (2010). A review of identification and monitoring methods for electric loads in commercial and residential buildings. In *2010 IEEE Energy Conversion Congress and Exposition* (pp. 4527–4533). IEEE.

Egarter, D., Bhuvana, V. P., & Elmenreich, W. (2014). PALDi: Online load disaggregation via particle filtering. *IEEE Transactions on Instrumentation and Measurement, 64*(2), 467–477.

Esa, N. F., Abdullah, M. P., & Hassan, M. Y. (2016). A review disaggregation method in Non-intrusive Appliance Load Monitoring. *Renewable and Sustainable Energy Reviews, 66,* 163–173.

Figueiredo, M., De Almeida, A., & Ribeiro, B. (2012). Home electrical signal disaggregation for non-intrusive load monitoring (NILM) systems. *Neurocomputing, 96,* 66–73.

Figueiredo, M., Ribeiro, B., & de Almeida, A. (2013). Electrical signal source separation via nonnegative tensor factorization using on site measurements in a smart home. *IEEE Transactions on Instrumentation and Measurement, 63*(2), 364–373.

Gabaldón, A., Ortiz-García, M., Molina, R., & Valero-Verdú, S. (2014). Disaggregation of the electric loads of small customers through the application of the Hilbert transform. *Energy Efficiency, 7*(4), 711–728.

Gohil, A., Modi, H., & Patel, S. K. (2013). 5G technology of mobile communication: A survey. In *2013 international conference on intelligent systems and signal processing (ISSP)* (pp. 288–292). IEEE.

Gulati, M., Ram, S. S., & Singh, A.(2014). An in depth study into using EMI signatures for appliance identification. In *Proceedings of the 1st ACM Conference on Embedded Systems for Energy-efficient Buildings* (pp. 70–79).

Hart, G. W. (1992). Nonintrusive appliance load monitoring. *Proceedings of the IEEE, 80*(12), 1870–1891.

Hassan, T., Javed, F., & Arshad, N. (2013). An empirical investigation of VI trajectory based load signatures for non-intrusive load monitoring. *IEEE Transactions on Smart Grid, 5*(2), 870–878.

Hastie, T., Rosset, S., Zhu, J., & Zou, H. (2009). Multi-class adaboost. *Statistics and its Interface, 2*(3), 349–360.

Himeur, Y., Alsalemi, A., Bensaali, F., & Amira, A. (2020). Robust event-based non-intrusive appliance recognition using multi-scale wavelet packet tree and ensemble bagging tree. *Applied Energy, 267,* 114877. https://doi.org/10.1016/j.apenergy.2020.114877.

Hinton, G. E., Osindero, S., & Teh, Y.-W. (2006). A fast learning algorithm for deep belief nets. *Neural Computation, 18*(7), 1527–1554.

Hock, D., Kappes, M., & Ghita, B. (2018). Non-intrusive appliance load monitoring using genetic algorithms. *IOP Conference Series: Materials Science and Engineering, 366,* 012003. https://doi.org/10.1088/1757-899x/366/1/012003.

Hou, R., Pan, M., Zhao, Y., & Yang, Y. (2019). Image anomaly detection for IoT equipment based on deep learning. *Journal of Visual Communication and Image Representation, 64,* 102599.

Hua, Y., Guo, J., & Zhao, H. (2015). Deep belief networks and deep learning. In *Proceedings of 2015 International Conference on Intelligent Computing and Internet of Things* (pp. 1–4). IEEE.

Huang, Y.-T., Shi, J., Xu, J.-M., & Xu, B. (2019a). Research advances and perspectives on the cocktail party problem and related auditory models. *Acta Automatica Sinica, 45*(2), 234–251.

Huang, Y., Zhan, J., Luo, C., Wang, L., Wang, N., Zheng, D., et al. (2019b). An electricity consumption model for synthesizing scalable electricity load curves. *Energy, 169,* 674–683.

Jakhar, K., & Hooda, N. (2018). Big data deep learning framework using Keras: A case study of pneumonia prediction. In *2018 4th International Conference on Computing Communication and Automation (ICCCA)* (pp. 1–5). IEEE.

Jia, M., Komeily, A., Wang, Y., & Srinivasan, R. S. (2019). Adopting Internet of Things for the development of smart buildings: A review of enabling technologies and applications. *Automation in Construction, 101,* 111–126.

Jia, Y., Shelhamer, E., Donahue, J., Karayev, S., Long, J., Girshick, R., Guadarrama, S., & Darrell, T. (2014). Caffe: Convolutional architecture for fast feature embedding. In *Proceedings of the 22nd ACM international conference on Multimedia* (pp. 675–678).

Jimenez, Y.., Duarte, C., Petit, J., & Carrillo, G. (2014). Feature extraction for nonintrusive load monitoring based on S-Transform. In *2014 Clemson university power systems conference* (pp. 1–5). IEEE.

Konar, A. (2006). Computational intelligence: Principles, techniques and applications. Springer Science & Business Media.

Kramer, O., Klingenberg, T., Sonnenschein, M., & Wilken, O. (2015). Non-intrusive appliance load monitoring with bagging classifiers. *Logic Journal of the IGPL, 23*(3), 359–368.

LeCun, Y., Bottou, L., Bengio, Y., & Haffner, P. (1998). Gradient-based learning applied to document recognition. *Proceedings of the IEEE, 86*(11), 2278–2324.

Leeb, S. B., Shaw, S. R., & Kirtley, J. L. (1995). Transient event detection in spectral envelope estimates for nonintrusive load monitoring. *IEEE Transactions on Power Delivery, 10*(3), 1200–1210. https://doi.org/10.1109/61.400897.

Li, D., & Dick, S. (2018). Residential household non-intrusive load monitoring via graph-based multi-label semi-supervised learning. *IEEE Transactions on Smart Grid, 10*(4), 4615–4627.

Li, L., Yang, L., Chen, H., Li, M., & Zhang, C. (2019). Multi-objective evolutionary algorithms applied to non-intrusive load monitoring. *Electric Power Systems Research, 177,* 105961. https://doi.org/10.1016/j.epsr.2019.105961.

Li, R., Li, F., & Smith, N. D. (2016a). Multi-resolution load profile clustering for smart metering data. *IEEE Transactions on Power Systems, 31*(6), 4473–4482.

Li, W., Fu, H., Yu, L., Gong, P., Feng, D., Li, C., et al. (2016b). Stacked Autoencoder-based deep learning for remote-sensing image classification: A case study of African land-cover mapping. *International Journal of Remote Sensing, 37*(23), 5632–5646.

Lin, Y.-H., & Hu, Y.-C. (2018). Electrical energy management based on a hybrid artificial neural network-particle swarm optimization-integrated two-stage non-intrusive load monitoring process in smart homes. *Processes, 6*(12), 236.

Liu, H., Wu, H., & Yu, C. (2019). A hybrid model for appliance classification based on time series features. *Energy and Buildings, 196,* 112–123.

Liu, H., Yu, C., Wu, H., Chen, C., & Wang, Z. (2020). An improved non-intrusive load disaggregation algorithm and its application. *Sustainable Cities and Society, 53,* 101918.

Liu, J., Zhang, L., Chen, X., & Niu, J. (2017). Facial landmark automatic identification from three dimensional (3D) data by using Hidden Markov Model (HMM). *International Journal of Industrial Ergonomics, 57,* 10–22.

Long, Z., Wen, B., Chen, X., Wu, J., & Zhang, J. (2020). Demand-side ubiquitous electric power internet of things: Architecture, functionalities and technologies. In *2020 Chinese Control And Decision Conference (CCDC)* (pp. 5586–5591). IEEE.

Lu, L., Lu, L., Park, S.-W., & Wang, B.-H. (2012). Electric load signature analysis for home energy monitoring system. *International Journal of Fuzzy Logic and Intelligent Systems, 12*(3), 193–197.

Luo, D., Norford, L. K., Shaw, S. R., & Leeb, S. B. (2002). Monitoring HVAC equipment electrical loads from a centralized location–methods and field test results/Discussion. *ASHRAE Transactions, 108,* 841.

LV, X., Ren, Z., Tang, B., Liu, H., Yang, R., & Wu, H. (2019). Hybrid load identification model based on grey wolf optimization algorithm. In *2019 25th International Conference on Automation and Computing (ICAC)* (pp 1–5). IEEE.

Maus, V., Câmara, G., Cartaxo, R., Sanchez, A., Ramos, F. M., & De Queiroz, G. R. (2016). A time-weighted dynamic time warping method for land-use and land-cover mapping. *IEEE Journal of Selected Topics in Applied Earth Observations and Remote Sensing, 9*(8), 3729–3739.

Meehan, P., McArdle, C., & Daniels, S. (2014). An efficient, scalable time-frequency method for tracking energy usage of domestic appliances using a two-step classification algorithm. *Energies, 7*(11), 7041–7066.

Mets, K., Depuydt, F., & Develder, C. (2015). Two-stage load pattern clustering using fast wavelet transformation. *IEEE Transactions on Smart Grid, 7*(5), 2250–2259.

Nguyen, M., Alshareef, S., Gilani, A., & Morsi, W. G. (2015). A novel feature extraction and classification algorithm based on power components using single-point monitoring for NILM. In *2015 IEEE 28th Canadian Conference on Electrical and Computer Engineering (CCECE)* (pp. 37–40). IEEE.

Nguyen, T. K., Dekneuvel, E., Jacquemod, G., Nicolle, B., Zammit, O., & Nguyen, V. C. (2017). Development of a real-time non-intrusive appliance load monitoring system: An application level model. *International Journal of Electrical Power & Energy Systems, 90,* 168–180.

Norford, L. K., & Leeb, S. B. (1996). Non-intrusive electrical load monitoring in commercial buildings based on steady-state and transient load-detection algorithms. *Energy and Buildings, 24*(1), 51–64.

Parson, O., Ghosh, S., Weal, M., & Rogers, A. (2014). An unsupervised training method for non-intrusive appliance load monitoring. *Artificial Intelligence, 217,* 1–19.

Patel, S. N., Robertson, T., Kientz, J. A., Reynolds, M. S., & Abowd, G. D. (2007). At the flick of a switch: Detecting and classifying unique electrical events on the residential power line (nominated for the best paper award). In *International Conference on Ubiquitous Computing* (pp. 271–288). Springer.

Patri, O. P., Panangadan, A. V., Chelmis, C., & Prasanna, V. K. (2014). Extracting discriminative features for event-based electricity disaggregation. In *2014 IEEE Conference on Technologies for Sustainability (SusTech)* (pp. 232–238). IEEE.

Piga, D., Cominola, A., Giuliani, M., Castelletti, A., & Rizzoli, A. E. (2016). Sparse Optimization for Automated Energy End Use Disaggregation. *IEEE Transactions on Control Systems Technology, 24*(3), 1044–1051. https://doi.org/10.1109/TCST.2015.2476777.

Popescu, F., Enache, F., Vizitiu, I.-C., & Ciotîrnae, P.(2014). Recurrence plot analysis for characterization of appliance load signature. In *2014 10th International Conference on Communications (COMM)* (pp. 1–4). IEEE.

Rashid, H., Singh, P., Stankovic, V., & Stankovic, L. (2019). Can non-intrusive load monitoring be used for identifying an appliance's anomalous behaviour? *Applied Energy, 238,* 796–805.

Rashid, K. M., & Louis, J. (2019). Times-series data augmentation and deep learning for construction equipment activity recognition. *Advanced Engineering Informatics, 42,* 100944.

Ren, Z., Tang, B., Wang, L., Liu, H., Li, Y., & Wu, H. (2019). Non-intrusive load identification method based on integrated intelligence strategy. In *2019 25th International Conference on Automation and Computing (ICAC)* (pp. 1–6) IEEE.

Rosdi, N. A. M., Nordin, F. H., & Ramasamy, A. K. (2014). Identification of electrical appliances using non-intrusive magnetic field and probabilistic neural network (pnn). In *2014 IEEE International Conference on Power and Energy (PECon)* (pp. 47–52). IEEE.

Shahriar, M. S., & Rahman, A. (2013). Energy disaggregation using ensemble of classifiers. In *IEEE 2013 Tencon-Spring* (pp. 161–164). IEEE.

Shaw, S. R., Abler, C., Lepard, R., Luo, D., Leeb, S. B., & Norford, L. K. (1998). Instrumentation for high performance nonintrusive electrical load monitoring. *Journal of Solar Energy Engineering, 120*(3), 224–229.

Su, S., Yan, Y., Lu, H.., Kangping, L., Yujing, S., Fei, W., Liming, L., & Hui, R. (2016). Non-intrusive load monitoring of air conditioning using low-resolution smart meter data. In *2016 IEEE International Conference on Power System Technology (POWERCON)* (pp. 1–5). IEEE.

Su, Y.-C., Lian, K.-L., & Chang, H.-H. (2011). Feature selection of non-intrusive load monitoring system using STFT and wavelet transform. In *2011 IEEE 8th international conference on e-business engineering* (pp. 293–298). IEEE.

Sun, L., Zhou, K., & Yang, S. (2020). An ensemble clustering based framework for household load profiling and driven factors identification. *Sustainable Cities and Society, 53,* 101958. https://doi.org/10.1016/j.scs.2019.101958.

Sun, X., Li, T., Chen, M., & Liu, L. (2019). Key technology applications and development prospects of ubiquitous electric internet of things. In *2019 IEEE Sustainable Power and Energy Conference (iSPEC)* (pp. 134–139). IEEE.

Tarutani, Y., Ohsita, Y., & Murata, M. (2016). Placement of virtual storages for distributed robust cloud storage. *IEICE Transactions on Communications, 99*(4), 885–893.

Teeraratkul, T., O'Neill, D., & Lall, S. (2018). Shape-based approach to household electric load curve clustering and prediction. *Ieee Transactions on Smart Grid, 9*(5), 5196–5206. https://doi.org/10.1109/Tsg.2017.2683461.

Tina, G. M., Amenta, V. A., Tomarchio, O., & Di Modica, G. (2014). Web interactive non intrusive load disaggregation system for active demand in smart grids. *EAI Endorsed Transactions on Energy Web 1*(3).

Tu, Y., Zhang, Z., Li, Y., Wang, C., & Xiao, Y. (2019). Research on the Internet of Things device recognition based on RF-fingerprinting. *IEEE Access, 7,* 37426–37431.

Varga, E. D., Beretka, S. F., Noce, C., & Sapienza, G. (2015). Robust real-time load profile encoding and classification framework for efficient power systems operation. *IEEE Transactions on Power Systems, 30*(4), 1897–1904. https://doi.org/10.1109/Tpwrs.2014.2354552.

Wang, A. L., Chen, B. X., Wang, C. G., & Hua, D. (2018). Non-intrusive load monitoring algorithm based on features of V-I trajectory. *Electric Power Systems Research, 157,* 134–144.

Wang, X., Lei, D., Yong, J., Zeng, L., & West, S. (2013). An online load identification algorithm for non-intrusive load monitoring in homes. In *2013 IEEE Eighth International Conference on Intelligent Sensors, Sensor Networks and Information Processing* (pp. 1–6) IEEE.

Wang, Y., Yan, J., Sun, Q., Li, J., & Yang, Z. (2019). A mobilenets convolutional neural network for GIS partial discharge pattern recognition in the ubiquitous power internet of things context: Optimization, comparison, and application. *IEEE Access, 7,* 150226–150236.

Wang, Z., & Zheng, G. (2011). Residential appliances identification and monitoring by a nonintrusive method. *IEEE Transactions on Smart Grid, 3*(1), 80–92.

Wild, B., Barsim, K. S., & Yang, B. (2015). A new unsupervised event detector for non-intrusive load monitoring. In *2015 IEEE global conference on signal and information processing (GlobalSIP)* (pp. 73–77). IEEE.

Wong, K.-Y., Chen, W., & Chen, K. J. (2010). Integrated voltage reference generator for GaN smart power chip technology. *IEEE Transactions on Electron Devices, 57*(4), 952–955.

Wortmann, F., & Flüchter, K. (2015). Internet of things. *Business & Information Systems Engineering, 57*(3), 221–224.

Wu, X., Gao, Y., & Jiao, D. (2019). Multi-label classification based on random forest algorithm for non-intrusive load monitoring system. *Processes, 7*(6), 337.

Wu, X., Qi, B., Han, L., Wang, Z., & Dong, C. (2017). Fast non-intrusive load identification algorithm for resident load based on template filtering. *Autom Electr Power Syst, 2,* 021.

Xia, F., Yang, L. T., Wang, L., & Vinel, A. (2012). Internet of things. *International Journal of Communication Systems, 25*(9), 1101.

Xia, M., Wang, K., Zhang, X., & Xu, Y. (2019). Non-intrusive load disaggregation based on deep dilated residual network. *Electric Power Systems Research, 170,* 277–285.

Xu, S., Lin, W., & Wang, J. (2016). Virtual machine placement algorithm based on peak workload characteristics. *Journal of Software, 27*(7), 1876–1887.

Yang, C. C., Soh, C. S., & Yap, V. V. (2014) Comparative study of event detection methods for non-intrusive appliance load monitoring.

Yang, C. C., Soh, C. S., & Yap, V. V. (2015). A systematic approach to ON-OFF event detection and clustering analysis of non-intrusive appliance load monitoring. *Frontiers in Energy, 9*(2), 231–237.

Yang, C. C., Soh, C. S., & Yap, V. V. (2019a). A systematic approach in load disaggregation utilizing a multi-stage classification algorithm for consumer electrical appliances classification. *Frontiers in Energy, 13*(2), 386–398.

Yang, D., Wang, D., Zhou, B., Chen, Q., Yang, Z., Xu, G., et al. (2019b). Key technologies and application prospects of ubiquitous power internet of things. *Power generation technology, 40*(2), 107–114.

Yang, T., Zhai, F., Zhao, Y., & Pen, H. (2019c). Explanation and prospect of ubiquitous electric power internet of things. *Automation of Electric Power Systems, 43*(13), 9–20.

Zhang, G., Li, Y., & Deng, X. (2020). K-Means clustering-based electrical equipment identification for smart building application. *Information, 11*(1), 27.

Zheng, C., Wang, S., Liu, Y., & Liu, C. (2019). A novel RNN based load modelling method with measurement data in active distribution system. *Electric Power Systems Research, 166,* 112–124.

Zheng, L., Chen, S., Xiang, S., & Hu, Y.(2012). Research of architecture and application of Internet of Things for smart grid. In *2012 International Conference on Computer Science and Service System* (pp. 938–941). IEEE.

Zoha, A., Gluhak, A., Imran, M. A., & Rajasegarar, S. (2012). Non-intrusive load monitoring approaches for disaggregated energy sensing: A survey. *Sensors, 12*(12), 16838–16866.

Chapter 2
Smart Non-intrusive Device Recognition Based on Physical Methods

2.1 Introduction

Smart non-intrusive device recognition methods can effectively save production costs and realize the sustainable development of resources. Due to the limitation of acquisition equipment, the data used in the experiment are generally voltage and current. To ensure the accuracy of equipment identification, it is very important to extract the feature information in the data effectively (Lin et al. 2020). Physical properties are very classical and well-studied. Physical features are usually overlapped, so this feature can help the classifier to realize accurate identification of equipment effectively (Li et al. 2019a). At present, in the research of device identification, device features are mainly divided into steady-state features and transient features. The extraction process of load features is mainly to analyze three kinds of features: physical features, harmonic features and voltage-current relationship features (Li et al. 2018). Transient load features mainly analyze the load changes in the process of electrical switching, and then extract the changes of different device features before and after the changes, such as power changes, impedance changes, harmonic changes, peak power stability ratio, and so on (Wöhrl and Brunelli 2019). The main content of this chapter is to study the physical features and application prospects of different classifiers in the field of device recognition.

2.2 Device Recognition Method Based on Decision Tree

2.2.1 Evaluation Criteria

For the binary classification problem, the classification results can be divided into four cases, including True Positive (TP), False Positive (FP), True Negative (TN), and False Negative (FN). The confusion matrix of the classification results is shown in Table 2.1. Four indices are used to evaluate the classification performance of the

H. Liu et al., *Smart Device Recognition*,
https://doi.org/10.1007/978-981-33-4925-4_2

Table 2.1 Confusion matrix of classification results

Actual class	Predicted results	
	Positive	Negative
Positive	TP (True positive)	FN (False negative)
Negative	FP (False positive)	TN (True negative)

model, including precision, recall, accuracy, and F1-score. For the sake of brevity, they are also calculated on the average of all classes. The results are named as overall precision, overall recall, overall accuracy, and F1-macro, respectively.

The calculation of precision, recall, accuracy, and F1-score are all based on the confusion matrix of classification results. The Precision effectively reflects the probability that all the sample values predicted to be true are really true. It can be obtained as follows (Deng et al. 2016a):

$$Precision_i = \frac{TP_i}{TP_i + FP_i} \tag{2.1}$$

And the overall precision is calculated on the average of all classes:

$$Precision = \frac{1}{N_c} \sum_{i=1}^{N_c} Precision_i \tag{2.2}$$

where N_c is the total number of classes.

The recall rate reflects the probability that true samples are predicted to be true samples. The recall is given as follows (Zhang et al. 2019b):

$$Recall_i = \frac{TP_i}{TP_i + FN_i} \tag{2.3}$$

And the overall precision is calculated on the average of all classes:

$$Recall = \frac{1}{N_c} \sum_{i=1}^{N_c} Recall_i \tag{2.4}$$

Accuracy is a very important index that is made up of the percentage of the sample that predicts the correct outcome. Accuracy is a classical evaluation index. It is calculated as follows (Lee 2017):

$$Accuracy = \frac{TP + TN}{TP + TN + FP + FN} \tag{2.5}$$

F1-Score represents the weighted harmonic average of Precision and Recall. Accuracy and recall values are usually inconsistent. When the numerical gap between the two is relatively large, the F1-Score method can effectively combine these two indicators to obtain a better evaluation effect. F1-Score is obtained as follows (Picon et al. 2019):

$$F1 - Score_i = \frac{2 \times Precision_i \times Recall_i}{Precision_i + Recall_i} \tag{2.6}$$

The overall F1-Score is also called F1-macro. It is calculated as the average F1-Score of all classes. Since the amount of data is not taken into account, each class is treated equally. The calculation method of F1-macro is given as follows (Bills et al. 2020):

$$F1 - macro = \frac{1}{N_c} \sum_{i=1}^{N_c} F1 - Score_i \tag{2.7}$$

2.2.2 Basic Definitions of Physical Features

The core idea of device recognition is to identify the type of electrical device by analyzing the physical features of different electrical devices. Therefore, the feature extraction method will greatly affect the recognition accuracy of subsequent classifiers. Feature extraction is a process of extracting key information from raw data. This chapter mainly needs to extract various physical features from the original voltage and current time series. Traditional physical definitions effectively reflected device information. The main physical features are defined as follows:

The original $U(t)$ and $I(t)$ signals of the PLAID data set are the main experimental subjects. The $U(t)$ and $I(t)$ are defined as follows:

$$U = \sqrt{\frac{1}{T} \int_0^T u^2(t)dt} \tag{2.8}$$

$$I = \sqrt{\frac{1}{T} \int_0^T i^2(t)dt} \tag{2.9}$$

Common physical definition features mainly include the following contents: active power, reactive power, apparent power, admittance, impedance, etc., whose definitions are shown as follows:

(1) *Power*

Active power and reactive power are important output characteristics of circuit systems. Reactive power is a characteristic produced by the interaction between the circuit and magnetic field to ensure the stability of the magnetic field. Active power can be divided into instantaneous power and average power, which can reflect the operating characteristics of equipment and define the change of circuit energy consumption. Besides, the apparent power is also one of the features of the circuit, which represents the total energy consumption of the whole circuit. The active power and reactive power can be obtained as follows:

$$P = \frac{1}{T}\int_0^T u(t)i(t)dt \tag{2.10}$$

$$Q = \sqrt{S^2 - P^2} \tag{2.11}$$

where S represents the apparent power.

(2) *Power factor*
The power factor is one of the important physical indicators. High reactive power will increase circuit loss and lead to energy waste. The power factor can be got as follows:

$$\lambda = \frac{P}{\sqrt{P^2 + Q^2}} \tag{2.12}$$

(3) *Impedance*
When a circuit contains resistance, inductance, capacitance and other circuit elements, a variety of elements can be created that block currently. When only the capacitance obstructs the current, the obstruction is called capacitive reactance. When only the inductance obstructs the current, the obstruction is called the inductance. When both capacitance and inductance are present, the obstruction is called reactance.
The definition of impedance is shown as follows:

$$Z = \frac{U}{I} \tag{2.13}$$

The resistance can be defined as follows:

$$R = Z\cos\alpha_Z \tag{2.14}$$

where $\alpha_Z = \arcsin(\frac{Q}{S})$.
The reactance is obtained as follows:

$$X = Z\sin\alpha_Z \tag{2.15}$$

(4) *Admittance*
The admittance is calculated as follows:

$$Y = \frac{I}{U} \tag{2.16}$$

Table 2.2 The variable representation of load feature vectors

Features	Variable
Current	I
Voltage	U
Apparent power	S
Reactive power	Q
Admittance	Y
Conductance	G
Resistance	R
Impedance	Z
Reactance	X
Active power	P
Power factor	λ
Susceptance	B
The real and imaginary parts of the 1–9th harmonic current	H_i

The conductance is given as follows:

$$G = Y \cos \alpha_Z \tag{2.17}$$

The susceptance is obtained as follows:

$$B = -Y \sin \alpha_Z \tag{2.18}$$

The real and imaginary parts of the 1–9th harmonic current is calculated as follows:

$$H_i = \begin{cases} I_{(i+1)/2} \cos \alpha_{(i+1)/2} (i = 1, 3, 5, 7, 9, 11, 15, 17) \\ I_{i/2} \sin \alpha_{i/2} (i = 2, 4, 6, 8, 10, 12, 14, 16, 18) \end{cases} \tag{2.19}$$

Based on the above theoretical, all physical features can be summarized in Table 2.2.

2.2.3 *Original Dataset*

PLAID is the abbreviation of Plug Load Appliance Identification Dataset. It includes the current and voltage data of 11 different devices, including air conditioner, compact fluorescent lamp, fridge, hairdryer, laptop, microwave, washing machine, bulb, vacuum, fan, and heater (Gao et al. 2015). PLAID is a set of high-resolution common data sets for load recognition research, which has over 200 information measures for different devices (Gao et al. 2014). Five kinds of electric appliances such as Hairdryer, Microwave, Fan, Fridge, and Heater are applied for classification and model recognition research in this chapter.

The emphasis of this chapter is to analyze the recognition ability of different classifiers. Before the experiment, it is important to process the raw U and I data of the PLAID data set and extract the physical features. The sample of measurement data in this chapter contains 1000 samples of 5 devices, including 200 sample points of the Hairdryer, 200 sample points of the Microwave, 200 sample points of the Fan, 200 sample points of the Fridge, and 200 sample points of Heater.

2.2.4 The Theoretical Basis of Decision Tree

The decision tree algorithm in machine learning is a classic algorithm series. It can be used as either a classification algorithm or a regression algorithm and is especially suitable for ensemble learning such as random forest (Friedl and Brodley 1997).

(1) *The basic principles of the ID3 (Iterative Dichotomiser 3) method*
ID3 algorithm originates from the concept learning system (CLS) (Jin et al. 2014). It takes the decreasing speed of information entropy as the criterion to select the test attribute, that is, the attribute with the highest information gain that has not been used to be divided at each node is selected as the criterion (Cheng et al. 1988). Then, the ID3 algorithm repeats the above steps. When the accuracy of ID3 reaches a certain requirement or the training times are completed, all parameters of ID3 are saved (Bartczuk and Rutkowska 2006).
Although the ID3 algorithm puts forward new ideas, there are still many places worth improving.
ID3 algorithm cannot effectively analyze continuous features such as time series features, which limits the application scope of the ID3 algorithm to some extent. (Phu et al. 2017).
Based on the characteristics of information enhancement and basic theory, the ID3 algorithm needs to establish decision tree nodes first (Mitra et al. 2001). Therefore, ID3 algorithm analysis features are mainly feature-based quantities, which makes ID3 unable to deeply analyze the deep implicit information of the feature data. (Kinney and Murphy 1987).
The ID3 algorithm has four major shortcomings (Li et al. 2012):

- It cannot handle continuous features.
- It is the feature that information gain as the standard tends to have more values.
- The problem of missing value processing.
- The problem of overfitting.

(2) *The basic principles of the C4.5 method*
Based on the above and case analysis, it can be known that there are many factors affecting the classification accuracy of ID3. To improve the shortcomings of ID3 algorithm, scholars put forward C4.5 algorithm (Muniyandi et al. 2012).
For the first problem, continuous features cannot be dealt with. The idea of C4.5 is to discretize continuous features (Lee et al. 2018). In the C4.5 algorithm, the model gets the result of the boundary point by solving the average value of two adjacent samples (Szarvas et al. 2006). It is expressed as (Ngoc et al. 2019):

$$T_i = \frac{c_i + c_{i+1}}{2} \tag{2.20}$$

where T_i is the ith partition point.
For these M_I points, by calculating the information gain of each point, the point with the maximum information gain is selected to realize binary discrete classification (Meng et al. 2020). If the features of the current node have continuously changing characteristics, then the attribute needs to participate in the generation and selection process of the child nodes to build the model (Polat and Güneş 2009).
For the second problem, the information gain ratio variable $I_R(X, Y)$ is introduced (Budiman et al. 2017). It is the ratio of information gain to feature entropy (Benkercha and Moulahoum 2018). It is shown as follows (Rajeswari et al. 2014):

$$I_R(B, C) = \frac{I(C, B)}{H_C(B)} \tag{2.21}$$

where B denotes the output of sample features; C represents the sample features. The expression of feature entropy $H_A(D)$ is shown as follows (Li et al. 2019b):

$$H_C(B) = -\sum_{i=1}^{n} \frac{|B_i|}{|B|} \log_2 \frac{|B_i|}{|B|} \tag{2.22}$$

where n represents the number of categories; B_i represents the number of samples corresponding to feature C.
The greater the number of features, the greater the feature entropy. It can correct for the fact that the information gain tends to be biassed towards more features.
For the third problem of missing value processing, two main problems need to be solved (Li et al. 2015):

- Select the attribute of partition when some features of the sample are missing.
- The partition attribute is selected and the processing of samples with missing features on this attribute is conducted.

For the first sub-problem, C4.5 solves this problem by dividing the data and assigning weights. The data is divided into two main parts. Some of them have

eigenvalues, and some of them do not. Then, the problems of the model are improved by the weighted gain (Ling et al. 2003).
For the second sub problem, the sample of the missing feature can be divided into all the child nodes simultaneously. The weight of the sample is distributed in proportion to the number of samples of each child node (Song et al. 2008).
For the fourth problem, C4.5 introduces the regularization coefficient for preliminary pruning. Although C4.5 has improved several major issues of the ID3 algorithm, there is still room for optimization (Asri et al. 2016).

(3) *The basic principles of the Classification and Regression Tree (CART) algorithm*
Since the decision tree algorithm is very easy to overfit, pruning is necessary for the generated decision tree. There are many pruning algorithms and they can effectively improve the performance of C4.5. Therefore, the C4.5 pruning method has a lot of room for optimization (Crawford 1989). The main ideas are given as follows (Rutkowski et al. 2014):

- Pre-pruning: when the decision tree is generated, it is decided whether to prune or not.
- After pruning: The master becomes the decision tree, and then pruning through cross verification.

The CART tree mainly adopts post-pruning and cross-validation to select the most suitable decision tree (García et al. 2019). It is found that the binary tree model can get better results than the multi-tree model in many classification experiments (Deng et al. 2020). Besides, the binary tree method can also provide great help to improve efficiency. (Shao and Lunetta 2012).
Due to the use of the entropy model, C4.5 contains a lot of time-consuming logarithm operations (Han et al. 2018).
These problems are solved in the CART method. Therefore, if ensemble learning is not considered at present, the CART algorithm is a relatively superior algorithm among ordinary decision tree algorithms (Waheed et al. 2006).
In general, the decision tree algorithm has advantages and disadvantages, which can be summarized as follows. Advantages of decision tree algorithm are given as follow (Hasanipanah et al. 2017):

- The model framework is easy to understand and the structure is clear, which makes the logic of the model very clear
- The model does not need to normalize the initial data to get better results
- The model can be used to analyze and identify both discrete and continuous features
- It has good fault tolerance and robustness to outliers.

Disadvantages of the decision tree algorithm are shown as follow (Zimmerman et al. 2016):

- When the minimum sample number and depth of the model are not set properly, the decision tree algorithm will fall into overfitting, which greatly affects the generalization effect and classification accuracy of the model.
- A small change in the sample will lead to a drastic change in the tree structure. This can be done through integration learning or something like that.
- Finding the optimal decision tree is hard. Heuristic algorithms are often used to find locally optimal solutions. It can be improved by ensemble learning.
- For some complex nonlinear problems, the decision tree algorithm is difficult to calculate for the optimal result. However, these problems can be solved using the neural network classification method.
- When the samples are unbalanced, the decision tree algorithm tends to favor a large number of samples, which leads to poor practicability of the model.

(4) *Experimental results of the decision tree algorithm*
Based on the above theoretical basis, this chapter compares and analyzes the classification results of ID3, C4.5, and CART. Table 2.3 shows the classification results of three kinds of algorithms. Figures 2.1, 2.2 and 2.3 show the confusion matrix of the decision tree algorithm. Based on the experimental results, the conclusions can be obtained:

Table 2.3 Classification performance indexes of decision tree methods

Model	Type of device	Accuracy	Recall	F1-Score
ID3	Hairdryer	0.9067	0.7667	0.7667
	Microwave	0.9017	0.7833	0.7611
	Fan	0.9056	0.7667	0.7645
	Fridge	0.9075	0.7750	0.7702
	Heater	0.9107	0.7767	0.7767
	Average	0.9064	0.7737	0.7678
C4.5	Hairdryer	0.9067	0.8167	0.7778
	Microwave	0.9250	0.8250	0.8148
	Fan	0.9211	0.8111	0.8044
	Fridge	0.9225	0.8125	0.8075
	Heater	0.9213	0.8033	0.8033
	Average	0.9193	0.8137	0.8016
CART	Hairdryer	0.9333	0.8167	0.8305
	Microwave	0.9367	0.8500	0.8430
	Fan	0.9389	0.8611	0.8493
	Fridge	0.9400	0.8625	0.8519
	Heater	0.9373	0.8433	0.8433
	Average	0.9372	0.8467	0.8436

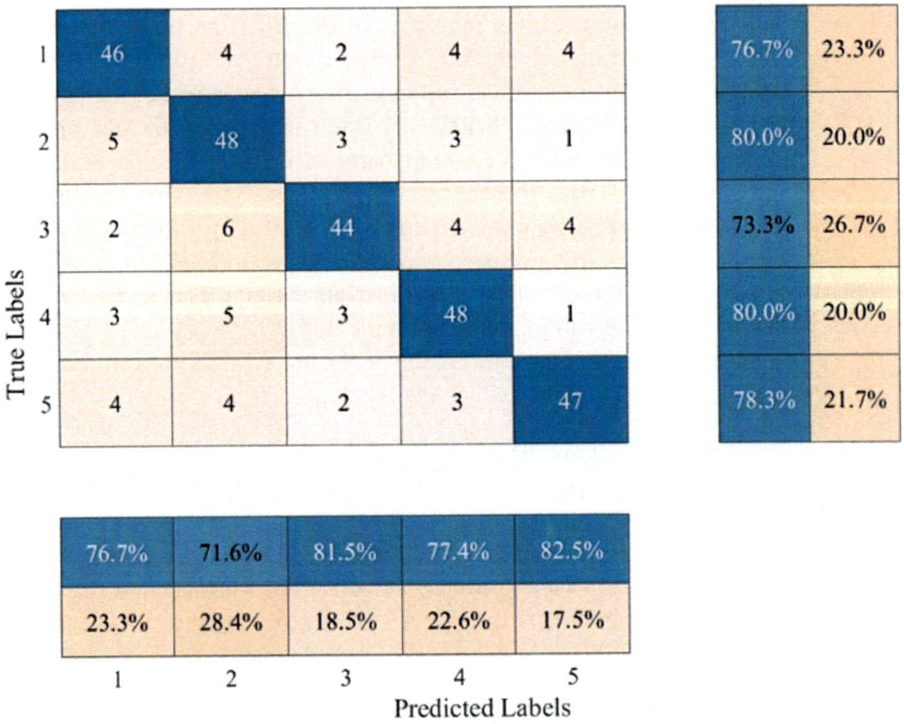

Fig. 2.1 Confusion matrix of the ID3 method

- All three kinds of decision tree algorithms can get more accurate classification results. It shows that the physical features can effectively help the decision tree to identify different electrical equipment. The possible reason is that physical features are important factors reflecting device features, and the decision tree algorithm has excellent feature sensitivity.
- Compared to the classification results of three kinds of decision tree algorithms, CART algorithm has the best result and ID3 algorithm has the worst result. This proves that CART algorithm has an excellent application prospect in the field of device recognition. The possible reason is that CART algorithm improves ID3 and C4.5 algorithms in multiple angles, which makes it useful and strong in feature analysis and device recognition.

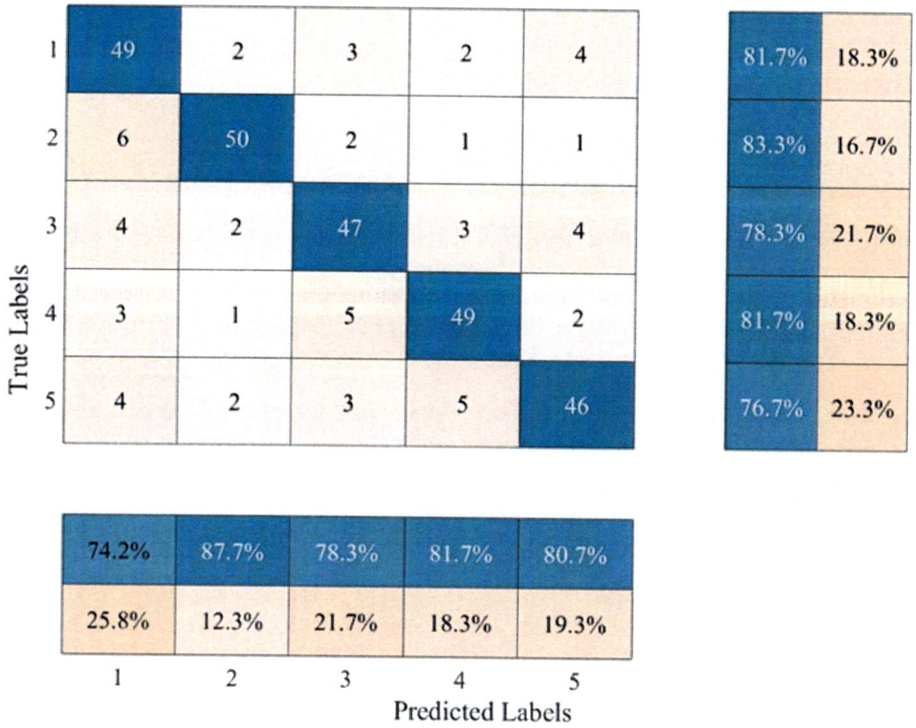

Fig. 2.2 Confusion matrix of the C4.5 method

2.3 Device Recognition Method Based on Template Matching Method

2.3.1 The Basic Content of the Template Matching Method

Template matching is often referred to as pattern classification. From the perspective of supervised classification of problem processing and method of problem-solving, template matching can be classified into two types: Supervised Classification and Unsupervised Classification (Choi and Kim 2002). The main difference is whether the category of each sample is known in advance (Cerciello et al. 2011). In general, supervised classification often requires the provision of a large sample of known categories (Sintorn et al. 2004). However, in practical problems, there are certain difficulties. Therefore, it is necessary to study the unsupervised classification. (Thanh et al. 2009). The template matching referred to in this chapter is mainly to identify and classify the specific patterns of speech waveform, text, picture, and other objects (Shin et al. 2013).

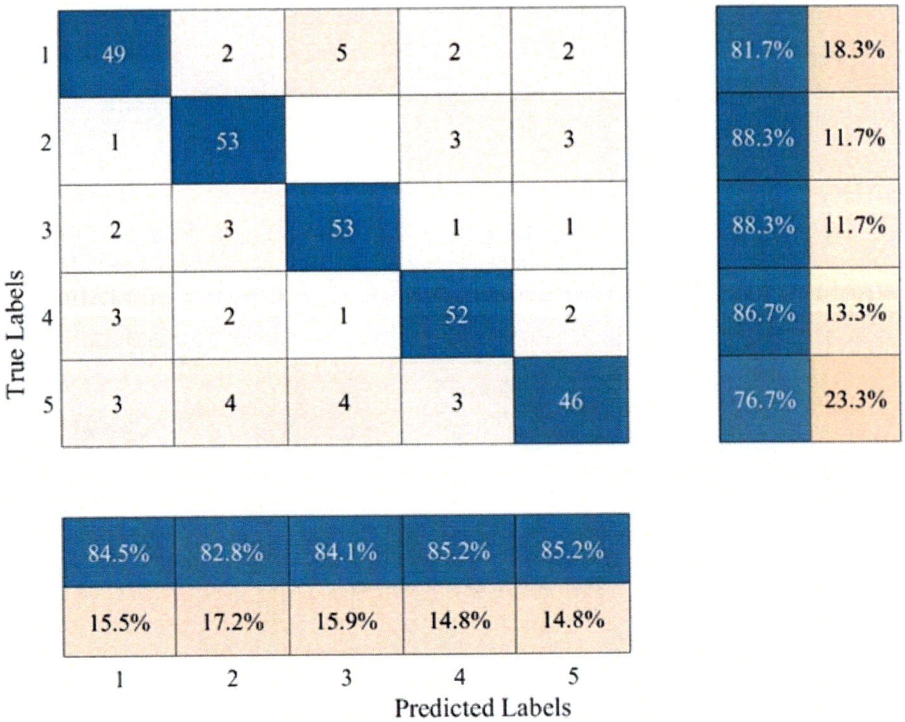

Fig. 2.3 Confusion matrix of the CART method

2.3.2 *Device Recognition Based on KNN Algorithm*

The KNN method is an efficient algorithm for nonparametric classification proposed in 1967 (Li and Guo 2007). In the process of learning, the model realizes device identification by the nearest neighbor method. The method builds templates by storing known training data. Then, the classification is achieved by comparing the minimum distance between the test data and the template (Du et al. 2009).

(1) *The basic idea of the KNN algorithm*
Based on the input and output mapping principle of supervised learning, this chapter transforms the extracted PLAID data set features into the input feature matrix form, expressed as $D = D(T_1, W_1; T_2, W_2; \ldots; T_n, W_n)$ (Shang et al. 2005). When the test machine is classified, the model analyzes the similarity degree of different features and training set templates through the test set and selects the minimum distance to realize the classification (Mejdoub and Amar 2013). By calculating the weighted distance between different samples, the KNN algorithm can effectively realize device identification (Trstenjak et al. 2014).
The steps of the device recognition method are introduced as follows (Haixiang et al. 2016):

- Extract physical feature information of different devices and divide training sets and test sets.
- Based on the basic principle of the algorithm, the similarity and distance between different devices are calculated as follows:

$$Sim(x_i, x_j) = \frac{\sum_{k=1}^{n} M_{ik} \times M_{jk}}{\sqrt{\sum_{k=1}^{n} M_{ik}^2 \sqrt{\sum_{k=1}^{n} M_{jk}^2}}} \tag{2.23}$$

where x_i represents the feature vector; n is the number of dimensions; x_j represents the center vector; M_k represents the k-th dimension of the vector.

- According to the calculated similarity results, K training samples similar to the test samples were obtained.
- The weights of different devices in k neighborhood in the test set are calculated as follows (Deng et al. 2016b):

$$P(X, Y) = \begin{cases} 1 \text{ when } \sum\limits_{b \in KNN} Sim(a, b) y(b, Y) - \beta \geq 0 \\ 0 \text{ else} \end{cases} \tag{2.24}$$

- Based on the weights calculated by KNN, the classification of each device is realized, and the classification result is the class with the highest weight.

KNN algorithm is an effective algorithm to analyze the nonlinear correlation between different kinds of data. When the distribution of sample categories cannot be expressed by the linear formula, KNN can achieve an excellent classification effect (Jiang et al. 2012). However, KNN also has some defects. On the one hand, the KNN algorithm is a lazy algorithm, which makes the computational efficiency of the model average. On the other hand, KNN is highly sensitive to features, which makes it easy to overfit (Cheng et al. 2014). In the process of classification, the KNN algorithm needs to analyze the correlation between different features, sample weights, and distances to ensure the accuracy of the model (Östermark 2009).
The advantages and disadvantages of the KNN algorithm are shown as follows:
The advantages of the KNN are given as follows (Saçlı et al. 2019):

- It is simple, easy to understand, and easy to implement, without parameter estimation and training.
- It is suitable for classifying rare events.
- KNN performs better than SVM in multiple classification problems.

Disadvantages of KNN (Li et al. 2016):

- It requires a lot of computation, has a large memory overhead, and scores slowly;
- Its interpretability is poor, and it cannot give rules like a decision tree.

Notes for KNN algorithm (Chen et al. 2020):

- The setting of k value: The size of the K value will affect the recognition accuracy of proximity distance. If the K value is improperly selected, noise data will be introduced and classification accuracy will be reduced.
- Choice of distance measurement: the dimension of input variables greatly affects the calculation of Euclidean distance. Therefore, selecting the appropriate data dimension and standardizing the variables can help the model to effectively calculate the Euclidean distance.
- Performance issues: KNN is a lazy algorithm, which leads to some consequences. Modeling is simple, but classifying test samples is expensive.

(2) *Steps of KNN algorithm*
The flow chart of the KNN method is shown in Fig. 2.4. The basic steps of KNN are shown as follow (Hmeidi et al. 2008):

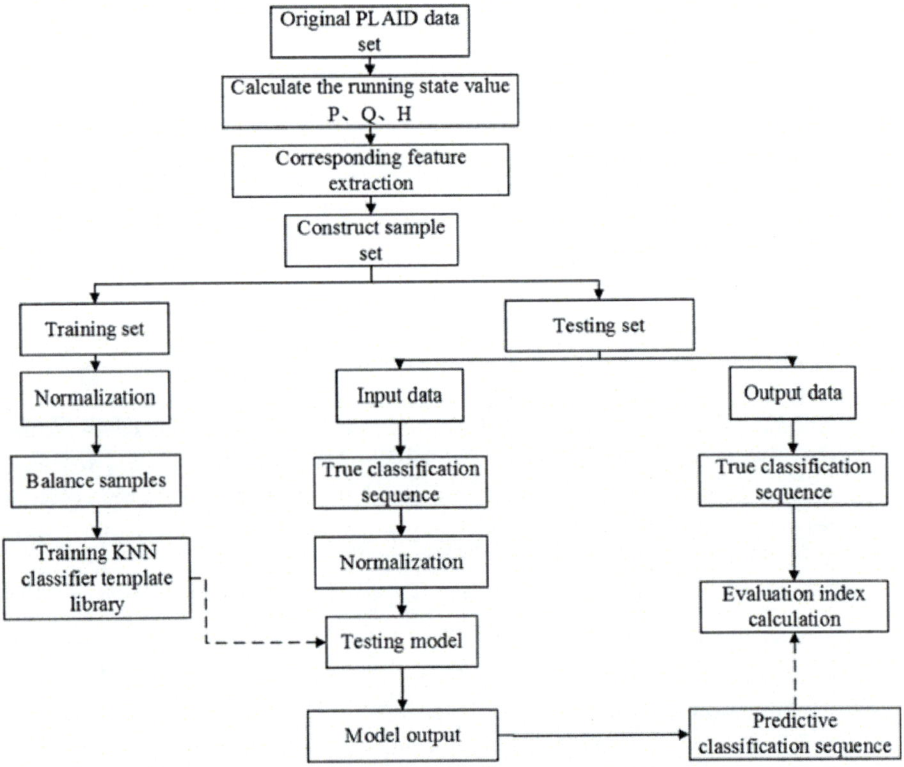

Fig. 2.4 Flow chart of the KNN method

Table 2.4 Classification performance indexes of the KNN method

Model	Type of device	Accuracy	Recall	F1-Score
KNN	Hairdryer	0.9433	0.8833	0.8618
	Microwave	0.9483	0.8917	0.8735
	Fan	0.9478	0.8722	0.8698
	Fridge	0.9467	0.8625	0.8661
	Heater	0.9480	0.8700	0.8700
	Average	0.9468	0.8759	0.8682

- Preprocess the original PLAID data set and extract the physical features
- Divide the training set and test set.
- Set the basic parameters of the KNN algorithm, such as K value.
- A queue with K values from large to small is constructed, which is used to store the relevant parameters of the training set of the model. K values are randomly selected from the training set as the initial nearest neighbor category, which will be used to calculate the distance between the test set and K values.
- The Euclidean distance is calculated and the obtained distance L is compared with the maximum distance L_{max}.
- Based on the calculation and comparison results, the model decides whether or not to discard the tuple.
- Store the modified training tuples in the priority queue of the model.
- When the model implements a full traversal, it computes most of the classes in the priority queue of the model and uses them as categories for test tuples.
- When the test tuple set is modeled, the error rate is calculated. Then, the K value was modified to compare the different classification results. Finally, choose the K value with the lowest error rate.

(3) *Experimental results of KNN*

Table 2.4 shows the classification results of the KNN method. Figure 2.5 shows the confusion matrix of the KNN model. Analyzing the experimental results, the conclusions can be obtained:

KNN algorithm can achieve good classification accuracy, but it also has certain feature sensitivity, which makes it easy to fall into overfitting and other phenomena. Therefore, optimizing the core parameters of the KNN algorithm can effectively improve the classification ability of KNN. In this chapter, the heuristic algorithm will be adopted to optimize K parameters of KNN to effectively improve model classification accuracy.

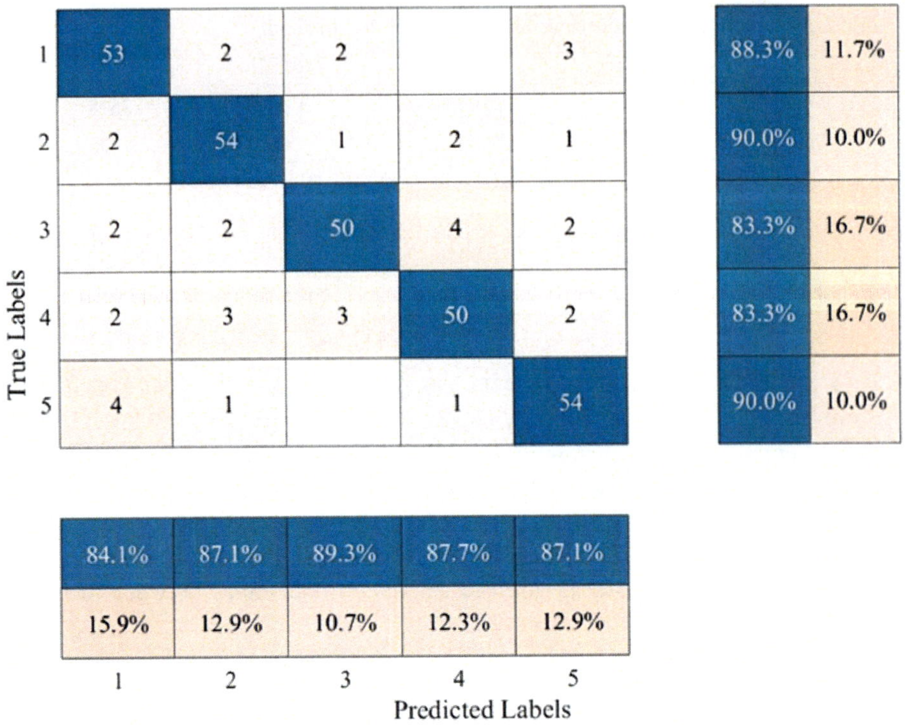

Fig. 2.5 Confusion matrix of the KNN method

2.3.3 *Device Recognition Based on DTW Algorithm*

Time series data is a classic way of data existence, and in most data mining work, it is an important task to calculate the similarity between time series. How to calculate the similarity of non-isometric time series is a very important field. DTW is used to solve this problem. DTW (Dynamic Time Warping) algorithm is a kind of dynamic programming algorithm that calculates the similarity of two-time series, especially sequences of different lengths (Chen et al. 2017b). It is mainly used in sequential data, such as speech recognition of isolated words, gesture recognition, data mining, and information retrieval. DTW algorithm is a dynamic programming algorithm in essence.

(1) *The basic idea of the DTW algorithm*
In this chapter, based on the time-series information of different electrical devices, the DTW algorithm is used to realize equipment identification and classification. In the process of the DTW algorithm, the similarity difference between feature vectors is generally calculated to achieve device classification, such as Euclidean distance and Markov distance (Orozco-Alzate et al. 2015). In this chapter, the chi-square test will be used as the core method of the DTW algorithm to build the classification model.

In the process of device identification, the DTW algorithm realizes device identification by calculating the minimum value of the point pair base distance sum. The calculation formula of DTW distance is shown as follows (Iwana et al. 2020):

$$DTW(X,\ Y) = \min\left\{\sum_{K=1}^{K} D_{base}|(w_k)\right\} \tag{2.25}$$

The optimal path selection is realized by constructing a cumulative distance matrix. The element $\gamma_{i,j}$ in matrix M is given as follows:

$$W_{i,j} = D_{base}(x_i, y_j) + \min(W_{i,j-1}, W_{i-1,j}, W_{i-1,j-1}) \tag{2.26}$$

where, $W_{i,j}$ is the DTW distance between series $X[1{:}j]$ and series $Y[1{:}j]$. The steps of DTW are shown as follows (Górecki and Łuczak 2014):

- Use power acquisition devices to collect power timing data of different devices.
- Based on the collected data, the device template library was established.
- Compare the data from the template library to the data from the test samples. Based on the DTW principle, the similarity distance is calculated.
- Compare the results of the similarity calculation and get the template matching results.

(2) *Experimental results of DTW*
Table 2.5 shows the classification results of the DTW method. Figure 2.6 shows the confusion matrix of the DTW model. Based on the experimental results, the conclusions can be obtained:
The classification accuracy of the DTW algorithm is not very high. The possible reason is that DTW is not good at identifying current and voltage series and related physical features. Besides, the improvement of parameters and feature input of DTW is not as good as the KNN and decision tree algorithm. Therefore, this chapter mainly adopts a heuristic algorithm to improve the core parameters of KNN and Decision Tree.

Table 2.5 Classification performance indexes of the DTW method

Model	Type of device	Accuracy	Recall	F1-Score
DTW	Hairdryer	0.8967	0.7500	0.7438
	Microwave	0.9100	0.7833	0.7769
	Fan	0.9144	0.7944	0.7879
	Fridge	0.9167	0.7958	0.7925
	Heater	0.9173	0.7933	0.7933
	Average	0.9110	0.7834	0.7789

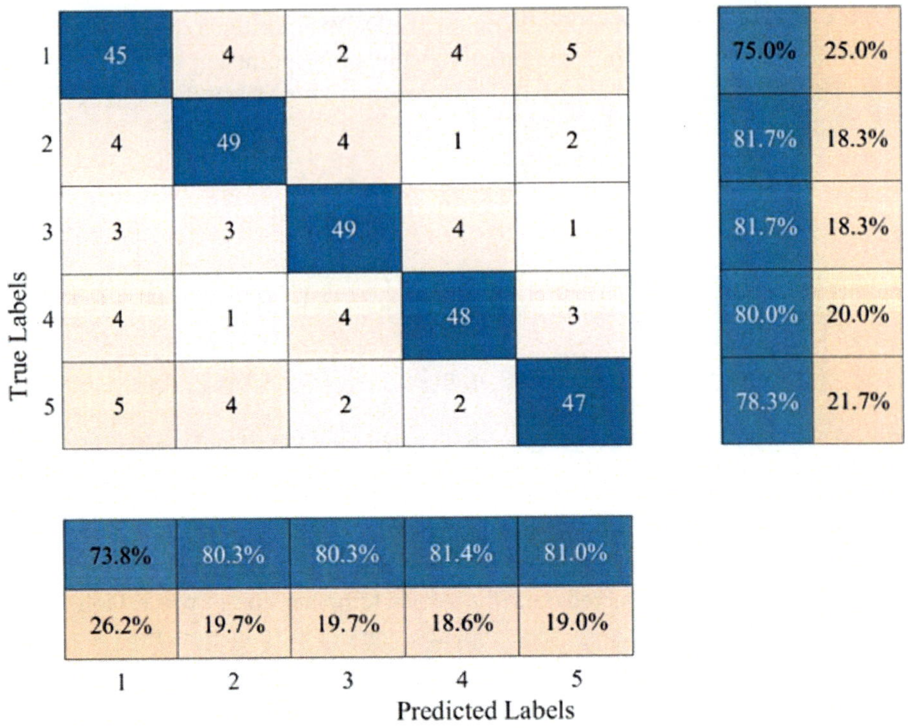

Fig. 2.6 Confusion matrix of the DTW method

2.4 Device Recognition Method Based On Current Decomposition

2.4.1 Introduction of the Current Decomposition Method

The theory of current decomposition is a research hotspot in device recognition. It is developed based on steady-state voltage and current decomposition methods of the device, such as Kirchhoff's law and Ohm's law (Singh and Verma 2008). These methods are based on the basic principles of time series. It can decompose the total signal into a combination of device signals (Wang et al. 2007). This chapter briefly introduces the principles of these methods and the features used. Detailed experiments and improvement strategies will be introduced in subsequent chapters.

The steps of the current decomposition method are shown as follows (Kumar et al. 2019):

- Preprocess the raw data. Then, based on the principle of physical feature extraction, the feature library of the Device is established.
- Extracting the total current sequence or the total power sequence of the equipment;
- Based on the feature base and the total time series data, the current decomposition model is established.

- Based on the results obtained in the previous step, the performance of different algorithms is compared and analyzed.

The input eigenvectors used in the most classical decomposition methods are extracted from the low-frequency sampling data. Based on the extracted features, the algorithm can be effectively applied to the existing power system (Yazdani et al. 2009). At the same time, this method has the advantage of high transmission efficiency. However, these features also have some common drawbacks (Kumar et al. 2017). Due to the low data sampling rate, features contain limited information. Besides, the inherent information and power features of high-frequency data are not fully utilized (Kompella et al. 2018). However, this method still has an excellent research prospect in the field of device recognition, because it can effectively help enterprises and society to save energy.

2.4.2 *Physical Features of Current Decomposition*

In the process of current decomposition, the features of the model must satisfy the following two conditions (Ji et al. 2017):

- The features should be superposition.
- The features are stable when devices are in a stable operating state.

(1) *Active power and reactive power.*
In the steady-state decomposition algorithm, the most classic feature time series feature is the power series, which includes active power P_a and reactive power P_r. P_a and P_r are both superpositions. The power series can directly reflect the energy consumption information of electrical equipment and the features of the variation of reactive voltage and current (Thahab and Asumadu 2017). However, when there are similar devices with active power series and reactive power series, it is difficult to realize decomposition due to the uniqueness of features (Cekic and Eren 2018).

(2) *Power phase angle θ_P and apparent power P_s.*
The combination of apparent power and the phase angle of the power series is another manifestation of power feature. This is called power vector P_v. Different from active power and reactive power, both phase Angle and apparent power belong to non-superposition eigenvectors (Harirchi and Simoes 2018). However, based on the principle of vector computation, power vectors can be superimposed. Therefore, as a feature, it has better feature recognition than active and reactive power. The power vector is defined as follows (Salas-Biedma et al. 2019):

$$P_v = P_s \angle \theta_p \tag{2.27}$$

(3) *Current series.*
Current series sampled at high frequencies are often used as features of decomposition algorithms. In a stable state, the current sequence of the equipment usually has a certain operation and sampling rule (Ray et al. 2017). They are stable and regular and do not produce abrupt changes in a stable operation state. Based on the Fourier transform method, the instantaneous current i_a of appliance a can be obtained as follows (Gerkšič et al. 2018):

$$i_a(t) = I_{a1}\cos(\omega t + \theta_{a1}) + I_{a2}\cos(2\omega t + \theta_{a2}) + \cdots + I_{am}\cos(m\omega t + \theta_{am}) \tag{2.28}$$

where I_{a1} represents the amplitude of the fundamental harmonic; θ_{a1} represents the initial phase angle; I_{am} represents the amplitude; θ_{am} represents the phase angle.
When n devices work in parallel, the different current signals i_z can be superposed by the linear regression formula. It is defined as follows (Džamić et al. 2019):

$$i_z(t) = \alpha_1 i_1(t) + \alpha_2 i_2(t) + \ldots + \alpha_n i_n(t) \tag{2.29}$$

where, $\alpha_1, \alpha_2, \ldots, \alpha_n$ are the current weight of n devices.
The equation is usually expressed as a matrix-vector, which can be seen as follow: (Neiman et al. 2019):

$$\begin{bmatrix} I_{z1}\angle\theta_{z1} \\ I_{z2}\angle\theta_{z2} \\ \vdots \\ I_{zm}\angle\theta_{zm} \end{bmatrix} = \begin{bmatrix} I_{11}\angle\theta_{11} & I_{21}\angle\theta_{21} & \cdots & I_{n1}\angle\theta_{n1} \\ I_{12}\angle\theta_{12} & I_{22}\angle\theta_{22} & \cdots & I_{n2}\angle\theta_{n2} \\ \vdots & \vdots & \cdots & \vdots \\ I_{1m}\angle\theta_{1m} & I_{2m}\angle\theta_{2m} & \cdots & I_{nm}\angle\theta_{nm} \end{bmatrix} \begin{bmatrix} \alpha_1 \\ \alpha_2 \\ \vdots \\ \alpha_n \end{bmatrix} \tag{2.30}$$

The feature information can be applied to several harmonics of electric current, which effectively solves the problem of the singleness of the feature. Therefore, this shows that the feature has excellent information and device identification. In the field of device identification, this feature has not been fully utilized by researchers (Chattopadhyaya et al. 2016). The phase information of a harmonic is mainly based on the calculation of its cosine value. Besides, the feature information obtained by superimposing all harmonics may be masked by the first harmonics (Porter et al. 2020).

(4) *Admittance.*
In parallel circuits, the admittance of each branch is usually superimposed in vector form. This feature includes the timing characteristics of the current sequence and voltage sequence of the device, which makes it more effective in assisting the decomposition algorithm to identify the device (Huang and Zhang 2019). Admittance is a transient variable, which is one of the main characteristic

variables still being explored and improved by researchers. The basic calculation of admittance is illustrated below (Tyuryukanov and Popov 2020):

$$\begin{cases} \dot{G}_t = \dot{I}_t / \dot{U}_t \\ |\dot{G}_t| = |\dot{I}_t| / |\dot{U}_t| \\ \theta_{G_t} = \theta_{I_t} - \theta_{U_t} \end{cases} \tag{2.31}$$

where $\dot{G}_t$ represents the admittance vector; $\dot{I}_t$ represents the relevant current; $\dot{U}_t$ represents voltage vectors; $|\dot{G}_t|$, $|\dot{I}_t|$, $|\dot{U}_t|$ represent the amplitudes of $\dot{G}_t$, $\dot{I}_t$, $\dot{U}_t$; θG_t, θI_t, θU_t represent the phase angle.

2.5 Experiment Analysis

2.5.1 Common Optimization Algorithms

To effectively optimize the accuracy of the model, scholars put forward a method to improve the model parameters by using the optimization algorithm(Liu et al. 2018). The heuristic algorithm is the most classical and practical optimization method. Based on the theory of population selection, the heuristic algorithm can effectively balance the global optimal and local optimal, which makes it effectively optimize the model parameters and get the optimal prediction results (Al-Musaylh et al. 2018).

Figure 2.7 is a flow chart of the classification model optimized heuristic algorithms (Liu et al. 2020b):

(1) *Genetic algorithm (GA)*

The genetic algorithm (GA) is the most classical heuristic optimization algorithm. The algorithm follows the principle of survival of the fittest (Liu et al. 2020a). The evolution of a population is realized by constantly eliminating the inferior individuals in the population and retaining the superior individuals in the population (Liu et al. 2015). Among them, the fitness function is the key to reflect the degree of fitness of each individual in the population. In the process of algorithm iteration, the fitness of each individual is calculated to obtain the quality of each individual (Liu et al. 2019). Then the low-quality individuals are eliminated, and finally, the optimal population individuals are selected (Büyükşahin and Ertekin 2019).

The basic steps of the genetic algorithm are shown as follows (Peng and Xiang 2019):

Step 1: Initialize the basic parameters of the population and randomly generate the population.

Step 2: Calculate the fitness (Loss function).

Step 3: Determine if the iteration completion condition is met. If the condition is not satisfied, step 4 is adopted; if it is satisfied, step 5 is adopted.

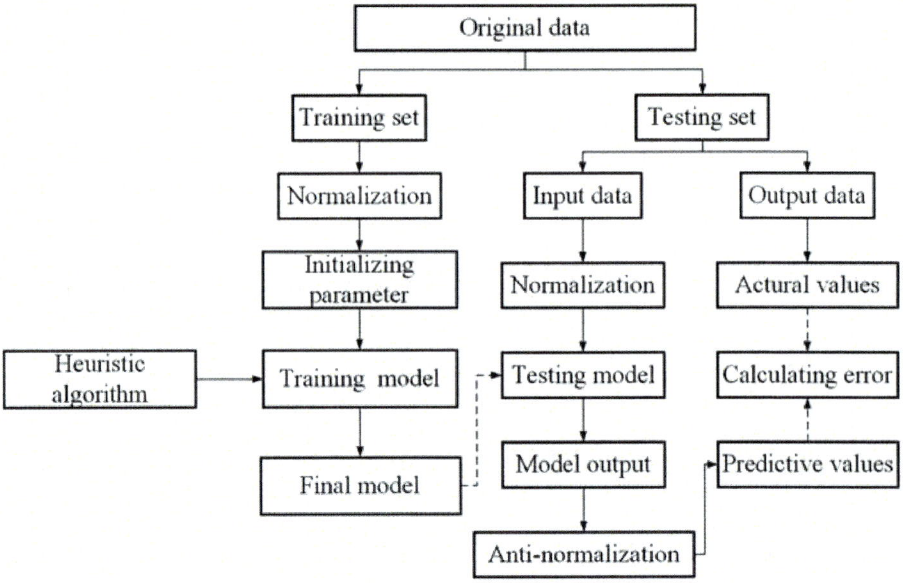

Fig. 2.7 Flow chart of the classification model

Step 4: Perform cross mutation and other operations on the population, update the individual information in the population, and then return to Step 2.
Step 5: When the end condition of the iteration is satisfied, the parameter of the individual with the best fitness in the population at this time is the final optimization goal.

(2) *Imperial competition algorithm (ICA)*
The imperial competition algorithm (ICA), which is inspired by the imperialist competition mechanism, is a new heuristic algorithm. Different from traditional heuristic algorithms based on the behavior of biological populations, ICA is a heuristic algorithm based on human social activities and mimics the mechanisms of colonial assimilation and imperial competition (Zhang et al. 2019a). Compared with other optimization algorithms, the ICA algorithm has the advantages of good optimization effect and fast solving speed.
The key to the ICA is the imperial competition mechanism. It is based on the colonization and assimilation between empires to realize the transmission of information between empires (Ardeh et al. 2017). It starts with an initial group and is optimized through assimilation, status reversal, imperial competition, and elimination. It starts with the initial group and effectively searches the region through a few specific steps. And then it will converge to or close to the optimal solution (Gordan et al. 2020).
By moving colonies to imperialist countries for local search, the ICA algorithm effectively guarantees the ability of local optimization. At the same time, the competitive activities of the empire made an effective breakthrough in the search area of the empire, which effectively solved the problem of falling into local

optimal(Liu and Dong 2020). Besides, based on the behavior of the empire merger, the convergence rate of the algorithm is improved effectively.
The imperial competition algorithm is mainly divided into the following parts (Yin and Gao 2020):

- Initialize the imperial (Chen et al. 2017a): Initializes the location and basic parameters of the national population in the search space. The power and power of a country are determined by the fitness function. Powerful groups of nations were chosen as imperialist states, while the rest were chosen as colonial states. Colonial countries were assigned to imperialist countries according to their position and fitness. The imperialist forces consist of one imperialist country and several colonial countries.
- Assimilation policy (Zhang et al. 2020): To effectively control the colonial countries, the imperialist countries would pass their parameters to the colonial countries. This process is called assimilation. During each ICA iteration, the colonial countries would move a certain distance randomly based on the spatial position of the imperialist countries. In the process of moving, the colonial countries may get a more adaptable position. At this point, it will replace the original imperialist state.
- Competition among imperialist countries (Khoshnoudian et al. 2017): Imperialist countries can greatly enhance their power by occupying other colonial powers. Therefore, in the iterative process of the ICA algorithm, there are mainly two important steps: (a) calculating the total strength of each empire. (b) The different empires competed with each other and then moved the colonial countries.
- The weakest empires fall (Ma et al. 2020a): An imperialist country will eventually perish if it loses all its colonies. As the iterative process continues, an imperialist country will eventually be left. This imperialist country represents the optimal outcome of the model.

2.5.2 Classification Results

(1) *Optimal classification model for decision trees (DT)*
The DT classification method has the advantages of easy to understand and good classification accuracy. Decision tree calculation does not require a large amount of prior user knowledge, which effectively alleviates the cost and difficulty of model calculation. Besides, in processing high-dimensional data, the decision tree algorithm has a certain capability of feature analysis (Harrison et al. 2018). However, the decision tree classification method is difficult to solve many difficult cases. For example, overfitting can seriously affect the classification accuracy of the decision tree algorithm. Decision trees tend to ignore the potential correlation between data, which greatly affects the selection of feature information (Sarker et al. 2020). To effectively improve the classification results of the

decision tree algorithm, scholars put forward the optimization algorithm to optimize the parameters of the Decision Tree, and some achievements have been made. To effectively improve the classification accuracy of the decision tree algorithm, the optimization algorithm can be used to optimize the parameters of the decision tree from the following perspectives (Chen et al. 2018):

- The minimum number of samples contained in different leaf nodes will affect the classification performance of the decision tree to some extent. Therefore, it is important to select the right number of nodes.
- The pruning method has some influence on the decision tree. Therefore, the choice of building function is very important.

The iteration range of the minimum sample number of leaf nodes is 10 to 100. To find the minimum sample number of optimal leaf nodes, the optimization algorithm can be used for adaptive selection (Yeo and Grant 2018). In this section, the classification performance of the decision tree is analyzed by pruning after optimization. After pruning, a complete decision tree is first generated. And then the nodes in the decision tree are evaluated for increased accuracy (Shaikhina et al. 2019).

Figure 2.8 shows the values of loss during the iterations. Figures 2.9 and 2.10 show the confusion matrix of the decision tree algorithms optimized by GA and ICA. Table 2.6 shows the classification results.

(2) *Optimal classification model for KNN*

At present, many factors will affect the results of the KNN algorithm, such as the selection of K value and the smoothness of sample size. When the number

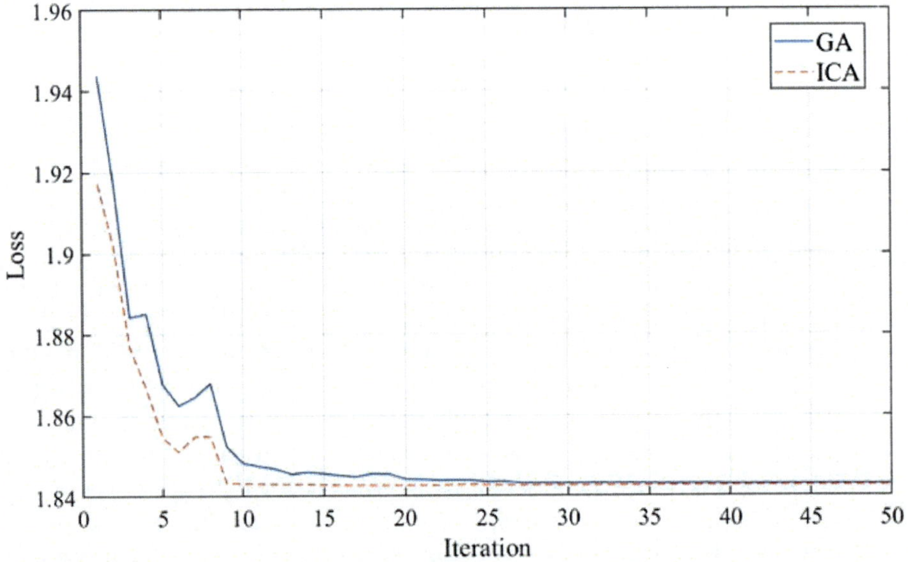

Fig. 2.8 Values of Loss during the iterations of GA and ICA

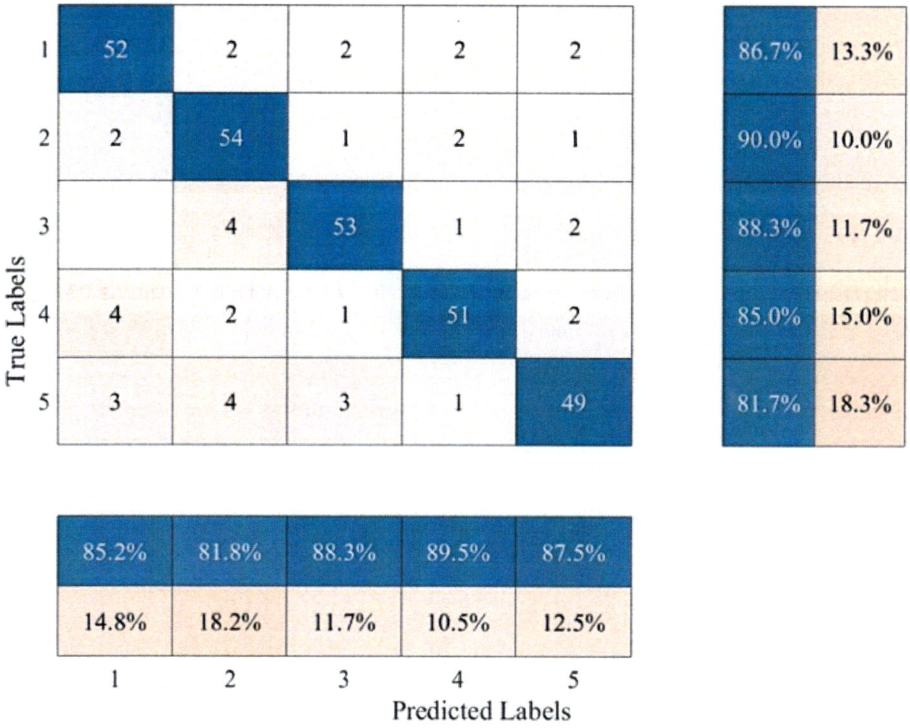

Fig. 2.9 Confusion matrix of the decision tree classification optimized by GA

of samples is unbalanced, sample coverage will occur, which greatly affects the accuracy of the model. (Harrou et al. 2020). Therefore, researchers usually adopt methods such as sample expansion and weighted balance to ensure the rationality of samples (Hu et al. 2020). Besides, the basic principle of KNN method is to get the best *K* samples in the training tuple, which has the closest European distance to the test sample (Haixiang et al. 2016). If the feature attributes of *K* samples are similar to those of the test samples, the classification results of the test samples are obtained (Nayak et al. 2020). In sum, the size of *K* value will significantly affect the device identification effect of KNN algorithm (Ma et al. 2020b). The ICA method and GA method are applied to find the best *K* value. The classification results of ICA-KNN and GA-KNN are shown as follows. Figures 2.11 and 2.12 show the confusion matrix of GA-KNN and ICA-KNN. Figure 2.13 shows the values of loss during the iterations. Table 2.7 shows the classification results.

Based on the above experimental results, the following conclusions can be obtained:

- The optimization algorithm can effectively improve the classification accuracy of KNN. It is proved that the heuristic algorithm has an excellent performance in optimizing model parameters.

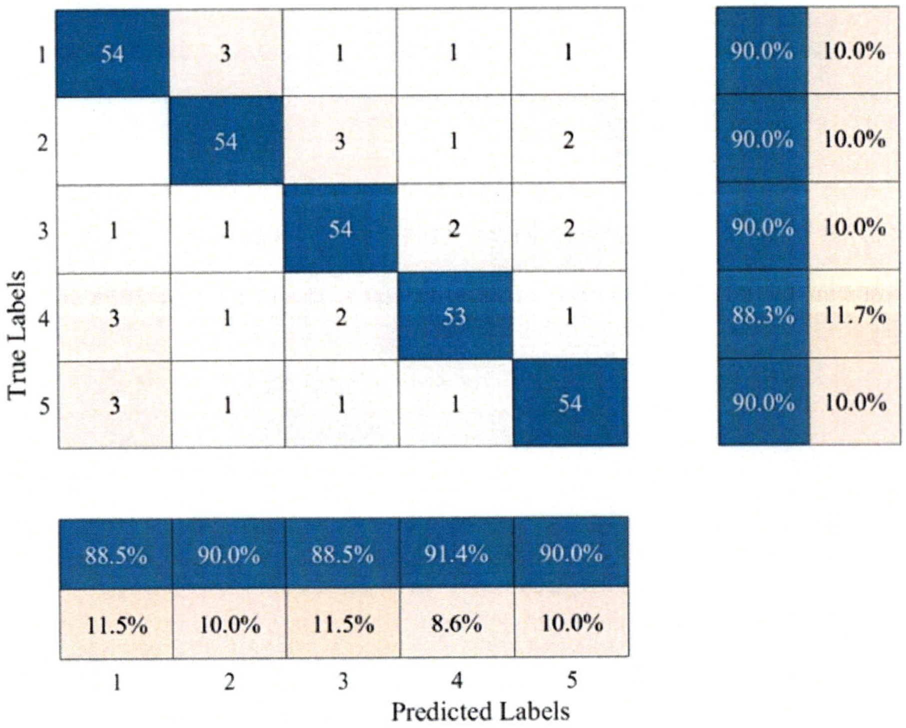

Fig. 2.10 Confusion matrix of the decision tree classification optimized by ICA

Table 2.6 Classification performance indexes of optimization decision tree model

Model	Type of device	Accuracy	Recall	F1-Score
GA	Hairdryer	0.9433	0.8667	0.8595
	Microwave	0.9417	0.8833	0.8583
	Fan	0.9456	0.8833	0.8665
	Fridge	0.9467	0.8750	0.8678
	Heater	0.9453	0.8633	0.8633
	Average	0.9445	0.8743	0.8631
ICA	Hairdryer	0.9567	0.9000	0.8926
	Microwave	0.9583	0.9000	0.8963
	Fan	0.9578	0.9000	0.8950
	Fridge	0.9583	0.8958	0.8958
	Heater	0.9587	0.8967	0.8967
	Average	0.9580	0.8985	0.8953

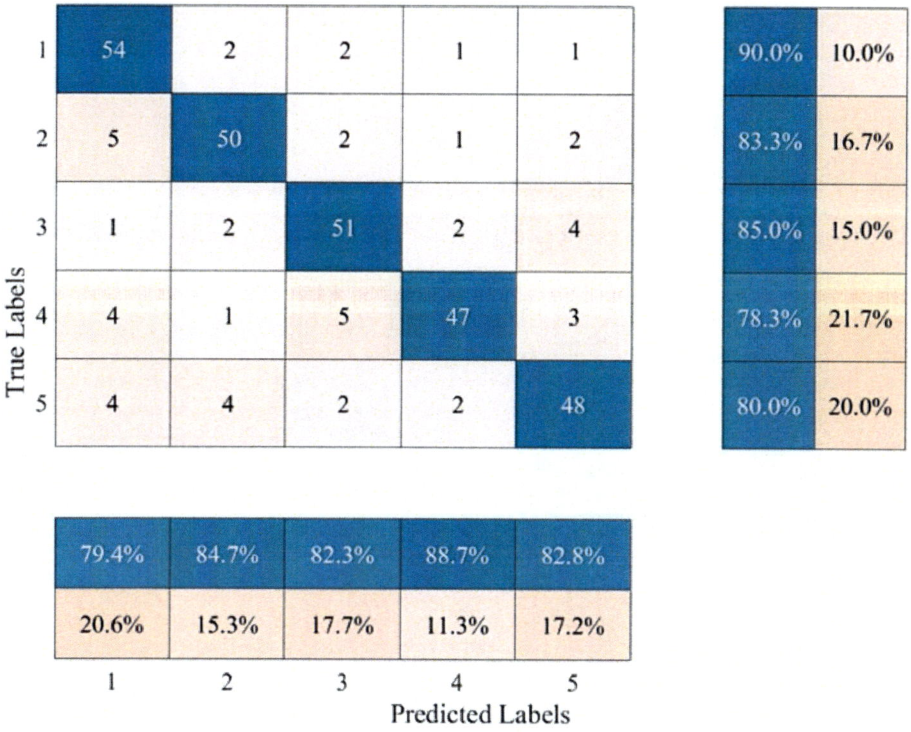

Fig. 2.11 Confusion matrix of the GA-KNN model

- The optimization performance of the ICA algorithm is better than that of GA algorithm. This proves the excellent capability of the ICA algorithm. On the one hand, the social heuristic algorithm can achieve better parameter optimization. On the other hand, ICA has a strong ability to avoid falling into local optimality.

2.5.3 Summary

This chapter mainly analyzes the physical characteristics and the application value of various classification algorithms in the field of equipment identification. Based on the traditional equipment load physical characteristics analysis principle, this chapter firstly extracts 33 physical characteristics of 5 devices in the PLAID data set. Then, based on the physical characteristics, this chapter studies the classification accuracy of the decision tree, KNN, and DTW algorithms. Finally, to improve the performance of the classifier, a heuristic algorithm is used to optimize the parameters of the decision tree and KNN. This chapter can be summarized from the following perspectives:

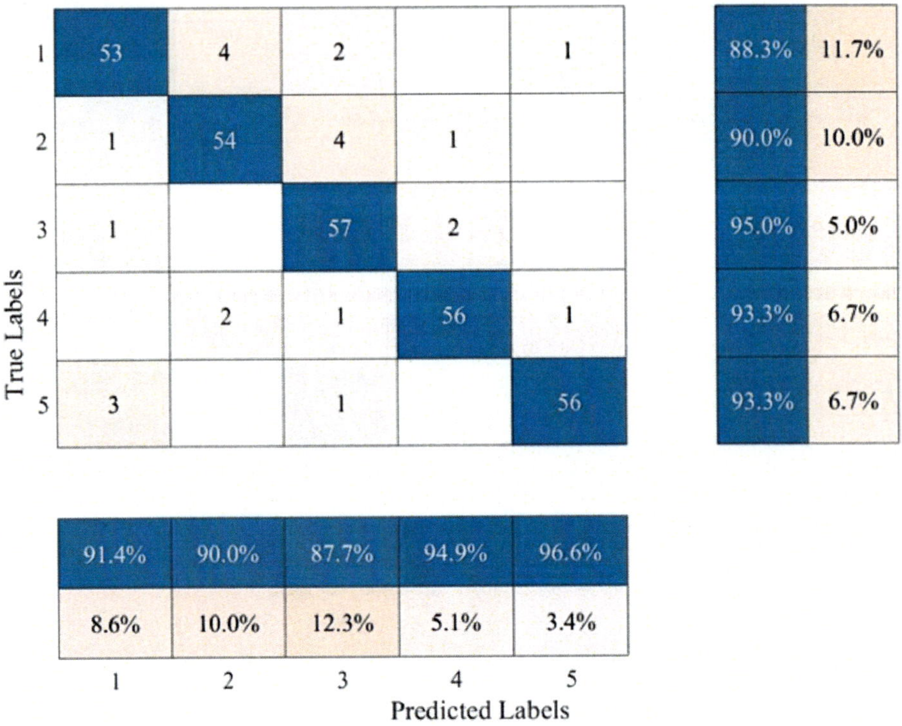

Fig. 2.12 Confusion matrix of the ICA-KNN model

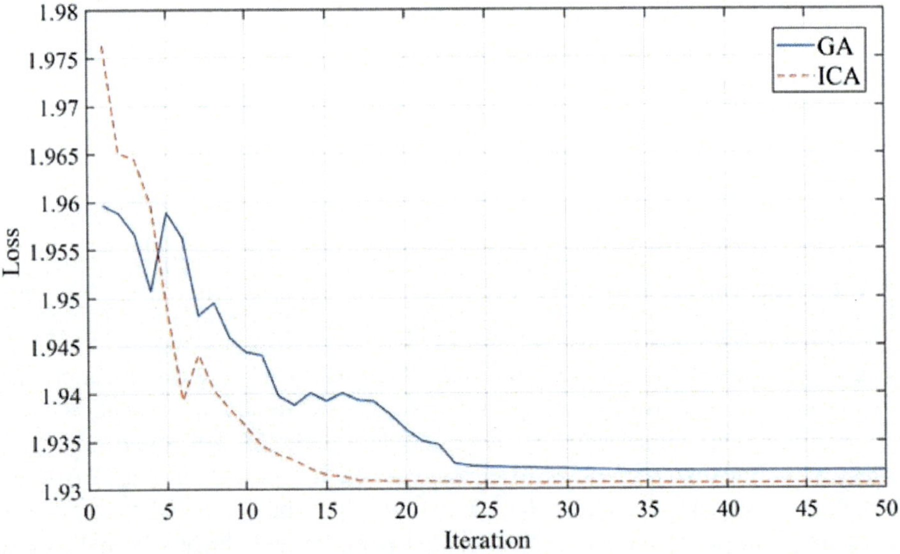

Fig. 2.13 Values of Loss during the iterations of GA and ICA

Table 2.7 Classification performance indexes of GA-KNN and ICA-KNN

Model	Type of device	Accuracy	Recall	F1-Score
GA-KNN	Hairdryer	0.9333	0.9000	0.8438
	Microwave	0.9350	0.8667	0.8421
	Fan	0.9344	0.8611	0.8401
	Fridge	0.9350	0.8417	0.8382
	Heater	0.9333	0.8333	0.8333
	Average	0.9342	0.8606	0.8395
ICA-KNN	Hairdryer	0.9600	0.8833	0.8983
	Microwave	0.9600	0.8917	0.8992
	Fan	0.9611	0.9111	0.9036
	Fridge	0.9650	0.9167	0.9129
	Heater	0.9680	0.9200	0.9200
	Average	0.9628	0.9046	0.9068

- The decision tree algorithm series can get better classification results. Among them, the classification performance of the CART algorithm is better than ID3 and C4.5 algorithms.
- KNN algorithm has excellent recognition performance in the field of classification. However, the size of K value affects the recognition accuracy of the model. So it's important to choose the right K value.
- DTW algorithm is a good modeling algorithm for time series analysis. However, if the original data features cannot be processed effectively, the classification accuracy is relatively poor.
- The heuristic algorithm can effectively optimize the parameters of the model and improve the classification accuracy. Besides, the optimization performance of the ICA algorithm based on social behavior is better than that of the genetic algorithm based on biological behavior.

References

Al-Musaylh, M. S., Deo, R. C., Adamowski, J. F., & Li, Y. (2018). Short-term electricity demand forecasting with MARS, SVR and ARIMA models using aggregated demand data in Queensland, Australia. *Advanced Engineering Informatics, 35,* 1–16.

Ardeh, M. A., Menhaj, M. B., Esmailian, E., & Zandhessami, H. (2017). EXPLICA: An explorative imperialist competitive algorithm based on the notion of explorers with an expansive retention policy. *Applied Soft Computing, 54,* 74–92.

Asri, H., Mousannif, H., Al Moatassime, H., & Noel, T. (2016). Using machine learning algorithms for breast cancer risk prediction and diagnosis. *Procedia Computer Science, 83,* 1064–1069.

Bartczuk, Ł., & Rutkowska, D. (2006) A new version of the Fuzzy-ID3 algorithm. In *International Conference on Artificial Intelligence and Soft Computing* (pp. 1060–1070). Springer.

Benkercha, R., & Moulahoum, S. (2018). Fault detection and diagnosis based on C4. 5 decision tree algorithm for grid connected PV system. *Solar Energy, 173,* 610–634.

Bills, M. V., Loh, A., Sosnowski, K., Nguyen, B. T., Ha, S. Y., Yim, U. H., et al. (2020). Handheld UV fluorescence spectrophotometer device for the classification and analysis of petroleum oil samples. *Biosensors & Bioelectronics, 159,* 112193. https://doi.org/10.1016/j.bios.2020.112193.

Budiman, E., Kridalaksana, A. H., & Wati, M. (2017). Performance of Decision Tree C4. 5 Algorithm in Student Academic Evaluation. In *International Conference on Computational Science and Technology* (pp. 380–389). Springer

Büyükşahin, Ü. Ç., & Ertekin, Ş. (2019). Improving forecasting accuracy of time series data using a new ARIMA-ANN hybrid method and empirical mode decomposition. *Neurocomputing, 361,* 151–163. https://doi.org/10.1016/j.neucom.2019.05.099.

Cekic, Y., & Eren, L. (2018). Broken rotor bar detection via four-band wavelet packet decomposition of motor current. *Electrical Engineering, 100*(3), 1957–1962.

Cerciello, T., Romano, M., Bifulco, P., Cesarelli, M., & Allen, R. (2011). Advanced template matching method for estimation of intervertebral kinematics of lumbar spine. *Medical Engineering & Physics, 33*(10), 1293–1302.

Chattopadhyaya, A., Chattopadhyay, S., Bera, J., & Sengupta, S. (2016). Wavelet decomposition based skewness and kurtosis analysis for assessment of stator current harmonics in a PWM–fed induction motor drive during single phasing condition. AMSE J, 1–14.

Chen, M.-H., Chen, S.-H., & Chang, P.-C. (2017a). Imperial competitive algorithm with policy learning for the traveling salesman problem. *Soft Computing, 21*(7), 1863–1875.

Chen, Y., Liu, X., Li, X., Liu, X., Yao, Y., Hu, G., et al. (2017b). Delineating urban functional areas with building-level social media data: A dynamic time warping (DTW) distance based k-medoids method. *Landscape and Urban Planning, 160,* 48–60. https://doi.org/10.1016/j.landurbplan.2016.12.001.

Chen, Y., Hu, X., Fan, W., Shen, L., Zhang, Z., Liu, X., et al. (2020). Fast density peak clustering for large scale data based on kNN. *Knowledge-Based Systems, 187,* 104824.

Cheng, D., Zhang, S., Deng, Z., Zhu, Y., & Zong, M. (2014). kNN algorithm with data-driven k value. In *International Conference on Advanced Data Mining and Applications* (pp. 499–512). Springer.

Chen, W., Zhang, S., Li, R., & Shahabi, H. (2018). Performance evaluation of the GIS-based data mining techniques of best-first decision tree, random forest, and naïve Bayes tree for landslide susceptibility modeling. *Science of the Total Environment, 644,* 1006–1018.

Cheng, J., Fayyad, U. M., Irani, K. B., & Qian, Z. (1988). Improved decision trees: a generalized version of id3. In *Machine Learning Proceedings* (pp. 100–106). Elsevier.

Choi, M.-S., & Kim, W.-Y. (2002). A novel two stage template matching method for rotation and illumination invariance. *Pattern Recognition, 35*(1), 119–129.

Crawford, S. L. (1989). Extensions to the CART algorithm. *International Journal of Man-Machine Studies, 31*(2), 197–217.

Deng, H., Diao, Y., Wu, W., Zhang, J., Ma, M., & Zhong, X. (2020). A high-speed D-CART online fault diagnosis algorithm for rotor systems. *Applied Intelligence, 50*(1), 29–41.

Deng, X., Liu, Q., Deng, Y., & Mahadevan, S. (2016a). An improved method to construct basic probability assignment based on the confusion matrix for classification problem. *Information Sciences, 340,* 250–261.

Deng, Z., Zhu, X., Cheng, D., Zong, M., & Zhang, S. (2016b). Efficient kNN classification algorithm for big data. *Neurocomputing, 195,* 143–148.

Du, P., Cao, S., & Li, Y. (2009). SubChlo: predicting protein subchloroplast locations with pseudo-amino acid composition and the evidence-theoretic K-nearest neighbor (ET-KNN) algorithm. *Journal of Theoretical Biology, 261*(2), 330–335.

Džamić, D., Aloise, D., & Mladenović, N. (2019). Ascent–descent variable neighborhood decomposition search for community detection by modularity maximization. *Annals of Operations Research, 272*(1–2), 273–287.

Friedl, M. A., & Brodley, C. E. (1997). Decision tree classification of land cover from remotely sensed data. *Remote Sensing of Environment, 61*(3), 399–409.

Gao, J., Giri, S., Kara, E. C., & Bergés, M. (2014). Plaid: a public dataset of high-resoultion electrical appliance measurements for load identification research: demo abstract. In *Proceedings of the 1st ACM Conference on Embedded Systems for Energy-Efficient Buildings* (pp. 198–199).

Gao, J., Kara, E. C., Giri, S., & Bergés, M. (2015). A feasibility study of automated plug-load identification from high-frequency measurements. In *2015 IEEE Global Conference on Signal and Information Processing (GlobalSIP)* (pp. 220–224) IEEE.

García, V. J., Márquez, C. O., Isenhart, T. M., Rodríguez, M., Crespo, S. D., & Cifuentes, A. G. (2019). Evaluating the conservation state of the páramo ecosystem: An object-based image analysis and CART algorithm approach for central Ecuador. *Heliyon, 5*(10), e02701.

Gerkšič, S., Pregelj, B., Perne, M., Ariola, M., De Tommasi, G., & Pironti, A. (2018). Model predictive control of ITER plasma current and shape using singular-value decomposition. *Fusion Engineering and Design, 129,* 158–163.

Gordan, M., Razak, H. A., Ismail, Z., Ghaedi, K., Tan, Z. X., & Ghayeb, H. H. (2020). A hybrid ANN-based imperial competitive algorithm methodology for structural damage identification of slab-on-girder bridge using data mining. *Applied Soft Computing, 88,* 106013.

Górecki, T., & Łuczak, M. (2014). Non-isometric transforms in time series classification using DTW. *Knowledge-Based Systems, 61,* 98–108. https://doi.org/10.1016/j.knosys.2014.02.011.

Haixiang, G., Yijing, L., Yanan, L., Xiao, L., & Jinling, L. (2016). BPSO-Adaboost-KNN ensemble learning algorithm for multi-class imbalanced data classification. *Engineering Applications of Artificial Intelligence, 49,* 176–193.

Han, J., Mao, K., Xu, T., Guo, J., Zuo, Z., & Gao, C. (2018). A soil moisture estimation framework based on the CART algorithm and its application in China. *Journal of Hydrology, 563,* 65–75.

Harirchi, F., & Simoes, M. G. (2018). Enhanced instantaneous power theory decomposition for power quality smart converter applications. *IEEE Transactions on Power Electronics, 33*(11), 9344–9359.

Harrison, P. A., Dunford, R., Barton, D. N., Kelemen, E., Martín-López, B., Norton, L., et al. (2018). Selecting methods for ecosystem service assessment: A decision tree approach. *Ecosystem Services, 29,* 481–498.

Harrou, F., Zeroual, A., & Sun, Y. (2020). Traffic congestion monitoring using an improved kNN strategy. *Measurement, 156,* 107534. https://doi.org/10.1016/j.measurement.2020.107534.

Hasanipanah, M., Faradonbeh, R. S., Amnieh, H. B., Armaghani, D. J., & Monjezi, M. (2017). Forecasting blast-induced ground vibration developing a CART model. *Engineering with Computers, 33*(2), 307–316.

Hmeidi, I., Hawashin, B., & El-Qawasmeh, E. (2008). Performance of KNN and SVM classifiers on full word Arabic articles. *Advanced Engineering Informatics, 22*(1), 106–111.

Hu, J., Peng, H., Wang, J., & Yu, W. (2020). kNN-P: A kNN classifier optimized by P systems. *Theoretical Computer Science, 817,* 55–65. https://doi.org/10.1016/j.tcs.2020.01.001.

Huang, X., & Zhang, X. (2019). Current-driving dissolution of nanoscale brittle precipitates produced by spinodal decomposition in FeCrAl alloys. *Journal of Alloys and Compounds, 805,* 26–32.

Iwana, B. K., Frinken, V., & Uchida, S. (2020). DTW-NN: A novel neural network for time series recognition using dynamic alignment between inputs and weights. *Knowledge-Based Systems, 188,* 104971. https://doi.org/10.1016/j.knosys.2019.104971.

Ji, T., Shi, M., Li, M., Zhang, L., & Wu, Q. (2017). Current transformer saturation detection using morphological gradient and morphological decomposition and its hardware implementation. *IEEE Transactions on Industrial Electronics, 64*(6), 4721–4729.

Jiang, S., Pang, G., Wu, M., & Kuang, L. (2012). An improved K-nearest-neighbor algorithm for text categorization. *Expert Systems with Applications, 39*(1), 1503–1509.

Jin, C., Li, F., & Li, Y. (2014). A generalized fuzzy ID3 algorithm using generalized information entropy. *Knowledge-Based Systems, 64,* 13–21.

Khoshnoudian, F., Talaei, S., & Fallahian, M. (2017). Structural damage detection using FRF data, 2D-PCA, artificial neural networks and imperialist competitive algorithm simultaneously. *International Journal of Structural Stability and Dynamics, 17*(07), 1750073.

Kinney, E. L., & Murphy, D. D. (1987). Comparison of the ID3 algorithm versus discriminant analysis for performing feature selection. *Computers and Biomedical Research, 20*(5), 467–476.

Kompella, K. D., Rao, M. V. G., & Rao, R. S. (2018). Bearing fault detection in a 3 phase induction motor using stator current frequency spectral subtraction with various wavelet decomposition techniques. *Ain Shams Engineering Journal, 9*(4), 2427–2439.

Kumar, A., Gupta, N., Gupta, V., & Babu, B. C. (2019). A Novel Orthogonal Current Decomposition Control for Grid-Connected Voltage Source Converter. *IEEE Transactions on Industry Applications, 55*(6), 7628–7641.

Kumar, P., Gupta, N., Niazi, K., & Swarnkar, A. (2017). Current decomposition method for loss allocation in distribution systems. *IET Generation, Transmission and Distribution, 11*(18), 4599–4607.

Lee, H. (2017). Framework and development of fault detection classification using IoT device and cloud environment. *Journal of Manufacturing Systems, 43,* 257–270. https://doi.org/10.1016/j.jmsy.2017.02.007.

Lee, S.-J., Xu, Z., Li, T., & Yang, Y. (2018). A novel bagging C4. 5 algorithm based on wrapper feature selection for supporting wise clinical decision making. *Journal of Biomedical Informatics, 78,* 144–155.

Li, Y., & Guo, L. (2007). An active learning based TCM-KNN algorithm for supervised network intrusion detection. *Computers & security, 26*(7–8), 459–467.

Li, Y., Jiang, D., & Li, F. (2012). The application of generating fuzzy ID3 algorithm in performance evaluation. *Procedia Engineering, 29,* 229–234.

Li, Y., Jiang, Z. L., Yao, L., Wang, X., Yiu, S.-M., & Huang, Z. (2019b) Outsourced privacy-preserving C4. 5 decision tree algorithm over horizontally and vertically partitioned dataset among multiple parties. *Cluster Computing, 22*(1), 1581–1593.

Li, N., Kong, H., Ma, Y., Gong, G., & Huai, W. (2016). Human performance modeling for manufacturing based on an improved KNN algorithm. *The International Journal of Advanced Manufacturing Technology, 84*(1–4), 473–483.

Li, L., Yang, L., Chen, H., Li, M., & Zhang, C. (2019a). Multi-objective evolutionary algorithms applied to non-intrusive load monitoring. *Electric Power Systems Research, 177,* 105961.

Li, Y., Yang, Q., Lai, S., & Li, B. (2015). A new speculative execution algorithm based on C4. 5 decision tree for Hadoop. In *International Conference of Young Computer Scientists, Engineers and Educators* (pp. 284–291). Springer.

Li, F., Zhang, W., Liu, H., & Zhang, M. (2018). Feature extraction of dichotomous equipment based on non-intrusive load monitoring and decomposition. In *International Conference on Algorithms and Architectures for Parallel Processing* (pp. 192–200). Springer.

Lin, J., Ding, X., Qu, D., & Li, H. (2020). Non-intrusive load monitoring and decomposition method based on decision tree. *Journal of Mathematics in Industry, 10*(1), 1.

Ling, C. X., Huang, J., & Zhang, H. (2003). AUC: a better measure than accuracy in comparing learning algorithms. In *Conference of the canadian society for computational studies of intelligence* (pp. 329–341). Springer.

Liu, H., & Dong, S. (2020). A novel hybrid ensemble model for hourly PM2. 5 forecasting using multiple neural networks: a case study in China. *Air Quality, Atmosphere & Health,* 1–10.

Liu, H., Duan, Z., Han, F-z, & Li, Y-f. (2018). Big multi-step wind speed forecasting model based on secondary decomposition, ensemble method and error correction algorithm. *Energy Conversion and Management, 156,* 525–541. https://doi.org/10.1016/j.enconman.2017.11.049.

Liu, H., Duan, Z., Wu, H., Li, Y., & Dong, S. (2019). Wind speed forecasting models based on data decomposition, feature selection and group method of data handling network. *Measurement, 148,* 106971.

Liu, H., Tian, H.-q, & Li, Y.-f. (2015). Comparison of new hybrid FEEMD-MLP, FEEMD-ANFIS, Wavelet Packet-MLP and Wavelet Packet-ANFIS for wind speed predictions. *Energy Conversion and Management, 89,* 1–11. https://doi.org/10.1016/j.enconman.2014.09.060.

Liu, H., Yu, C., Wu, H., Duan, Z., & Yan, G. (2020a). A new hybrid ensemble deep reinforcement learning model for wind speed short term forecasting. *Energy, 202,* 117794. https://doi.org/10.1016/j.energy.2020.117794.

Liu, H., Yu, C., Yu, C., Chen, C., & Wu, H. (2020b). A novel axle temperature forecasting method based on decomposition, reinforcement learning optimization and neural network. *Advanced Engineering Informatics, 44,* 101089. https://doi.org/10.1016/j.aei.2020.101089.

Ma, L., Xiao, L., Meng, Z., & Huang, X. (2020a). Robust adaptive fault reconfiguration for micro-gas turbine based on optimized T–S fuzzy model and nonsingular TSMO. *International Journal of Fuzzy Systems*, 1–19.

Ma, Z-f, Tian, H-p, Liu, Z-c, & Zhang, Z-w. (2020b). A new incomplete pattern belief classification method with multiple estimations based on KNN. *Applied Soft Computing, 90,* 106175. https://doi.org/10.1016/j.asoc.2020.106175.

Mejdoub, M., & Amar, C. B. (2013). Classification improvement of local feature vectors over the KNN algorithm. *Multimedia Tools and Applications, 64*(1), 197–218.

Meng, X., Zhang, P., Xu, Y., & Xie, H. (2020) Construction of decision tree based on C4. 5 algorithm for online voltage stability assessment. *International Journal of Electrical Power & Energy Systems, 118,* 105793.

Mitra, P., Mitra, S., & Pal, S. K. (2001). Evolutionary modular MLP with rough sets and ID3 algorithm for staging of cervical cancer. *Neural Computing and Applications, 10*(1), 67–76.

Muniyandi, A. P., Rajeswari, R., & Rajaram, R. (2012). Network anomaly detection by cascading k-Means clustering and C4. 5 decision tree algorithm. *Procedia Engineering, 30,* 174–182.

Nayak, S., Panda, M., & Palai, G. (2020). Realization of optical ADDER circuit using photonic structure and KNN algorithm. *Optik, 212,* 164675. https://doi.org/10.1016/j.ijleo.2020.164675.

Neiman, A., Semin, V., Meisner, L., & Ostapenko, M. (2019). Structural decomposition and phase changes in TiNi surface layer modified by low-energy high-current pulsed electron beam. *Journal of Alloys and Compounds, 803,* 721–729.

Ngoc, P. V., Ngoc, C. V. T., Ngoc, T. V. T., & Duy, D. N. (2019) A C4. 5 algorithm for english emotional classification. *Evolving Systems, 10*(3), 425–451.

Orozco-Alzate, M., Castro-Cabrera, P. A., Bicego, M., & Londoño-Bonilla, J. M. (2015). The DTW-based representation space for seismic pattern classification. *Computers & Geosciences, 85,* 86–95. https://doi.org/10.1016/j.cageo.2015.06.007.

Östermark, R. (2009). A fuzzy vector valued KNN-algorithm for automatic outlier detection. *Applied Soft Computing, 9*(4), 1263–1272.

Peng, Y., & Xiang, W. (2019). Short-term traffic volume prediction using GA-BP based on wavelet denoising and phase space reconstruction. *Physica A: Statistical Mechanics and its Applications*:123913. doi:https://doi.org/10.1016/j.physa.2019.123913

Phu, V. N., Tran, V. T. N., Chau, V. T. N., Dat, N. D., & Duy, K. L. D. (2017). A decision tree using ID3 algorithm for English semantic analysis. *International Journal of Speech Technology, 20*(3), 593–613.

Picon, A., Alvarez-Gila, A., Seitz, M., Ortiz-Barredo, A., Echazarra, J., & Johannes, A. (2019). Deep convolutional neural networks for mobile capture device-based crop disease classification in the wild. *Computers and Electronics in Agriculture, 161,* 280–290. https://doi.org/10.1016/j.compag.2018.04.002.

Polat, K., & Güneş, S. (2009). A novel hybrid intelligent method based on C4. 5 decision tree classifier and one-against-all approach for multi-class classification problems. *Expert Systems with Applications, 36*(2), 1587–1592.

Porter, M. A., Williams, J., Broeg, M., Corzine, K., & Weatherford, T. (2020). Current and temperature measurement via spectral decomposition of light emission from a GaN Power Diode. In *2020 IEEE Applied Power Electronics Conference and Exposition (APEC)* (pp. 640–646). IEEE.

Rajeswari, C., Sathiyabhama, B., Devendiran, S., & Manivannan, K. (2014). A gear fault identification using wavelet transform, rough set based GA, ANN and C4. 5 algorithm. *Procedia Engineering, 97,* 1831–1841.

Ray, N. J., Styrov, V. V., & Karpov, E. G. (2017). Interfacial contributions of H2O2 decomposition-induced reaction current on mesoporous Pt/TiO2 systems. *Chemical Physics Letters, 689,* 111–115.

Rutkowski, L., Jaworski, M., Pietruczuk, L., & Duda, P. (2014). The CART decision tree for mining data streams. *Information Sciences, 266,* 1–15.

Saçlı, B., Aydınalp, C., Cansız, G., Joof, S., Yilmaz, T., Çayören, M., et al. (2019). Microwave dielectric property based classification of renal calculi: Application of a kNN algorithm. *Computers in Biology and Medicine, 112,* 103366.

Salas-Biedma,, P., Gonzalez-Prieto, I., & Duran, M. J. (2019). Current imbalance detection method based on vector space decomposition approach for five-phase induction motor drives. In *IECON 2019–45th Annual Conference of the IEEE Industrial Electronics Society* (pp. 975–980). IEEE.

Sarker, I. H., Colman, A., Han, J., Khan, A. I., Abushark, Y. B., & Salah, K. (2020). Behavdt: a behavioral decision tree learning to build user-centric context-aware predictive model. *Mobile Networks and Applications, 25*(3), 1151–1161.

Shaikhina, T., Lowe, D., Daga, S., Briggs, D., Higgins, R., & Khovanova, N. (2019). Decision tree and random forest models for outcome prediction in antibody incompatible kidney transplantation. *Biomedical Signal Processing and Control, 52,* 456–462.

Shang, W., Huang, H., Zhu, H., Lin, Y., Wang, Z., & Qu, Y. (2005) An improved kNN algorithm–fuzzy kNN. In *International Conference on Computational and Information Science* (pp. 741–746). Springer.

Shao, Y., & Lunetta, R. S. (2012). Comparison of support vector machine, neural network, and CART algorithms for the land-cover classification using limited training data points. *ISPRS Journal of Photogrammetry and Remote Sensing, 70,* 78–87.

Shin, K., Yang, H., Lee, S.-K., & Lee, Y.-S. (2013). Group delay based location template matching method for the identification of the impact location on a plate. *Journal of Sound and Vibration, 332*(8), 2111–2117.

Singh, B., & Verma, V. (2008). Selective compensation of power-quality problems through active power filter by current decomposition. *IEEE Transactions on Power Delivery, 23*(2), 792–799.

Sintorn, I.-M., Homman-Loudiyi, M., Söderberg-Nauclér, C., & Borgefors, G. (2004). A refined circular template matching method for classification of human cytomegalovirus capsids in TEM images. *Computer Methods and Programs in Biomedicine, 76*(2), 95–102.

Song, Q., Shepperd, M., Chen, X., & Liu, J. (2008). Can k-NN imputation improve the performance of C4. 5 with small software project data sets? A comparative evaluation. *Journal of Systems and Software, 81*(12), 2361–2370.

Szarvas, G., Farkas, R., & Kocsor, A. (2006). A multilingual named entity recognition system using boosting and c4. 5 decision tree learning algorithms. In *International Conference on Discovery Science* (pp. 267–278). Springer,

Thahab, R. T., & Asumadu, J. A. (2017). Current decomposition based on a double/multiple synochrouns reference frames and Fryze-Buchholz-Depenbrock theory for a non-islanded micro-grid with a finite control set-model predictive controller: A comparative approach. In *2017 IEEE 3rd International Future Energy Electronics Conference and ECCE Asia (IFEEC 2017-ECCE Asia)* (pp. 356–363). IEEE.

Thanh, N. D., Li, W., & Ogunbona, P. (2009). An improved template matching method for object detection. In *Asian Conference on Computer Vision*. Springer.

Trstenjak, B., Mikac, S., & Donko, D. (2014). KNN with TF-IDF based framework for text categorization. *Procedia Engineering, 69,* 1356–1364.

Tyuryukanov, I., & Popov, M. (2020) D-Decomposition based robust discrete-time current regulator for grid-connected VSI. In *2020 IEEE 29th International Symposium on Industrial Electronics (ISIE)* (pp. 100–107). IEEE.

Waheed, T., Bonnell, R., Prasher, S. O., & Paulet, E. (2006). Measuring performance in precision agriculture: CART—A decision tree approach. *Agricultural Water Management, 84*(1–2), 173–185.

Wang, K., Lu, J., & Zhuang, L. (2007). A Current-decomposition study of the borohydride oxidation reaction at Ni electrodes. *The Journal of Physical Chemistry C, 111*(20), 7456–7462.

Wöhrl, H., & Brunelli, D. (2019) Non-intrusive load monitoring on the edge of the network: A smart measurement node. In *International Conference on Applications in Electronics Pervading Industry, Environment and Society* (pp. 477–482). Springer.

Yazdani, D., Bakhshai, A., & Jain, P. K. (2009). A three-phase adaptive notch filter-based approach to harmonic/reactive current extraction and harmonic decomposition. *IEEE Transactions on Power Electronics, 25*(4), 914–923.

Yeo, B., & Grant, D. (2018). Predicting service industry performance using decision tree analysis. *International Journal of Information Management, 38*(1), 288–300.

Yin, Z., & Gao, Q. (2020). A novel imperialist competitive algorithm for scheme configuration rules mining of product service system. *Arabian Journal for Science and Engineering*, 1–13

Zhang, D., Fang, X., Wu, H., Zhang, X., He, S., & Liu, C. (2019a) Recognition classification based on Hu moment invariants and imperial competitive algorithm for axis trajectory of magnetic bearing-rotor system. *Personal and Ubiquitous Computing*, 1–10.

Zhang, Y., Hu, X., & Wu, C. (2020). Improved imperialist competitive algorithms for rebalancing multi-objective two-sided assembly lines with space and resource constraints. *International Journal of Production Research, 58*(12), 3589–3617.

Zhang, Z., Wang, Z., Gan, C., & Zhang, P. (2019b). A double auction scheme of resource allocation with social ties and sentiment classification for Device-to-Device communications. *Computer Networks, 155*, 62–71. https://doi.org/10.1016/j.comnet.2019.03.018.

Zimmerman, R. K., Balasubramani, G., Nowalk, M. P., Eng, H., Urbanski, L., Jackson, M. L., et al. (2016). Classification and Regression Tree (CART) analysis to predict influenza in primary care patients. *BMC Infectious Diseases, 16*(1), 503.

Chapter 3
Smart Non-intrusive Device Recognition Based on Intelligent Single-Label Classification Methods

3.1 Introduction

Smart device identification is a research hotspot in non-intrusive load monitoring. Non-intrusive load monitoring is an important part of the Internet of Things in Power Systems (IOTIPS). Thus, the smart device identification is a crucial technique in realizing the IOTIPS (Xia and Mei 2015). The intelligent device identification can be classified as the identification of independent operation devices and the identification of multiple devices running simultaneously. The identification research of independent operation devices only collects the operation data of devices running independently and realizes the classification of device types. On the one hand, the research can deepen people's understanding of different devices and provide a reference for device identification under the condition of simultaneous operation of multiple devices. On the other hand, the device identification under the condition of the device operation can also be used in the smart socket to realize the intelligent perception of a single electrical device on the socket (Tabatabaei et al. 2017). On the occasion of multiple devices running at the same time, the on-off state's identification of single operation devices can be realized. The equipment identification under the condition of simultaneous operation of multiple devices is based on the running state of actual electrical equipment and can be used for intelligent equipment identification in the actual scenes of home, office, factory, and so on.

This chapter will analyze the characteristics of various features and analyze the matching relationship between different features and different classifiers. The potential possibility of improving model performance by optimizing feature extraction parameters and classifier parameters was analyzed. This chapter is expanded according to the following chapters: Sect. 3.2 focuses on device identification based on Support Vector Machines (SVM) and the model performances with traditional physical features, harmonic features, and the Wavelet Packet Decomposition features are also analyzed; Sect. 3.3 discusses device recognition based on Extreme Learning Machine (ELM), the device recognition ability of models using various features is tested; Sect. 3.4 discusses device recognition based on Artificial Neural Networks

H. Liu et al., *Smart Device Recognition*,
https://doi.org/10.1007/978-981-33-4925-4_3

(ANN), also named Multilayer Perceptron (MLP). Based on the experimental results from Sect. 3.2 to Sect. 3.4, Sect. 3.5 fully discusses the performances of classifiers with their appropriate features. All performance evaluations will be validated based on single device operation data and multiple device simultaneous operation data.

Non-Intrusive Load Monitoring (NILM), also known as energy decomposition, refers to decomposing global energy dissipation through non-intrusive methods and analyzing specific conditions of energy consumption (Parson et al. 2012). Through the analysis of the proportion of various energy dissipation methods, the consumer energy consumption behavior can be improved by well energy management (Figueiredo et al. 2012).

Among them, most of the data sets about NILM are obtained by using low-frequency sampling signals, but there are few data sets obtained by using high-frequency sampling signals. Although this has promoted the rapid development of low-frequency NILM technology, it has also led to the slow development of high-frequency NILM technology.

Using whole-house data for NILM algorithm modeling can make the load monitoring model well trained. However, since the whole house data does not provide the usage status of individual electrical appliances, it's not adapted for the training of supervised models. On the other hand, the individual device data can adapt to the training of the model and it has the ability to achieve non-intrusive device recognition when using features of currents at the time of the electric switch.

In the chapter, the COOLL public data set is utilized to evaluate the performance of device recognition models. It is a dataset of plug-level electrical measurements. It is obtained by measuring the turn-on current and voltage of 42 independent electrical appliances. The sampling frequency is 100 kHz and there are 12 different types of devices, namely Fan, Drill, Grinder, Hedge trimmer, Hair-dryer, Lamp, Planer, Paint stripper, Router, Saw, Vacuum cleaner, and Sander, each with a certain number of samples.

3.2 Device Recognition Method Based on Support Vector Machine

3.2.1 Feature Extraction

(1) *The traditional physical quantities*

The physical quantities are frequently used as features in traditional electric device recognition. For data-based non-intrusive device recognition, the physical quantities such as active power, impedance, and other quantities, are also the most distinctive characteristics. According to Sect. 2.2.2, several physical features are calculated for each sample and used in the experiments. The physical features used in the chapter are given in Table 3.1.

Table 3.1 The adopted physical features

Feature name	Abbreviation	Feature name	Abbreviation
Original current	I	Reactance	X
Reactive power	Q	Impedance	Z
Active power	P	Susceptance	B
Power factor	λ	Conductance	G
Resistance	R	Admittance	Y

(2) *The theoretical basis of the harmonics*

In the power supply system, the harmonics are a kind of non-sinusoidal electric signal with the Fourier decomposition. After decomposition, the current signal obtains a fundamental wave signal consistent with its fundamental frequency. In addition, the decomposition result also includes some components greater than the fundamental frequency, which is called harmonics.

Generally speaking, the load on the grid can be divided into linear load and non-linear load. In normal work, the voltage is 220 V, frequency 50 Hz sinusoidal alternating current. When the waveform of the load current is also a 50 Hz sine waveform, this type of load is called a linear load. When the phase of the load current waveform is consistent with the voltage, it is called a resistive load, such as common incandescent lamps, electric stoves, and electric water heaters. When the load current phase leads the voltage, this type of load is called a capacitive load. There are relatively few pure capacitive loads. For example, the reactive power compensation cabinets commonly used in general power distribution rooms are pure capacitive loads. When the load current phase lags behind the voltage waveform, this type of load is called an inductive load. Common inductive loads such as motors, transformers, inductive ballasts, etc.

The characteristic of the above linear load is whether its current phase leads or lags behind the voltage waveform. The current waveform is always a sine wave without distortion. The relationship between voltage and current conforms to Ohm's law, which is the key to linear loads. The non-linear load means that its current waveform is a non-sine wave, or a sine wave having greater distortion relative, such as the currents of computers, color TVs, induction cookers, power frequency UPS, inverters, electroplating equipment, etc. The load current waveform is not a sine wave and each type of devices has its own unique current form.

In the frequency domain, the differences of various current forms are embedded in its harmonic amplitudes and phases. Thus, by analyzing the harmonic features of each device with machine learning models, the non-intrusive device recognition is able to be realized. The harmonic amplitudes and phase angles can be obtained through the Fast Flourier Transformation. After analyzing the harmonic features of all samples, the even harmonics are found useless in recognizing various devices. As shown in Fig. 3.1, the first ten harmonics of various device samples are analyzed. It is obvious that for all of the devices,

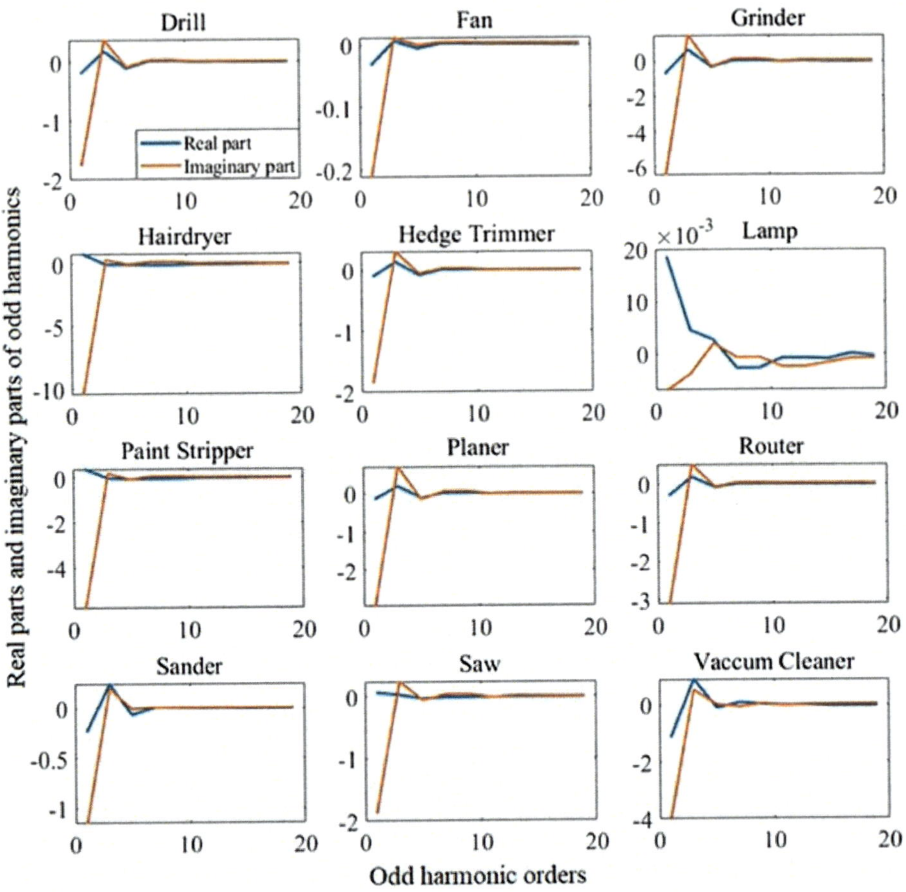

Fig. 3.1 The odd harmonics of various devices

the harmonics of higher orders are very closed to zero. That means the high-order harmonic features would be useless in device recognition. However, the first five harmonics of different devices are distinctive, which can be used as the harmonic features for device recognition. In the chapter, the phase angles and amplitudes of the first 5 odd current harmonics are taken as the harmonic features of devices.

(3) *The theoretical basis of the Wavelet Packet Decomposition (WPD)*
For a continuous periodic signal, Fourier transform can decompose it into a combination of trigonometric function signals with different frequencies. However, for abrupt signals, a large number of trigonometric function signals are required for linear combinations. In order to analyze non-stationary signals, wavelet transform methods are often used. Wavelet transform takes a set of orthogonal and rapidly decayed wavelet function bases for signal fitting. However, this method only decomposes the low-frequency part of the signal when processing the signal, and does not process the high-frequency part of the

signal, that is, this method has a poor resolution for the high-frequency part. The WPD is a more accurate time-frequency analysis method extended on the basis of the wavelet transform. This method can analyze the signal characteristics in more detail. The defect of the wavelet transform improves the time-frequency resolution and has more excellent application effects (Cui et al. 2006).

Wavelet packet analysis is actually developed and improved on the basis of wavelet analysis (Peng and Lu 2016). It is more flexible than wavelet analysis. The wavelet packet decomposition tree can be adjusted according to different actual needs. It is more purposeful and makes signal decomposition more effective. It is meticulous and has a wider practical value. From the point of view of the filter, the WPD method is based on the discrete wavelet transform and can further subdivide its high-frequency detail signal. The wavelet packet decomposition is equivalent to a low-pass filter and 2^n-1 band-pass filters. Figure 3.2 shows the complete decomposition diagram of the three-layer wavelet packet of the signal.

In the dyadic wavelet packet transform, when the wavelet packet is decomposed at various levels, the scaling function of the adjacent series and the wavelet function also have a recurrence relationship. In the wavelet packet transform, the father wavelet $\phi(t)$ is $\mu_0^0(t)$ and the mother wavelet $\psi(t)$ is $\mu_1^0(t)$. The superscript represents the decomposition level of the wavelet packet, and the subscript represents the position of the wavelet packet at its level.

Then the above recurrence relationship can be expressed as follows:

$$\gamma_{2n}(t) = \sum_k h_k \gamma_n(2t - k) \tag{3.1}$$

$$\gamma_{2n+1}(t) = \sum_k g_k \gamma_n(2t - k) \tag{3.2}$$

where h_k and g_k are defined as wavelet transform, and is called wavelet packet. The WPD method applied to the initial data in the COOLL dataset, the wavelet features of samples are extracted. The wavelet features are also added to the feature database. Figure 3.3 gives an example of the WPD feature extraction procedure.

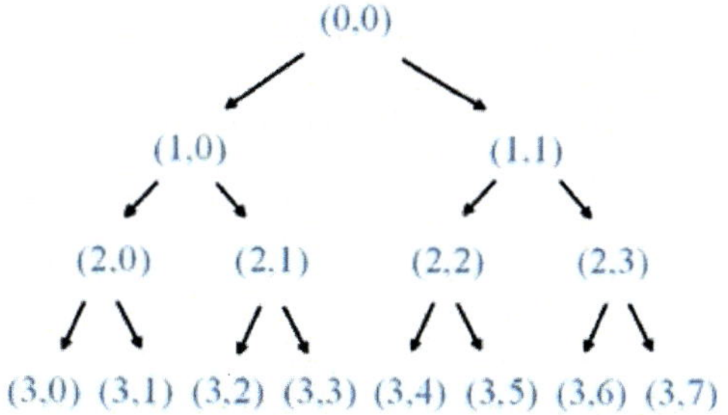

Fig. 3.2 The complete decomposition diagram of the three-layer wavelet packet

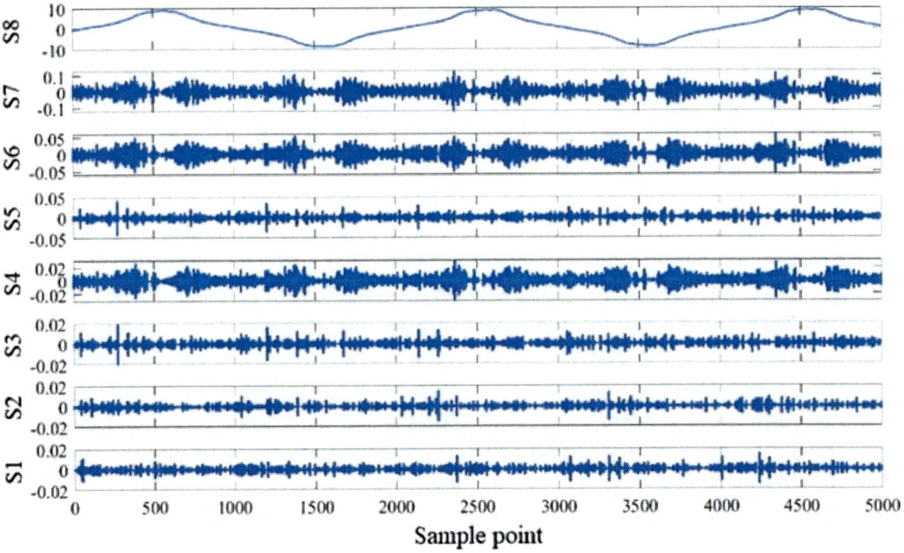

Fig. 3.3 The example of WPD feature extraction

3.2.2 *Steps of the Model Based on SVM*

Support vector machines (SVM) is a model for dealing with two classification problems (Kandaswamy et al. 2011). For a linearly separable input sample, a hyperplane can be used to divide the data set into two parts. There are countless such interfaces, but there is only one interface with the largest interval. The learning strategy of SVM is to find the interface with the maximum interval, so as to achieve more accurate classification. In the equation, is the normal vector of the hyperplane, and a is an offset parameter. The interface with the largest interval shall meet:

$$\min_{\omega,b} \frac{1}{2}\|\beta\|^2, \quad s.t.\ y_i[\beta x_i + a] - 1 \geq 0\ (i = 1, 2, \cdots, m) \tag{3.3}$$

Because in practical application problems, the sample data set is not always linearly separable. When the data is more complex, it is difficult to accurately classify each data with the maximum separation interface. In order to solve the possible misclassification in SVM, the concept of "soft interval" is introduced. Allow some data points not to meet the constraints: $y_i[\beta x_i + a] - 1 \geq 0$, and use hinge loss to rewrite the original optimization problem as below:

$$\min_{\omega,b} \frac{1}{2}\|\beta\|^2 + C\sum_{i=1}^{m} \varepsilon_i, \quad s.t.\ y_i[\beta x_i + a] - 1 \geq 1 - \ \varepsilon_i\ (i = 1, 2, \cdots, m) \tag{3.4}$$

$$\varepsilon_i = \max(0,\ 1 - y_i(\beta x_i + a)) \tag{3.5}$$

where ε_i is the "slack variable". Each sample has a corresponding slack variable, which represents the degree to which the sample does not meet the constraints. And in the equation, C is the penalty parameter (Huang et al. 2020).

In real life, some classification problems are not simply linearly separable. Compared with linear classification problems, nonlinear classification problems need to realize spatial mapping between low and high dimensional space. Non-linear SVM can achieve this non-linear transformation by using a kernel function. The definition of the kernel function $K(x, z)$ is given as follows:

$$K(x, z) = \psi(x)\psi(z) \tag{3.6}$$

where, $\psi(x)$ refers to the relationship between the feature space and the input space.

3.2.3 Performance Evaluation

(1) *Performance metrics*
Several classification performance metrics have already been introduced in Sect. 2.2.1 and the calculating methods have also been clarified. In the chapter, the metrics, named Recall (R), Precision (P), and F1-score, are selected for model evaluation and validation. Sometimes, there are non-numeric values of these measures, which is caused by the extremely bad results. In this case, the metric values are replaced by NaN.

(2) *The SVM model performance on the training set and test set*
Tables 3.2 and 3.3 list the identification results of each appliance. Figures 3.4 and 3.5 show the confusion matrixes of the SVM model performed on the training

Table 3.2 The performance of the physical-SVM model in the training process

Appliance	TP	FN	FP	TN	P	R	F1-score
Drill	84	0	0	504	1	1	1
Fan	27	1	0	560	1	0.964	0.981
Grinder	28	0	0	560	1	1	1
Hair dryer	56	0	0	532	1	1	1
Hedge trimmer	16	26	0	546	1	0.380	0.551
Lamp	56	0	1	531	0.982	1	0.99115
Paint tripper	14	0	0	574	1	1	1
Planer	13	1	0	574	1	0.928	0.962
Router	14	0	1	573	0.933	1	0.965
Sander	42	0	0	546	1	1	1
Saw	107	5	26	450	0.804	0.955	0.873
Vacuum cleaner	98	0	5	485	0.951	1	0.975

Table 3.3 The performance of the physical-SVM model in the testing process

Appliance	TP	FN	FP	TN	P	R	F1-score
Drill	36	0	0	216	1	1	1
Fan	12	0	0	240	1	1	1
Grinder	12	0	0	240	1	1	1
Hair dryer	24	0	0	228	1	1	1
Hedge trimmer	5	13	0	234	1	0.277	0.434
Lamp	24	0	0	228	1	1	1
Paint tripper	6	0	0	246	1	1	1
Planer	6	0	0	246	1	1	1
Router	6	0	0	246	1	1	1
Sander	18	0	0	234	1	1	1
Saw	48	0	13	191	0.786	1	0.880
Vacuum cleaner	42	0	0	210	1	1	1

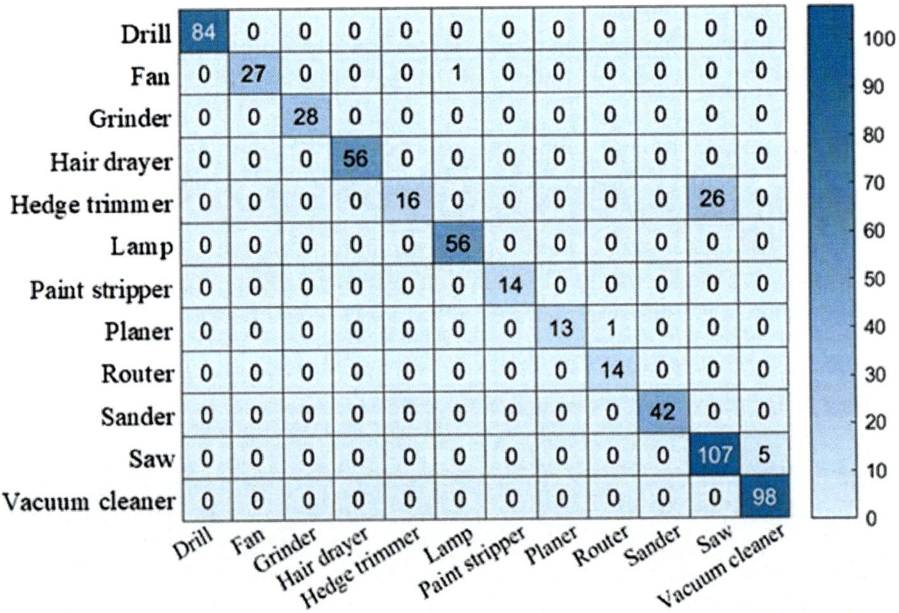

Fig. 3.4 Confusion matrix of the physical-SVM model in the training process

set and test set. By analyzing the results of SVM with physical features, it can be found that the SVM model with physical features performs well in the device recognition process. Nearly all of the devices are recognized correctly. Besides, according to the metrics in Tables 3.2 and 3.3, it can be concluded that the model performance of hedge trimmer is the worst in both of the training process and

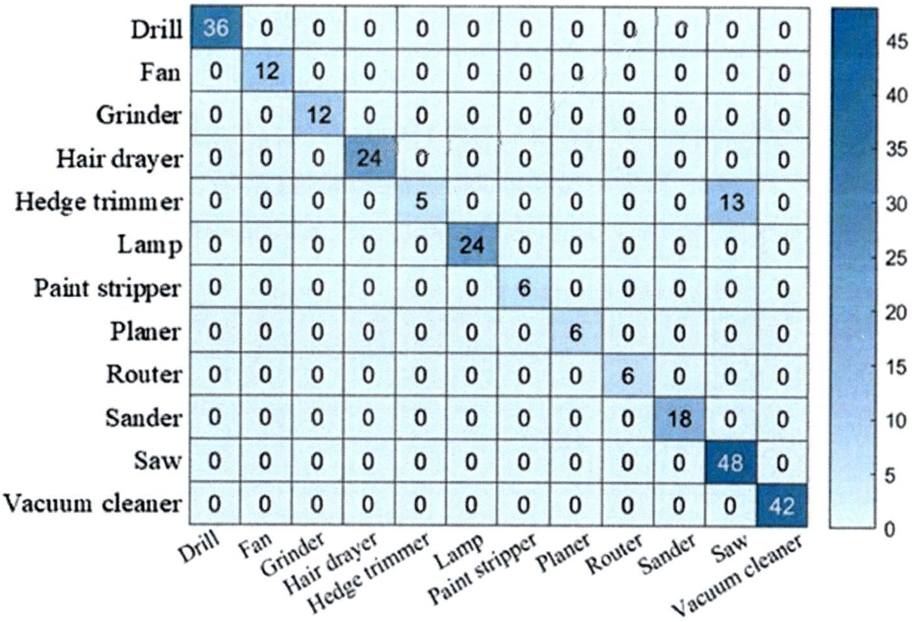

Fig. 3.5 Confusion matrix of the physical-SVM model in the testing process

testing process, and the performances of the SVM model using physical features on the test set and training set are similar which indicts the stationary of SVM.

(3) *The SVM model performance based on different features*

Comparison experiments are also conducted in the section in order to analyze the effect of different features on the performance of SVM. The physical features, harmonic features, and the WPD features are analyzed respectively. The device recognition results are given in Figs. 3.6, 3.7 and 3.8 in the form of the TP, FN, FP, and TN indexes. Table 3.4 shows the Precision, Recall, and F1-score values of model performance with various features. Analyzing the device recognition results, some conclusions are acquired as follows: (a) among all the features, the SVM model with physical features performs better than SVM using the other features, which means the physical features contain crucial and distinctive information of devices; (b) the WPD features are useless for SVM based device recognition model due to the extremely inaccurate results; (c) considering that the recognition results of hedge trimmer and saw are always unsatisfying no matter which features are selected, the model based on SVM cannot identify the samples of hedge trimmer and saw; (d) for the other devices, the SVM device recognition model using physical features is able to achieve high accuracy classification.

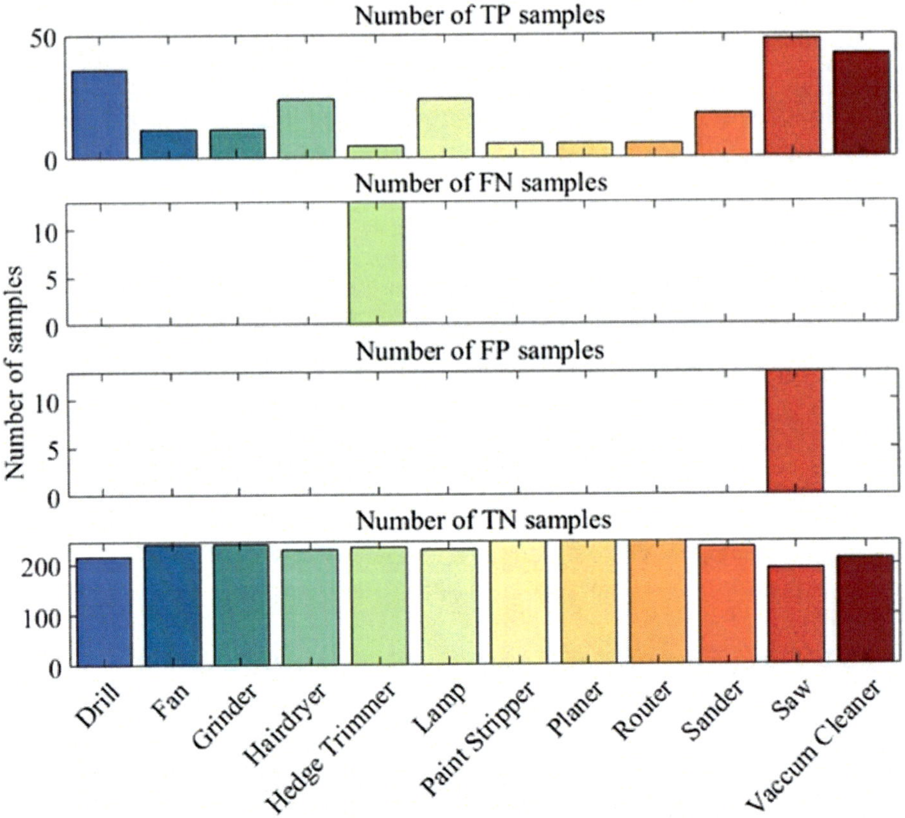

Fig. 3.6 The TP, FN, FP, TN results obtained by SVM with physical features

3.3 Device Recognition Method Based on Extreme Learning Machine

3.3.1 Data Process and Feature Extraction

The ELM-based device recognition model is also evaluated on the COOLL dataset. The methods of raw data processing and feature extraction are illustrated thoroughly in Sect. 3.2.1. The physical features, harmonic features, and WPD features are also utilized in this section to validate the effectiveness of the ELM-based device recognition model.

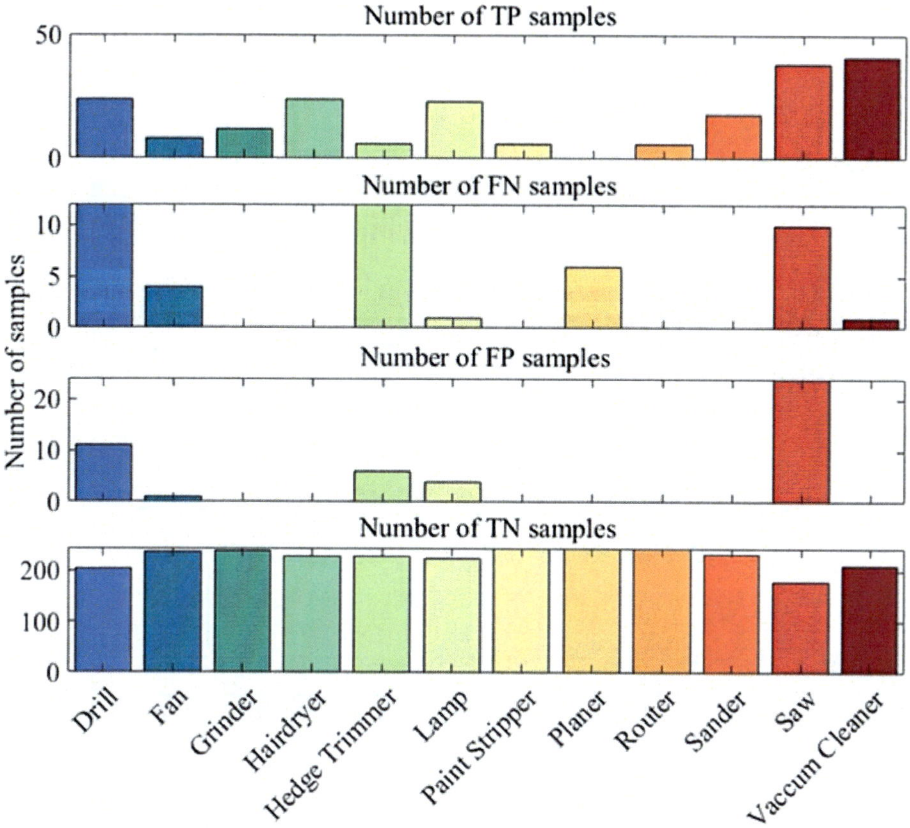

Fig. 3.7 The TP, FN, FP, TN results obtained by SVM with harmonic features

3.3.2 *Steps of the Model Based on Extreme Learning Machine*

The Extreme Learning Machine (ELM) is based on the Feedforward Neuron Network (FNN). The ELM network structure is composed of an output layer, an input layer, and a hidden layer. The weights and biases of the ELM hidden and input layer are randomly selected. The weight values of the output layer are optimized by constructing a minimized loss function. It is composed of a training error term and a regular term of the weight norm of the output layer. The essence is to solve a Moore-Penrose (MP) generalized inverse matrix.

Suppose a certain training set, $\{t_i, c_i \,|t_i \in R^m,\ c_i \in R^d,\ i = 1, 2, 3, \cdots, N\}$, which c_i represents the i-th data, t_i denotes the corresponding label.

The input of ELM is a training sample set, and the connection form of the hidden and input layers is fully connection form (Zhao et al. 2014). The output function of the hidden layer is denoted by $H(c)$. The hidden layer output $H(c)$ calculation formula is given below (Kang et al. 2020):

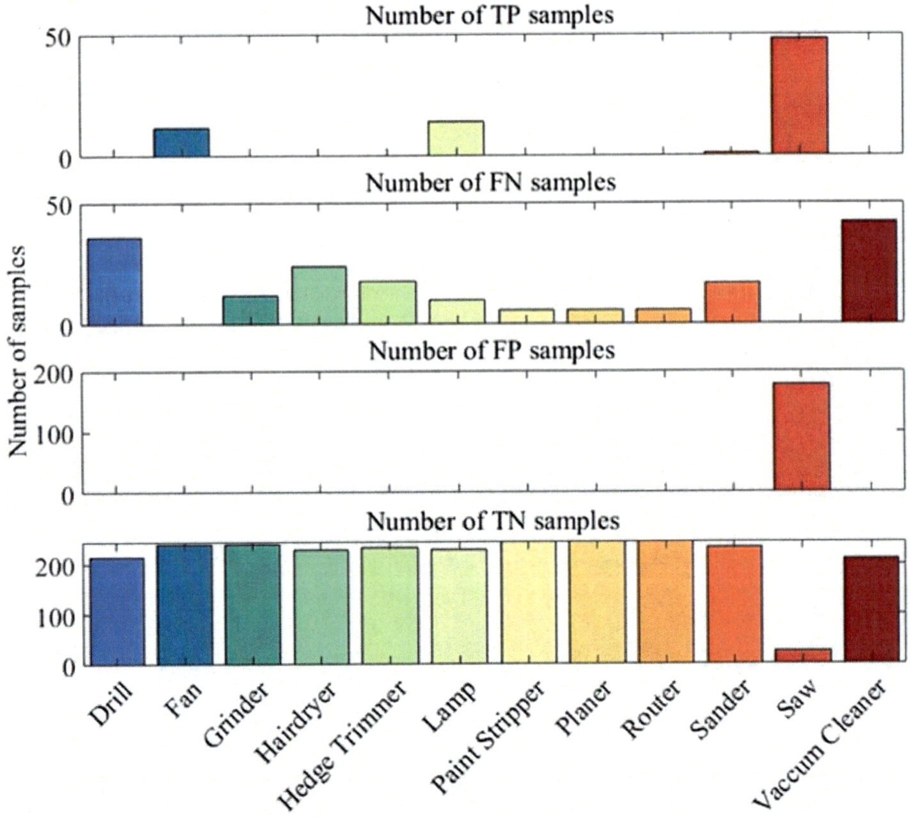

Fig. 3.8 The TP, FN, FP, TN results obtained by SVM with WPD features

$$H(c) = [h_1(c), h_2(c), \cdots, h_L(c)] \tag{3.7}$$

where L represents the quantity of the hidden layer nodes and $h(c)$ is one of the outputs, $h_i(c)$ represents output of the i-th hidden layer node (Fan 2009).

The input of the sample set is multiplied by the corresponding weight, plus the deviation, and then passes through a nonlinear function to be the outputs of the hidden layer nodes (Kongsorot et al. 2019). $H(c)$ is a nonlinear mapping. The output function of the hidden layer node is as given as follows:

$$h_i(c) = g(\omega_i c + b_i) \tag{3.8}$$

where $g(\omega_i c + b_i)$ represents the activation function (Oliehoek 2012). After the hidden layer is output, it enters the final output layer. According to the ELM network structure diagram, the output of ELM $f_{L(c)}$ is expressed as:

$$f_{L(c)} = \sum_{i=1}^{L} \alpha_i h_i(c) = \alpha H(c), \quad \alpha = [\alpha_1, \alpha_2, \cdots, \alpha_L]^T \tag{3.9}$$

Table 3.4 Classification results of SVM with different features

Appliance	Physical features			Harmonic features			WPD features		
	P	R	F1-score	P	R	F1-score	P	R	F1-score
Drill	1	1	1	0.685	0.667	0.676	NaN	0	NaN
Fan	1	1	1	0.889	0.667	0.762	1	1	1
Grinder	1	1	1	1	1	1	NaN	0	NaN
Hair dryer	1	1	1	1	1	1	NaN	0	NaN
Hedge trimmer	1	0.278	0.435	0.5	0.333	0.4	NaN	0	NaN
Lamp	1	1	1	0.852	0.958	0.902	1	0.583	0.737
Paint tripper	1	1	1	1	1	1	NaN	0	NaN
Planer	1	1	1	NaN	0	NaN	NaN	0	NaN
Router	1	1	1	1	1	1	NaN	0	NaN
Sander	1	1	1	1	1	1	1	0.056	0.105
Saw	0.787	1	0.881	0.613	0.792	0.691	0.213	1	0.352
Vacuum cleaner	1	1	1	1	0.976	0.988	NaN	0	NaN

The NaN value indicates that the metric cannot be computed because the classification results of that device are so inaccurate that the TN value is 0 as well as FP or FN

where is a weight vector of the output and hidden layers.

The ELM training process needs to randomize the hidden layer parameters and select an appropriate activation function. Construct the minimum error function as Eq. 3.10 and optimize the weight distribution of output layer.

$$\min\|\alpha H(c)\|^2, \quad \alpha \in R^{L\times M} \tag{3.10}$$

in which, H is the hidden layer output matrix, T is the label matrix. According to the knowledge of matrix theory, the optimal solution can be obtained as $\alpha^* = H^*T$, where H^* denotes the Moore-Penrose generalized inverse matrix of the matrix H. Compared with MLP, ELM has faster learning speed, fewer training parameters, and stronger generalization ability.

3.3.3 *Performance Evaluation*

(1) *The ELM model performance on training set and testing set*
The ELM-based device recognition model is evaluated on both of the training set and test set. The experiment results of the model using harmonic features are given in Tables 3.5 and 3.6 for comparing the performance of ELM on training set and test set. Besides, the confusion matrixes are shown in Figs. 3.9 and 3.10.

Table 3.5 The performance of the harmonic-ELM model in the training process

Appliance	TP	FN	FP	TN	P	R	F1-score
Drill	38	46	0	504	1	0.452	0.623
Fan	0	28	0	560	NaN	0	NaN
Grinder	28	0	0	560	1	1	1
Hair drayer	56	0	0	532	1	1	1
Hedge trimmer	27	15	30	516	0.474	0.643	0.545
Lamp	56	0	45	487	0.554	1	0.713
Paint tripper	14	0	0	574	1	1	1
Planer	0	14	0	574	NaN	0	NaN
Router	0	14	0	574	NaN	0	NaN
Sander	42	0	14	532	0.750	1	0.857
Saw	95	17	45	431	0.679	0.848	0.754
Vacuum cleaner	98	0	0	490	1	1	1

Table 3.6 The performance of the harmonic-ELM model in the testing process

Appliance	TP	FN	FP	TN	P	R	F1-score
Drill	21	15	1	215	0.955	0.583	0.724
Fan	0	12	0	240	NaN	0	NaN
Grinder	12	0	0	240	1	1	1
Hair drayer	24	0	0	228	1	1	1
Hedge trimmer	5	13	9	225	0.357	0.278	0.313
Lamp	24	0	15	213	0.615	1	0.762
Paint tripper	6	0	0	246	1	1	1
Planer	0	6	0	246	NaN	0	NaN
Router	0	6	0	246	NaN	0	NaN
Sander	18	0	6	228	0.75	1	0.857
Saw	44	4	26	178	0.629	0.917	0.746
Vacuum cleaner	41	1	0	210	1	0.976	0.988

Analyzing the results, it can be concluded that: (a) for most of the devices, the ELM model has similar recognition accuracy of the training set and testing set, but the model performance on hedge trimmer training samples is apparently better than that on the test samples; (b) the identification results of fan, planer, and router are the worst among all devices whether in testing set or training set, which indicates that the harmonic features of these devices are not distinctive enough; (c) the samples of grinder, hairdryer, and paint tripper are all correctly classified, pointing that the harmonic features of these devices are quite effective for ELM.

(2) *The ELM model performance based on different features*
The ELM device recognition models based on physical features, harmonic features, and WPD features are all conducted on the testing set to analyze the

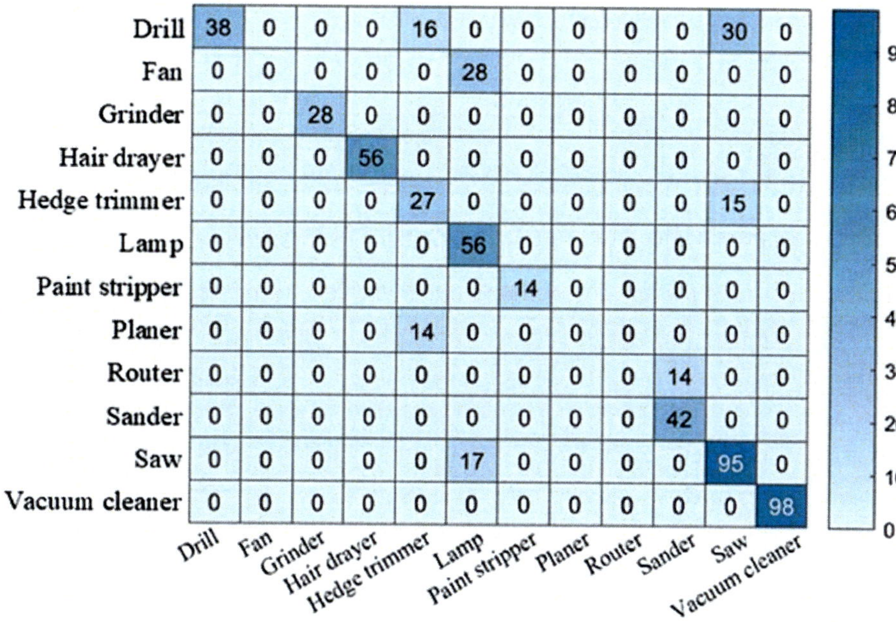

Fig. 3.9 Confusion matrix of the harmonic-ELM model in the training process

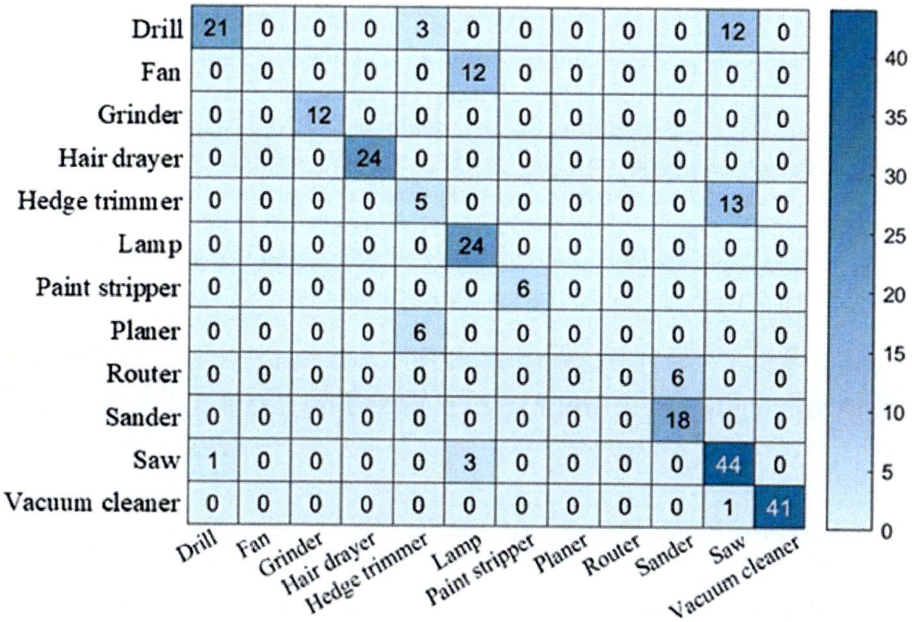

Fig. 3.10 Confusion matrix of the harmonic-ELM model in the testing process

Table 3.7 Classification results of ELM with different features

Appliance	Physical features			Harmonic features			WPD features		
	P	R	F1-score	P	R	F1-score	P	R	F1-score
Drill	1	1	1	0.685	0.667	0.676	NaN	0	NaN
Fan	1	0.5	0.667	0.889	0.667	0.762	1	1	1
Grinder	1	1	1	1	1	1	NaN	0	NaN
Hair dryer	1	1	1	1	1	1	NaN	0	NaN
Hedge trimmer	1	0.389	0.56	0.5	0.333	0.4	NaN	0	NaN
Lamp	0.8	1	0.889	0.852	0.958	0.902	1	0.583	0.737
Paint tripper	1	1	1	1	1	1	NaN	0	NaN
Planer	1	1	1	NaN	0	NaN	NaN	0	NaN
Router	1	1	1	1	1	1	NaN	0	NaN
Sander	1	1	1	1	1	1	1	0.056	0.105
Saw	0.813	1	0.897	0.613	0.792	0.691	0.213	1	0.352
Vacuum cleaner	1	1	1	1	0.976	0.988	NaN	0	NaN

impact of features on the ELM model. The results of ELM device recognition models with various features are shown in Table 3.7 and Figs. 3.11, 3.12 and 3.13. After analyzing the results, it can be summed up that (a) as the model with physical features performs the best among all ELM-based device recognition models, physical features are definitely effective in reaching an accurate recognition result; (b) since that most of the devices are wrongly recognized by ELM with WPD features, the WPD features is considered nearly useless for ELM-based device recognition models; (c) on the occasion that suitable features are selected, the ELM model has the ability to achieve accurate device recognition.

3.4 Device Recognition Method Based on Artificial Neural Network

3.4.1 Data Process and Feature Extraction

The dataset used in the section is COOLL public dataset, which is introduced in Sect. 3.1. Besides, the data process and feature extraction methods are detailed illustrated in Sect. 3.2.1. By applying those methods, the raw data from the COOLL dataset is well prepared for device recognition.

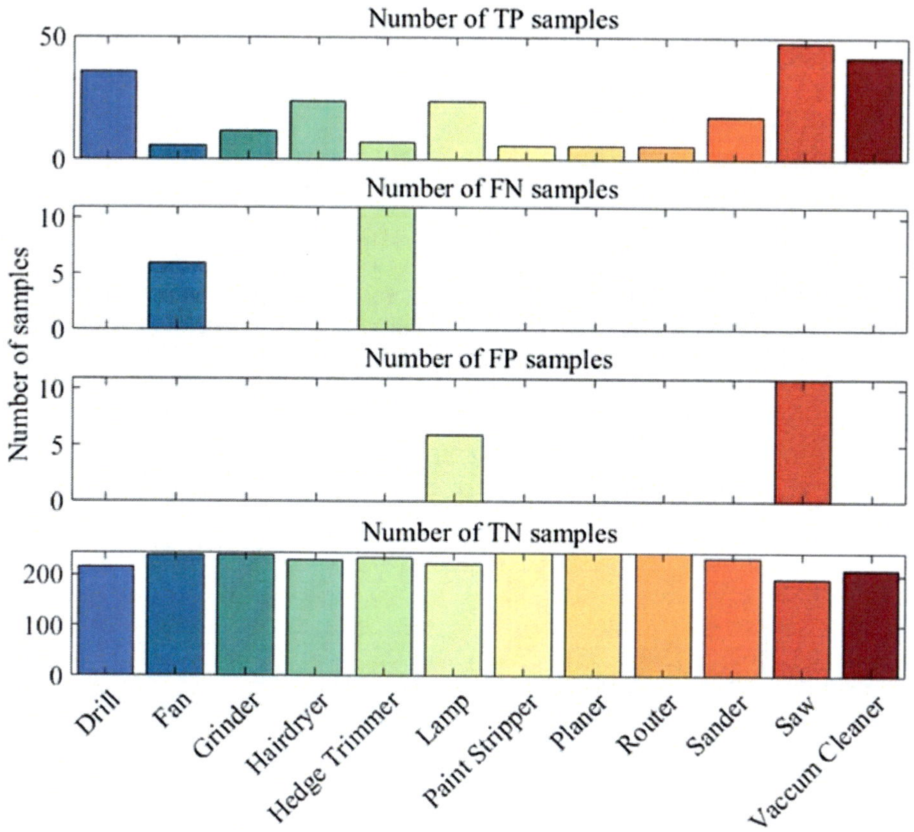

Fig. 3.11 The TP, FN, FP, TN results obtained by ELM with physical features

3.4.2 *Steps of the Multi-layer Perceptron Based Model*

Multi-layer Perceptron (MLP) is also called Artificial Neural Network (ANN) (Hu and Weng 2009). Different from a single-layer perceptron, a multi-layer perceptron has the input, output, and hidden layers. The output of each perceptron in each layer is linearly combined according to a certain weight, and then the bias is added to form the input of each perceptron in the next layer. The weight value and bias value of the MLP will be continuously updated during the training process (Gardner et al. 1998). According to the error obtained in the MLP forward propagation process, the weight value and the deviation value in the model are corrected to find the hyperplane with the best classification effect. The gradient descent method is usually used to correct the weight value and the deviation value. The principle of the gradient descent method is to calculate the gradient of loss function for all internal variables and perform backpropagation (Chtioui et al. 2015). The weight values and deviation values of each layer in MLP are internal variables, which are adjusted according to the steepest descent direction of the curved surface spanned by the loss function.

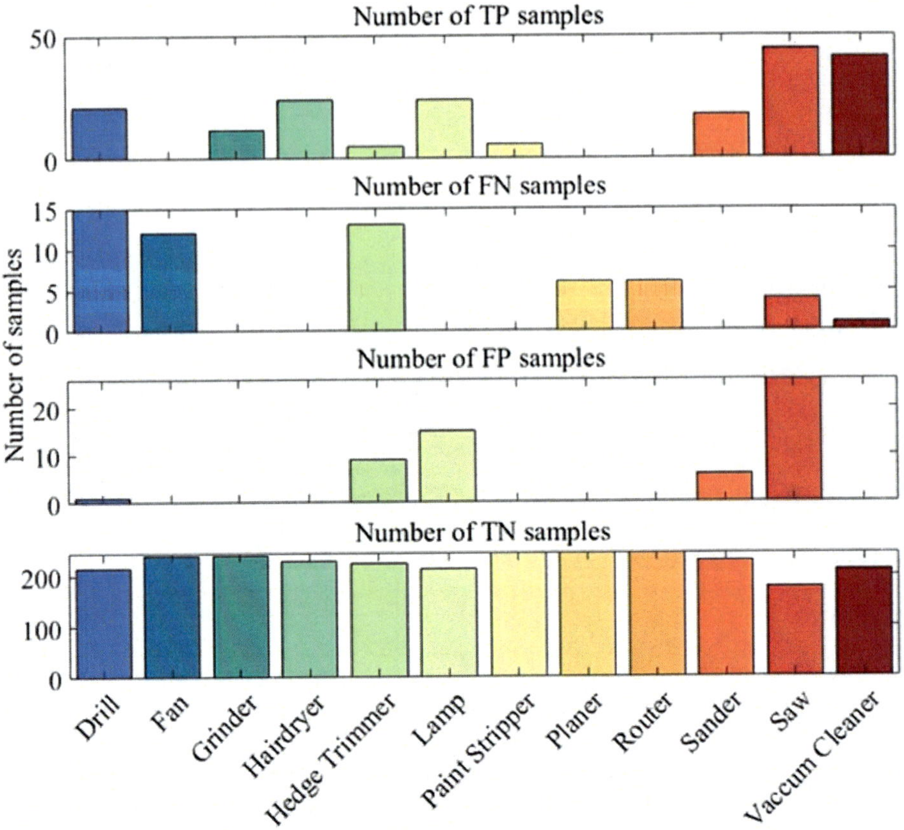

Fig. 3.12 The TP, FN, FP, TN results obtained by ELM with harmonic features

Although MLP can solve complex regression or classification problems after training, it also has shortcomings. The selection of the quantity of the hidden layer nodes has an impact on the comprehensive performance of the multilayer perceptron (Gomm and Yu 2000; Parlos et al. 2000). However, the selection of the quantity of the hidden layer nodes does not have the same rule. Besides, the learning speed of MLP is slow and the calculation is complicated.

3.4.3 Performance Evaluation

(1) *The MLP model performance on the training set and testing set*
The MLP-based device recognition method is evaluated on the training set and testing set respectively. In the section, results of the model with WPD features are displayed to analyze the performance of MLP on the training set and testing set. As shown in Tables 3.8 and 3.9, Figs. 3.14, and 3.15, the results and confusion matrixes of model performance on the training set and testing set are analyzed.

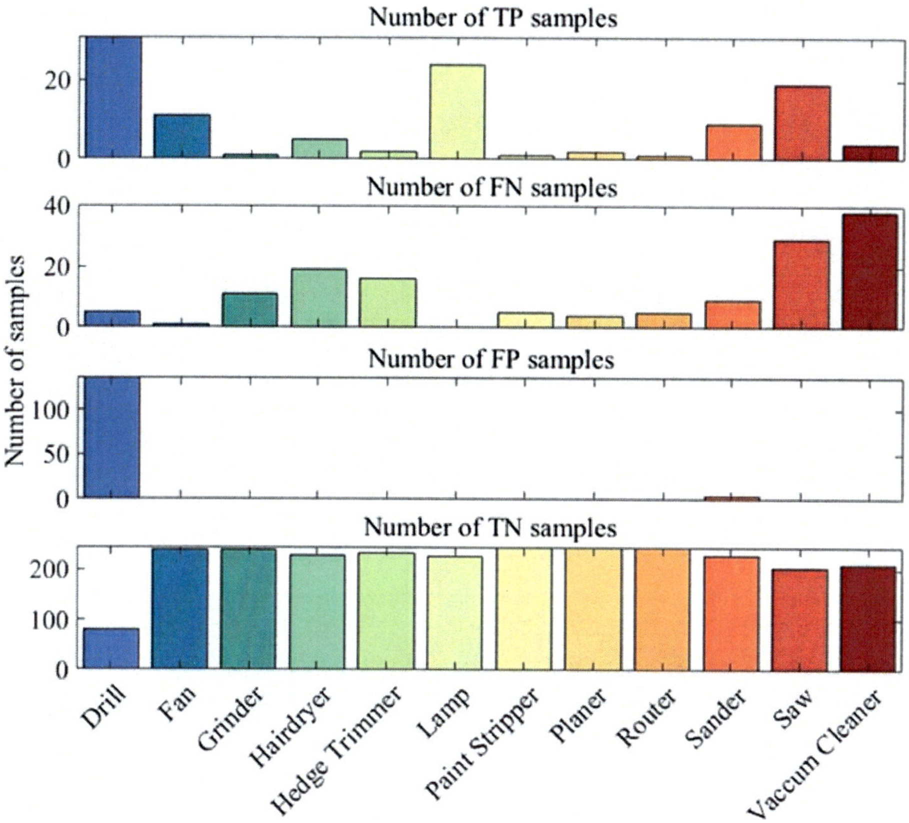

Fig. 3.13 The TP, FN, FP, TN results obtained by ELM with WPD features

Table 3.8 The performance of the WPD-MLP model in the training process

Appliance	TP	FN	FP	TN	P	R	F1-score
Drill	45	39	10	494	0.818	0.535	0.647
Fan	0	28	0	560	NaN	0	NaN
Grinder	27	1	0	560	1	0.964	0.981
Hair dryer	56	0	0	532	1	1	1
Hedge trimmer	25	17	2	544	0.925	0.595	0.724
Lamp	56	0	35	497	0.615	1	0.761
Paint tripper	14	0	0	574	1	1	1
Planer	14	0	0	574	1	1	1
Router	14	0	0	574	1	1	1
Sander	42	0	6	540	0.875	1	0.933
Saw	98	14	46	430	0.680	0.875	0.765
Vacuum cleaner	98	0	0	490	1	1	1

Table 3.9 The performance of the WPD-MLP model in the testing process

Appliance	TP	FN	FP	TN	P	R	F1-score
Drill	13	23	7	209	0.65	0.361	0.464
Fan	0	12	0	240	NaN	0	NaN
Grinder	12	0	0	240	1	1	1
Hair dryer	24	0	0	228	1	1	1
Hedge trimmer	9	9	2	232	0.818	0.5	0.62
Lamp	24	0	17	211	0.585	1	0.738
Paint tripper	6	0	0	246	1	1	1
Planer	6	0	0	246	1	1	1
Router	6	0	2	244	0.75	1	0.857
Sander	18	0	4	230	0.818	1	0.9
Saw	39	9	21	183	0.65	0.8125	0.722
Vacuum cleaner	42	0	0	210	1	1	1

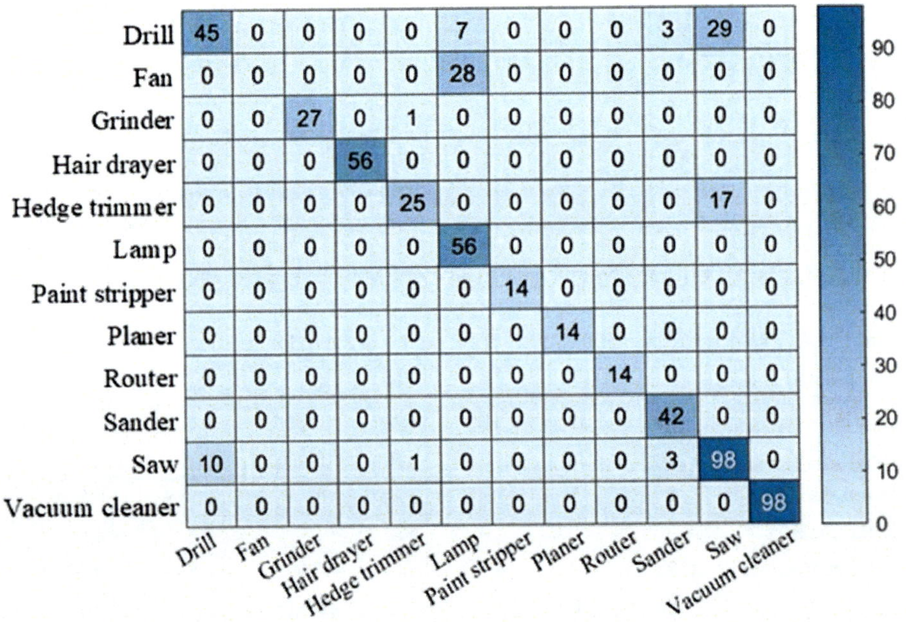

Fig. 3.14 Confusion matrix of the WPD-MLP model in the training process

It can be found that (a) compared to the SVM method and the ELM method, the MLP model performs better with WPD features; (b) the results of the MLP model performed on the training set are better than that performed on the testing set; (c) the recognition of samples of the drill, fan, and hedge trimmer is worse

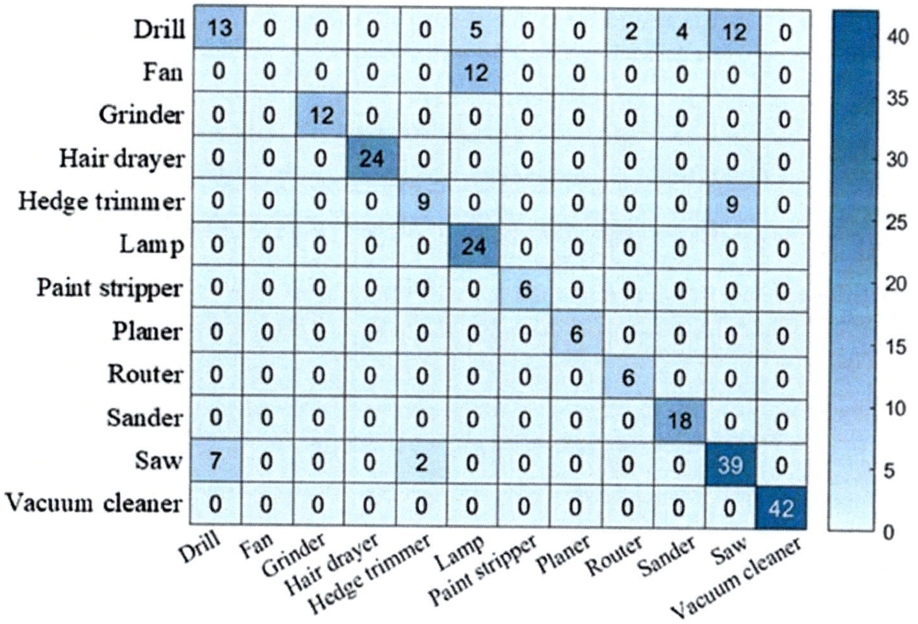

Fig. 3.15 Confusion matrix of the WPD-MLP model in the training process

than the other devices, which may be caused by the feature dispersion of these devices.

(2) *The MLP model performance based on different features*
In industrial applications, the characteristics of industrial data are very important for data analysis. It is of great significance to conduct the MLP-based device recognition model with different features and analyze the effect of features on MLP. The precision, recall, and F1-score of the results are given in Table 3.10 and the TP, FN, FP, and TN results of different features are shown in Figs. 3.16, 3.17 and 3.18. After discussing the results, it can be concluded that (1) as the accuracy of MLP using physical features is too low, the physical features are not suitable for the MLP-based device recognition model; (2) the metrics of the model using harmonic features is the best and from Fig. 3.17 it can be seen that nearly all of the samples are correctly classified; (3) the performance of MLP using WPD features are also satisfying, which indicates that the WPD features is appropriate to the MLP-based device recognition model.

3.5 Experiment Analysis

In the section, the results of SVM, ELM, and MLP models using the features which can apparently enhance their recognition accuracy are analyzed together. As listed in Table 3.11, the values of performance metrics of SVM, ELM, and MLP model are

Table 3.10 Classification results of MLP with different features

Appliance	Physical features			Harmonic features			WPD features		
	P	R	F1-score	P	R	F1-score	P	R	F1-score
Drill	1	1	1	0.968	0.861	0.911	0.804	0.916	0.857
Fan	1	0.5	0.666	1	1	1	0.631	1	0.774
Grinder	NaN	0	NaN	1	1	1	1	0.916	0.956
Hairdryer	0.8	1	0.888	1	1	1	1	1	1
Hedge trimmer	NaN	0	NaN	1	0.944	0.971	1	0.667	0.8
Lamp	0.454	0.208	0.285	1	1	1	0.944	0.708	0.809
Paint tripper	NaN	0	NaN	1	1	1	1	1	1
Planer	NaN	0	NaN	0.857	1	0.923	0.857	1	0.923
Router	NaN	0	NaN	1	1	1	1	1	1
Sander	NaN	0	NaN	1	1	1	1	1	1
Saw	0.389	0.916	0.546	0.903	0.979	0.94	0.833	0.833	0.833
Vacuum cleaner	0.75	1	0.857	1	1	1	1	1	1

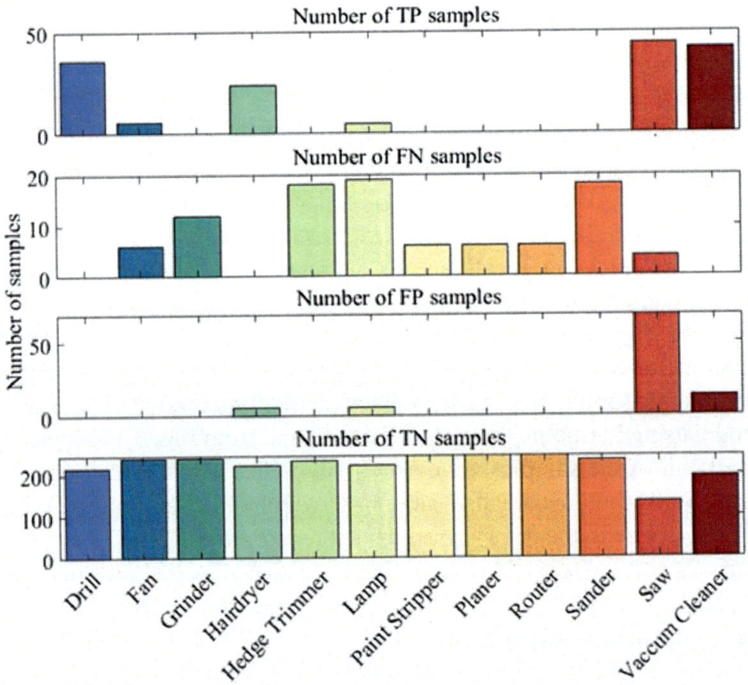

Fig. 3.16 The TP, FN, FP, TN results obtained by MLP with physical features

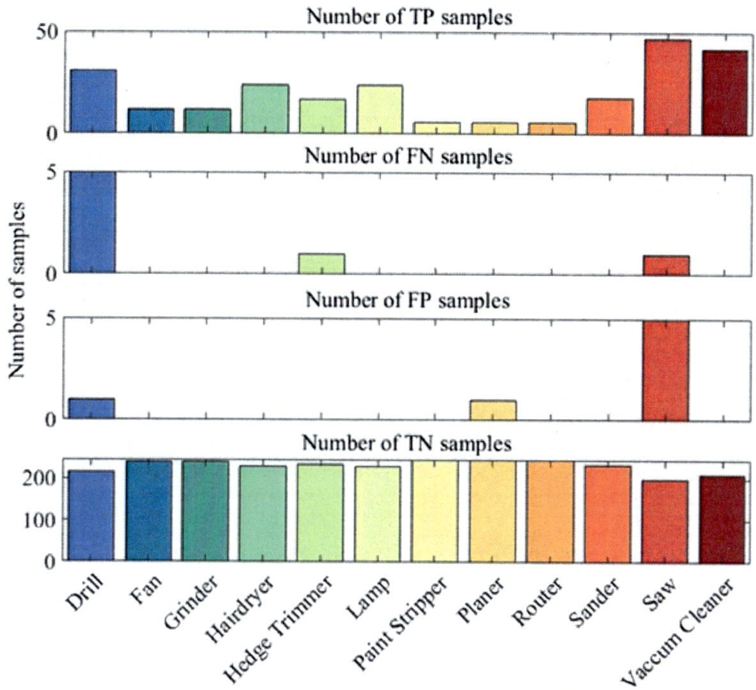

Fig. 3.17 The TP, FN, FP, TN results obtained by MLP with harmonic features

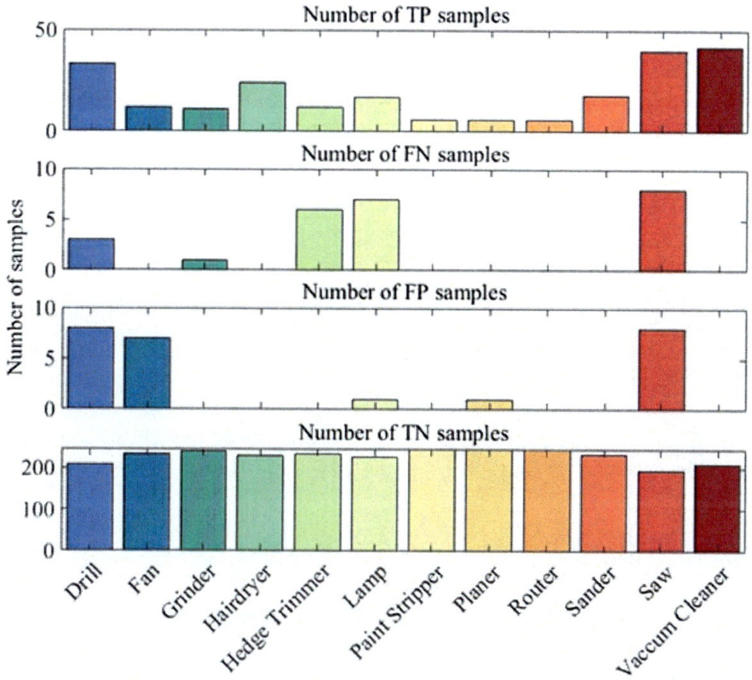

Fig. 3.18 The TP, FN, FP, TN results obtained by MLP with the WPD feature

Table 3.11 Classification results of SVM, ELM, and MLP with appropriate features

Appliance	Physical-SVM			Physical-ELM			Harmonic-MLP		
	P	R	F1-score	P	R	F1-score	P	R	F1-score
Drill	1	1	1	1	1	1	0.968	0.861	0.911
Fan	1	1	1	1	0.5	0.667	1	1	1
Grinder	1	1	1	1	1	1	1	1	1
Hair dryer	1	1	1	1	1	1	1	1	1
Hedge trimmer	1	0.278	0.435	1	0.389	0.56	1	0.944	0.971
Lamp	1	1	1	0.8	1	0.889	1	1	1
Paint tripper	1	1	1	1	1	1	1	1	1
Planer	1	1	1	1	1	1	0.857	1	0.923
Router	1	1	1	1	1	1	1	1	1
Sander	1	1	1	1	1	1	1	1	1
Saw	0.787	1	0.881	0.813	1	0.897	0.903	0.979	0.94
Vacuum cleaner	1	1	1	1	1	1	1	1	1

similar, which indicates that the best performances of the three device recognition models are approximate. However, for some specific devices, only certain models are able to perform well. In summary, some crucial conclusions are drawn as below:

(a) All of the three models have the ability to achieve high accuracy device identification on the occasion that appropriate features are selected.
(b) As for the hedge trimmer, only the MLP-based device recognition model has a satisfying performance; as for fan, the ELM model is out of work.
(c) For further improving the accuracy of recognition models, it may be effective to find a way to combine the advantages of different models.

References

Chtioui, Y., Bertrand, D., Devaux, M. F., Barba, D. (2015). Comparison of multilayer perceptron and probabilistic neural networks in artificial vision. Application to the discrimination of seeds. *Journal of Chemometrics, 11*(2), 111–129.

Cui, G., Cao, X., Wang, Y., Cao, L., & Yang, C. (2006). Wavelet packet decomposition-based fuzzy clustering algorithm for gene expression data. In: APCCAS 2006 – 2006 IEEE Asia Pacific Conference on Circuits and Systems, 2006/12.

Fan, X. (2009). Research and application of genetic algorithm-based optimized radial basis neural network model parameter design. In *International Conference on Electronic Measurement & Instruments*, 2009/10.

Figueiredo, M., De Almeida, A., & Ribeiro, B. (2012). Home electrical signal disaggregation for non-intrusive load monitoring (NILM) systems. *Neurocomputing, 96* (Complete), 66–73.

Gardner, J. W., Craven, M., Dow, C., & Hines, E. L. (1998). The prediction of bacteria type and culture growth phase by an electronic nose with a multi-layer perceptron network. *Measurement Science & Technology, 9*(1), 120.

Gomm, J. B., & Yu, D. L. (2000). Order and delay selection for neural network modelling by identification of linearized models. *International Journal of Systems Science, 31*(10), 1273–1283.

Hu, X., & Weng, Q. (2009). Estimating impervious surfaces from medium spatial resolution imagery using the self-organizing map and multi-layer perceptron neural networks. *Remote Sensing of Environment, 113*(10), 2089–2102.

Huang, P., Sang, G., Miao, Q., Ding, Y., & Jia, M. (2020). Soft measurement of ball mill load based on multi-classifier ensemble modelling and multi-sensor fusion with improved evidence combination. *Measurement Science and Technology.*

Kandaswamy, K. K., Pugalenthi, G., Hazrati, M. K., Kalies, K. U., & Martinetz, T. (2011). BLProt: prediction of bioluminescent proteins based on support vector machine and relieff feature selection. *BMC Bioinformatics, 12*(1), 345.

Kang, X., Lv, Z., Chen, Z., & Zhao, Y. (2020). Prediction of ammonia absorption in ionic liquids based on extreme learning machine modelling and a novel molecular descriptor SEP. *Environmental Research, 189,* 109951.

Kongsorot, Y., Horata, P., Musikawan, P., & Sunat, K. (2019). Kernel extreme learning machine based on fuzzy set theory for multi-label classification. *International Journal of Machine Learning and Cybernetics, 10*(5), 979–989.

Oliehoek, F. A. (2012). *Adaptation, Learning, and Optimization.* Germany: Springer.

Parlos, A. G., Rais, O. T., & Atiya, A. F. (2000). Multi-step-ahead prediction using dynamic recurrent neural networks. *Neural Networks the Official Journal of the International Neural Network Society, 13*(7), 765–786.

Parson, O., Ghosh, S., Weal, M., & Rogers. (2012). A non-intrusive load monitoring using prior models of general appliance types. In *Proceedings of theTwenty-Sixth Conference on Artificial Intelligence (AAAI-12),* 2012/07.

Peng, Y., & Lu, B.-L. (2016). Discriminative manifold extreme learning machine and applications to image and EEG signal classification. *Neurocomputing, 174* (JAN.22PT.A), 265–277.

Tabatabaei, S. M., Dick, S., & Xu, WJIToSG. (2017). Towards non-intrusive load monitoring via multi-label classification. *IEEE Transactions on Smart Grid* (99), 1–1.

Xia, C., & Mei, L. (2015). Empirical study on the demonstration project construction of IOTIPS. Paper presented at the 2015 International Conference on Industrial Technology and Management Science, 2015/11.

Zhao, Z., Chen, Z., Chen, Y., & Wang, S. (2014). A class incremental extreme learning machine for activity recognition. *Cognitive Computation, 6*(3), 423–431.

Chapter 4
Smart Non-intrusive Device Recognition Based on Intelligent Multi-label Classification Methods

4.1 Introduction

4.1.1 Background

As explained in Chap. 1, non-intrusive device recognition is an important part of Non-intrusive Load Monitoring and the Electric Internet of Things. Recently, multi-label device recognition has attracted many researchers. Unlike the methods which analyzing the opening and closing events of the devices, the multilabel device recognition methods identify the on and off states of devices during the steady operation of devices (Buddhahai et al. 2018). Besides, the method only requires samples of main circuit signals while with no need for the data of device operating signals. For multilabel non-intrusive device recognition, the data updating and model updating steps in practical application are more convenient due to no requirement for intrusive data collection (Tabatabaei et al. 2017).

In the multilabel identification domain, some scholars have proposed several effective multilabel classification models. By applying the Support Vector Machine to the multilabel classification field, the Ranking Support Vector Machine (Ranking SVM) is proposed (Hwanjo and Sungchul 2012). It is a method that can map the samples to a higher dimension space and conduct classification in the space. The Multilabel K-Nearest Neighbors (MLKNN) model is modified to realize the application of the K-Nearest Neighbors method in multilabel classification (Zhang and Zhou 2007). In contrast to other methods, MLKNN has great advantages in efficiency, which makes the method more feasible in practical application. The Backpropagation Neural Networks Multilabel Learning (BPMLL) (Zhang and Zhou 2006) model is derived from the Backpropagation Neural Networks (BPNN). The BPMLL model has the merits of neural networks that it can study the nonlinear relationship between the inputs and outputs. Many other multilabel classification methods have also been studied in recent years, such as Multilabel Decision Tree (Zhang and Zhou 2014), Calibrated Label Ranking (Zhang and Zhou 2014), RAndom k-labELsets (RAkEL)

H. Liu et al., *Smart Device Recognition*,
https://doi.org/10.1007/978-981-33-4925-4_4

method, Clustering-Based for MultiLabel Classification, and Ensemble of Classifier Chains (Moyano et al. 2018).

The chapter mainly introduces the application procedure of these multilabel models in non-intrusive device recognition. The most commonly used models, including Ranking SVM, MLKNN, and BPMLL, are evaluated on a popular public dataset named REDD dataset (Kolter and Johnson 2011). Moreover, some comparison experiments are conducted and the comparative analysis is also given in the experiment part.

4.1.2 Dataset Used in the Chapter

In the chapter, all of the multilabel non-intrusive device recognition methods are carried out on the public dataset, named REDD dataset (Kolter and Johnson 2011). The REDD dataset is a relatively commonly used public dataset in non-intrusive load monitoring. And there are both low-frequency data and high-frequency data included. Data of the main circuits and device circuits obtained from houses 1 to 5 are all sampled at a low frequency, but for house 3 and house 5 the high-frequency data is also collected. The current and voltage data of the five houses are collected for several months. Besides, the high-frequency current and voltage signals of main circuits in House 3 and House 5 are sampled at 15 kHz, the low-frequency power data of main circuits in all of five houses are recorded at 1 Hz, and the power data all of the devices in the five houses are recorded at one point every 3 s. The high-frequency data is stored using a special data form. Only one cycle data of the current is stored if several continuous current cycles are similar and then the number of the cycles is recorded. By utilizing the storing method, the data storage in need can be obviously reduced. Moreover, the timestamps of the data are also recorded. The detailed information of device categories in each house, the sampling rate, and data forms are introduced in Table 4.1. In the chapter, the data from House 3 and House 5 are used in the experiment.

4.2 Device Recognition Method Based on Ranking Support Vector Machine

The multilabel classification model is now a popular method in automatic speech recognition, pattern recognition, and many other aspects due to its low computational burden and well performance. In recent years, the method has been applicated to the field of device recognition. Generally, most of the device recognition problems are regarded as single-label classification issues. In the training part, those single-label classification models require the separate electric signal of each device which is obtained by intrusive load monitoring. However, the multilabel classification method

Table 4.1 Information of the REDD dataset

House number	Frequency	Device types
1	1 Hz for mains 1/3 Hz for devices	Furnace, Dishwasher, Smoke Alarms, Washer Dryer, Bathroom GFI, Disposal, Kitchen Outlets, Refrigerator, Lighting, Electronics, Microwave
2	1 Hz for mains 1/3 Hz for devices	Bathroom GFI, Washer Dryer, Kitchen Outlets, Dishwasher, Oven, Microwave, Refrigerator, Electric Heat, Lighting, Stove
3	15 kHz for mains 1/3 Hz for devices	Dishwasher, Disposal, Furnace, Washer Dryer, Refrigerator, Bathroom GFI, Kitchen Outlets, Lighting, Microwave, Electric Heat, Electronics, Outdoor Outlets
4	1 Hz for mains 1/3 Hz for devices	Bathroom GFI, Smoke Alarms, Washer Dryer, Kitchen Outlets, Furnace, Stove, Dishwasher, Disposal, Lighting, Air Conditioning
5	15 kHz for mains 1/3 Hz for devices	Washer Dryer, Dishwasher, Disposal, Kitchen Outlets, Refrigerator, Microwave, Lighting, Stove

is able to get the on-off state of each device by directly analyzing the total electric signal of all devices, making data collecting more convenient.

In the multilabel classification domain, the ranking support vector machine is now a ripe method having been applied to many aspects. And in the intelligent device recognition field, researchers find that it performs well in telling the on-off states of devices. In this segment, the application solution of the multilabel classification methods on non-intrusive device recognition is classified in detail and the SVM-based ranking approach is also introduced.

4.2.1 Model Framework

As introduced in Fig. 4.1, how the Rank-SVM-based multilabel model works in intelligent device recognition is introduced briefly by a general block diagram. Besides, the multilabel device recognition steps can be illustrated as follows:

(a) Obtain voltage and current data from the REDD dataset, mark the total voltage and current of the main circuit with the state labels of all devices in the circuit, extract the electrical and harmonic features, and then build the feature library.
(b) Fuse or reconstruct the electrical features and harmonic features to make full use of the information contained in them.
(c) Divide all samples into a training set and a test set, train the Ranking SVM multilabel device recognition model by means of modified cross-validation and compare the performances of different kernels.

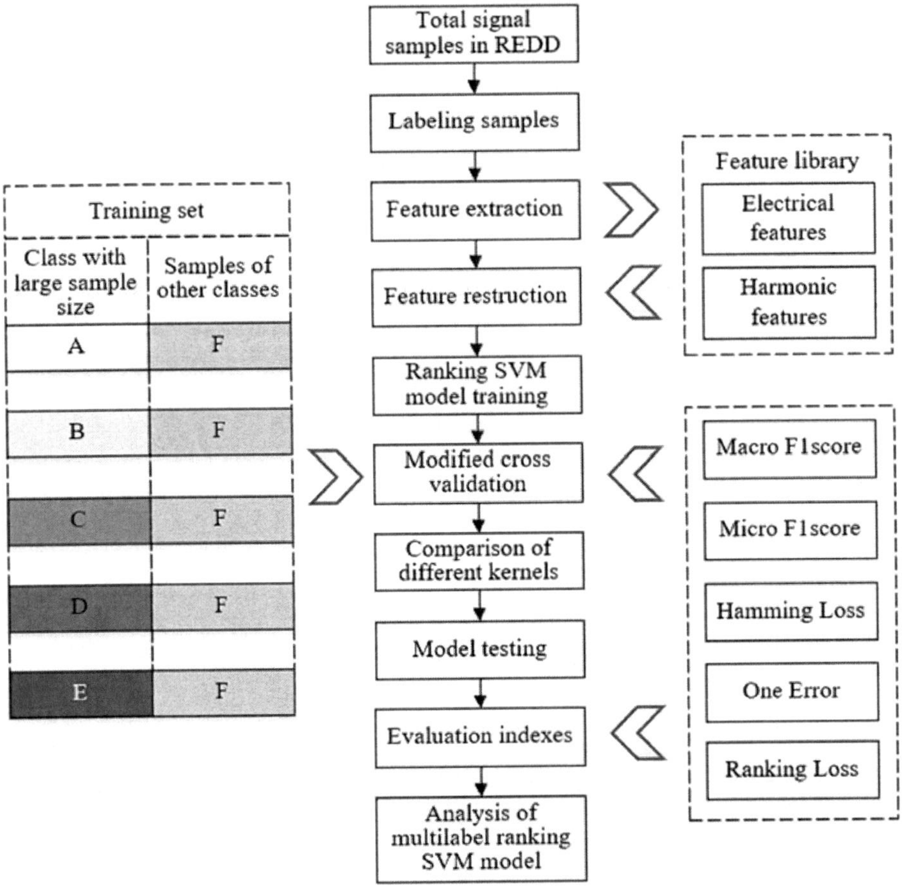

Fig. 4.1 The framework of multilabel ranking SVM classification model

(d) Perform the selected model on the test set, calculate the evaluation indexes, and then analyze the accuracy and application value of the Ranking SVM model in the field of intelligent device recognition.

4.2.2 Data Labeling

Specific and correct device operating data is the key to smart device recognition. Thus, it is crucial to ensure the exactitude of data labeling, especially for data with several labels. Generally, most of the datasets in terms of non-intrusive device recognition is not suited for multilabel learning. In the datasets, either there are only samples of devices operating independently, or the main circuit electrical signals are contained but with no labels. In the section, we provide a multilabel marking method to label the main circuit current and voltage correctly. Several labels which represent the

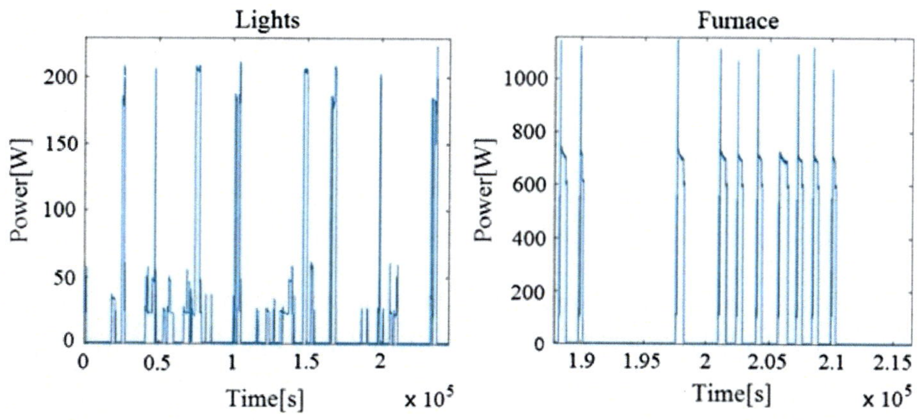

Fig. 4.2 The active power profile of two devices

operating states of devices connected to the circuit are tabbed on each sample of the main circuit signal. The procedure of data labeling is clarified in detail as follows:

(1) *Determination of the state of devices*
Generally, there are only time tags on the main circuit signals in most of the non-intrusive device recognition datasets, which is difficult for multilabel learning. For example, the main circuit current and voltage signals provided by the REDD dataset are sampled at 15 kHz for nearly four months (Kolter and Johnson 2011) with no device labels, but the device signals are also sampled simultaneously. Thus, it is feasible to obtain the device labels of total signals by determining the state of devices with device signals.
In the REDD dataset, the power of device signals is sampled once every 3 s, and by analyzing the power signals of devices we can determine whether a device is on or off. As shown in Fig. 4.2, the power signals of lights and microwaves are analyzed. The active power of light always changes within a range of 140–220 W while the active power of the furnace is in the range of 600–1200 W.
After power analysis of all devices, the power value of 80 W is utilized as the threshold to determine the operating states of devices:

$$L_i(t) = \begin{cases} 1, & P_i(t) > 80 \\ 0, & P_i(t) \leq 80 \end{cases} \tag{4.1}$$

in which $L_i(t)$ denotes the state of the device i at time t, and $P_i(t)$ is the power of the device.

(2) *Labeling main circuit signals*
Since the sampling rates of mains and device circuits are 15 kHz and 1/3 Hz (Kolter and Johnson 2011), a power point of the device signal corresponds to quantities of cycles of the main circuit current. The specific corresponding relation is given as follows:

$$1\ power\ point\ of\ devices = 45000\ current\ points\ of\ mains \\ = 180\ current\ cycles\ of\ mains \quad (4.2)$$

Thus, an algorithm based on the above relationship is designed for labeling each cycle of the main circuit current. Figure 4.3 introduces the method of the main circuit current labeling algorithm. In the figure, N is the number of labels. The detailed process of the algorithm is clarified as follows:
Step 1. Determine the state of the device i at time Td_i by the method introduced in the last section. Td_i is equal to the time tags of the device.
Step 2. The high-frequency current data in many datasets, like REDD, are stored by voltage cycles. Besides, for continuous voltage cycles which have approximate current waveforms, only one voltage period of data is logged as well as the number of voltage cycles n. Therefore, when time Td_i is within the range of $[Tm, Tm + n/f_b]$, the corresponding current is labeled with L_i. Tm is the time tags of the main circuit current and f_b is the grid fundamental frequency.
Step 3. If Td_i meets the condition but $n < 1.5\, f_b$, the next several continuous voltage cycles which have approximate current waveforms are also tagged with

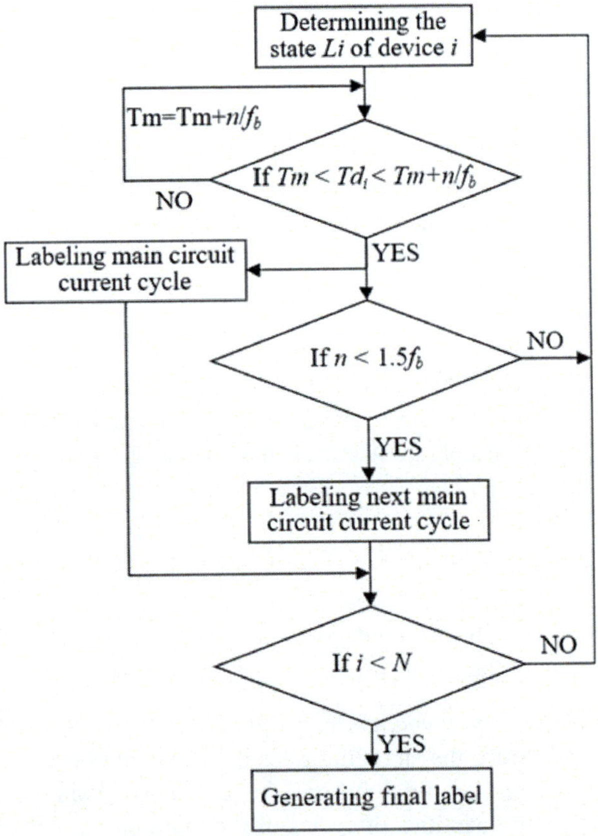

Fig. 4.3 Process of the main circuit current labeling algorithm

the label L_i. That means the power point may correspond to two different types of current cycles.
Step 4. Repeat step 1 to step 3 until the states of all devices are tagged on the main circuit current. Finally, the label of the current cycle is generated as below:

$$Lm = \{L_1, L_2, \ldots, L_i, \ldots\} \tag{4.3}$$

4.2.3 Feature Extraction and Reconstruction

As introduced in the last section, the low-frequency power data of House 3 and House 5 are utilized to label the high-frequency currents and voltages data of the main circuits in these two houses. In the feature extraction section, only the high-frequency data is used. Several electric features and harmonic features of the main circuits in House 3 and House 5 are calculated based on the high-frequency current data and voltage data. However, it is hard for the multilabel classifiers to perform well using all these features, considering that some features may be noisy for the classification and too many features may cause overfitting and low training efficiency. Thus, feature fusion is required to extract valuable information from the features and achieve feature dimension reduction. The feature extraction and feature fusion methods are elaborately introduced in the section.

(1) *Electric features extraction*
In the power system, there are several universal electric quantities, such as apparent power, reactive power, active power, power factor, impedance and so on (He et al. 2012). The calculation methods are introduced in Sect. 2.2.2.
By applying the feature extraction method to all samples of the main circuit signals, the feature space is constructed. As shown in Fig. 4.4, the electric features of the main circuit signals are presented by time series. It can be found that the time series of some features are similar, which indicates that some features are related.

(2) *Harmonic features extraction*
Since the features, like active power and impedance, are calculated using at least a period of voltage data and current data, the frequency of these features will not exceed 60 Hz, usually are 1 Hz. Besides, only harmonic features extracted from high sampling rate data contain enough useful information for device recognition. Thus, only the current data sampled at a high frequency is suitable for harmonic feature extraction (Bouhouras et al. 2019).
To get harmonic features, firstly Fast Flourier Transform (FFT) is applied to the current data. Considering that the data form of the current signal in REDD is exactly a voltage cycle, the spectrum leakage problem does not exist in the process of FFT. The fundamental frequency of current and voltage signals in REDD is 60 Hz, but generally, there are small variations (within 0.2 Hz) in the power grid (Liu et al. 2018). Therefore, the harmonic current amplitude and

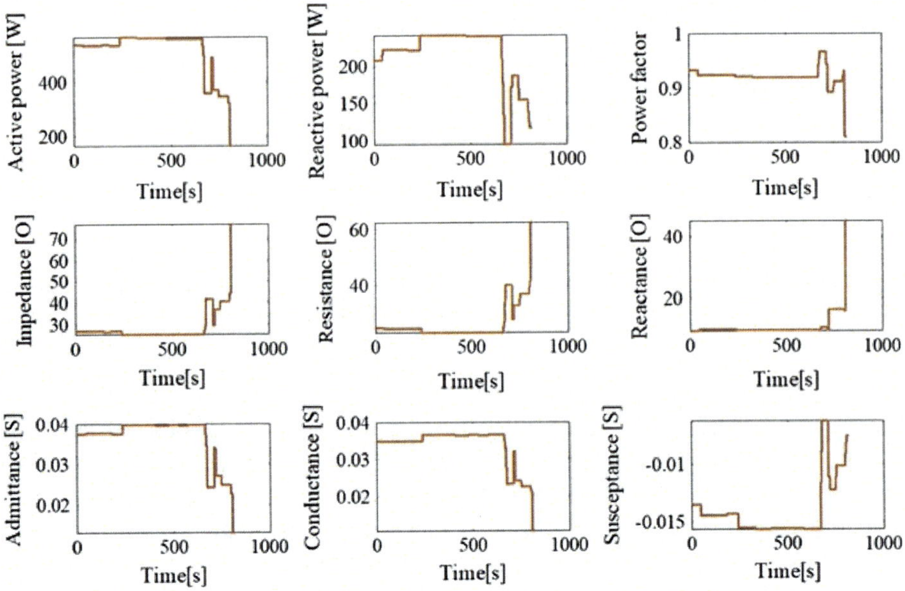

Fig. 4.4 Time series of the main circuit signal features

phase at integer multiples of the fundamental frequency might not be the exact values corresponding to the harmonic order. To get specific harmonic features, a search scope of ±15 Hz around integer multiples of the fundamental frequency is defined. By finding the maximum amplitude value in the search scope, the correct frequency of the harmonic is determined, and then the harmonic amplitude and phase are obtained. Figure 4.5 shows the amplitudes and phase angles of the first 6 odd harmonics. From the picture, it can be concluded that the harmonic amplitudes of higher orders are more unstable and the phase angle variation trends of different harmonics are various. It indicates that each order of harmonic contains some unique information that may be significant to device recognition.

In the section, the first 6 odd harmonic features are extracted and added to the feature space. All of the selected features are listed in Table 4.2 and the number of features is 23. A feature dimension reduction method is in need.

(3) *Feature fusion*

High dimensional characteristics usually contain more information than low dimensional characteristics. However, it brings several problems, such as increasing the computational burden, overfitting, and noisy features. A feature fusion method is conducted in the section so as to reconstruct the feature and reduce the feature dimension.

In the process of feature dimension reduction, the loss of feature information is inevitable, because part of feature data is eliminated or fused through certain methods. Therefore, it is very important to reduce feature information loss

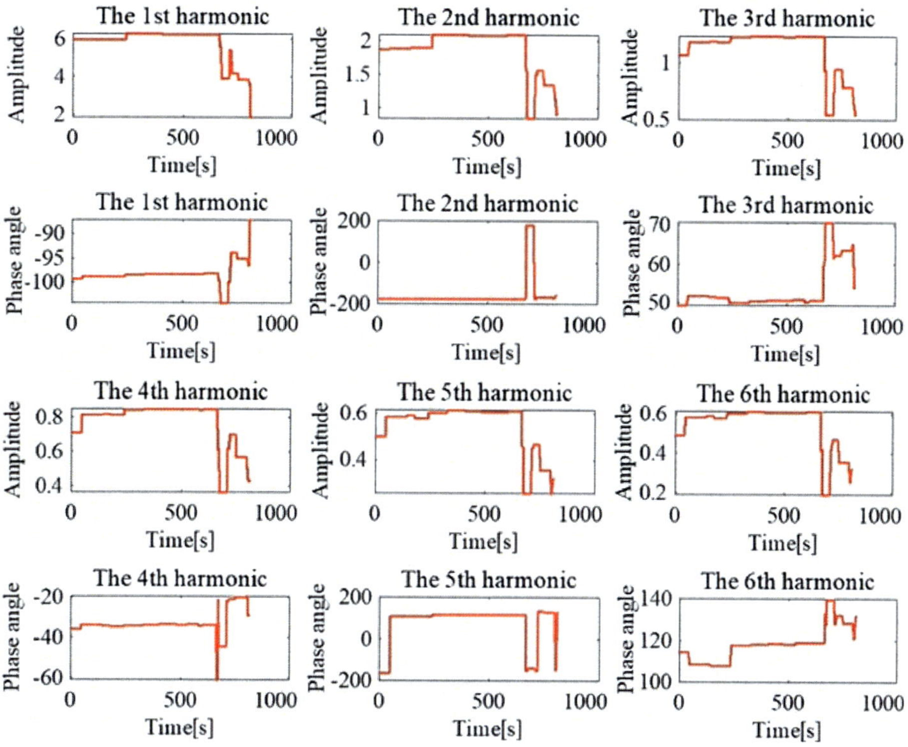

Fig. 4.5 Amplitudes and phase angles of the first 6 odd harmonics

Table 4.2 The feature space of main circuit signals

Feature name	Variable name	Feature name	Variable name
Original voltage	U	Reactance	X
Original current	I	Impedance	Z
Reactive power	Q	Resistance	R
Active power	P	Conductance	G
Power factor	λ	Admittance	Y
Amplitudes of the first 6 odd harmonics	A_i	Electrical susceptance	B
Amplitudes of the first 6 odd harmonics	φ_i		

during dimension reduction. The principal component analysis (PCA) method achieves feature fusion by extracting feature principal components. It has been widely used in many aspects, such as feature dimension reduction (Machlev et al. 2020), lossy data compression (Zhang et al. 2020), classification (Mahmoudi et al. 2020), and image compression (Sivasathya and Joans 2012). The PCA model tries to find several principal components that contain most of the information while the other components contain a small amount of information.

In this way, it can minimize the loss of feature information, fuse the feature information into several principal components, and realize feature dimension reduction. The specific PCA feature fusion method is clarified as follows:
Step 1. Normalize the data and get standardized feature variables.

$$\hat{c}_{ij} = \frac{c_{ij} - \bar{c}_j}{s_j} \tag{4.4}$$

where $\hat{c}_{ij}$ and c_{ij} are the *j*th standardized feature value and initial feature value of sample *i*, $\bar{c}_j$ and s_j are the mean value and standard variation of the *j*th feature.
Step 2. Calculate the correlation matrix R of all features as follows:

$$\begin{cases} R = \left(r_{kj}\right)_{M \times M} \\ r_{ij} = \frac{\sum_{i=1}^{N} \hat{c}_{ik}\hat{c}_{ij}}{N - 1} \end{cases} \tag{4.5}$$

where r_{kj} is the correlation coefficient of feature *k* and *j*, *M* is the quantity of features, and *N* is the quantity of samples.
Step 3. Calculate the eigenvalues and its corresponding eigenvectors, and then sort them in descending order of eigenvalues. The principal components can be expressed as below:

$$\begin{cases} cp_1 = \mu_{11}\hat{c}_1 + \mu_{21}\hat{c}_2 + \cdots + \mu_{M1}\hat{c}_M \\ cp_2 = \mu_{12}\hat{c}_1 + \mu_{22}\hat{c}_2 + \cdots + \mu_{M2}\hat{c}_M \\ \qquad \cdots\cdots \\ cp_n = \mu_{1n}\hat{c}_1 + \mu_{2n}\hat{c}_2 + \cdots + \mu_{Mn}\hat{c}_M \end{cases} \tag{4.6}$$

where *n* is the principal component size, cp_1 is the first principal component, $(\mu_{11}, \mu_{11}, \ldots, \mu_{M1})$ is the first eigenvector.
Step 4. Calculate the cumulative contribution rate of all eigenvalues, and if the cumulative contributive rate of the *p*-th principal component is higher than the threshold ε_1, the first *p* principal components are chosen as the new features. The calculation equation of the cumulative contribution rate is given as follows:

$$\alpha_p = \frac{\sum_{j=1}^{p} \lambda_j}{\sum_{j=1}^{n} \lambda_j} \tag{4.7}$$

By applying the PCA method to the initial feature space, 6 principal components whose cumulated contribution rate exceeds 0.95 (Fig. 4.6) are selected as the new features. Besides, the multilabel models will be trained and tested on the new feature data.

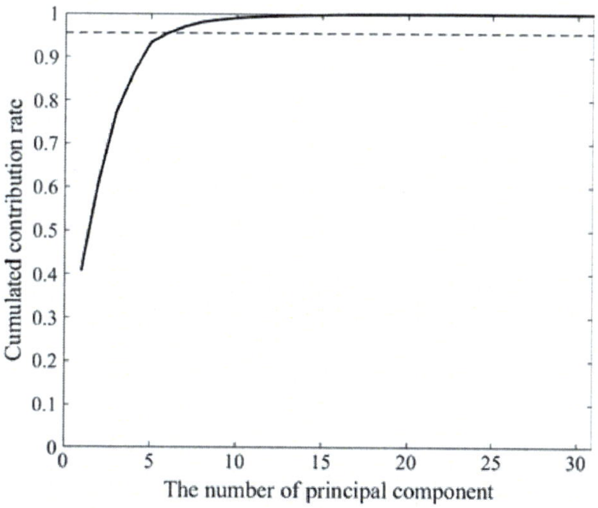

Fig. 4.6 The cumulated contribution rate of all principal component

4.2.4 *The Basic Theory of the Ranking Support Vector Machine*

In recent years, with the further study of support vector machine (SVM), the application of SVM in dichotomy and multi-classification has become very common. It realizes the classification of samples mainly by finding a hyperplane on the training samples that can minimize the classification error (Jayadeva et al. 2019). Kernel SVM is improved based on the Linear SVM classifier. The sample data is mapped to the high dimensional space by kernel function, and a hyperplane with the best classification effect is found in the high dimensional space. The Kernel SVM solves the problem of low dimensional spatial data overlap and linear inseparability (Kouziokas 2020). At present, the SVM model has been applicated in several fields, such as medical diagnosis, load recognition, text classification, speech recognition, and image recognition, due to its stability and satisfactory classification effect.

Some scholars tried to extend the SVM model to the field of multi-label classification. *Abe* proposed a fuzzy support vector machine to realize multi-label classification using the membership degree of data samples (Abe 2015). It is calculated by defining a region with associated membership function. *Feng* et al. put up with a multi-label SVM model based on tag correlation, which reduced the number of optional combinations of multi-label classification. The method improved the efficiency of multilabel classification by calculating tag correlation. Ranking SVM is a relatively stable multi-label classification model (Zhang and Zhou 2014), which is an algorithm adapted multi-label classification model. The linear Ranking SVM applies the SVM model to the sorting task to improve the rank-based multi-label classification algorithm. However, for the complex non-invasive device identification problem, the linear Ranking SVM has a poor classification effect. In addition, there are few studies on how to improve the application of multilabel classification in NILM, and the performance of current multilabel models in device recognition needs to be improved.

This section studies the application and improvement of the nonlinear Ranking SVM model in multi-label device identification, and proposes the application method of the model in intelligent non-invasive device identification, to improve the stability and accuracy of device identification.

The nonlinear Ranking SVM model is also called the Kernel Ranking SVM. The nonlinear Ranking SVM is to map the input data into a high-dimensional space by kernel function and replace the ranking of linear Ranking SVM by its ranking in the high-dimensional space. Besides, the rest steps of nonlinear Ranking SVM basically follow the linear Ranking SVM metho

(1) *The basic theory of Ranking SVM model*
Linear Ranking SVM is the basis of nonlinear Ranking SVM. The latter method only improves the hyperplane determination part of the former one. The specific principle of the multi-label linear Ranking SVM model is clarified as follows:

(a) Basic theory of Linear Ranking SVM
Above all, the basic principle of Linear Ranking SVM should be introduced. The key point of Linear Ranking SVM is to get the ranking function $f(x)$

$$f(x) = w^T x \tag{4.8}$$

so that for any sample pair there exists $f(x_i) > f(x_j) = w^T x_i > w^T x_j$. Besides, w is the ranking vector, x represents the sample.
In order to get the interval between two data points, the distance is defined as follows (Hwanjo and Sungchul 2012):

$$d = \min \frac{w^T (x_i - x_j)}{\|w\|} \tag{4.9}$$

By finding a ranking vector that maximizes the value of the interval between two data points, the distance can be maximized.

$$\max_w d = \max_w \min \frac{w^T (x_i - x_j)}{\|w\|} \tag{4.10}$$

Besides, how to transform the ranking problem into the classification problem is the key to implement the Ranking SVM algorithm. The L1 cost function is adopted to solve the problem:

$$f(x) = \frac{1}{2}\|w\|^2 + \beta \sum \max(0, 1 - w^T (x_i - x_j)) \tag{4.11}$$

where β is the regular parameter. By transforming the sorting function with independent variable x into a cost function with the independent variable $(x_i - x_j)$, the ranking problem finally becomes a classification issue that can be solved by the SVM model.

(b) Basic theory of Non-linear Ranking SVM
The difference between nonlinear Ranking SVM and linear Ranking SVM is that the former first transforms the sample space to space with a high dimension by kernel functions and then selects the hyperplane. The method can greatly improve the effectiveness of the classification and make the nonlinear Ranking SVM more suitable for dealing with complex multi-label classification problems such as non-intrusive device recognition.
The key of the kernel Ranking SVM is to map sample x to the regenerative kernel Hibbert space according to the kernel function mapping rule. The mapping relationship is shown below:

$$f_{\kappa}\left(x_i, x_j\right) = \left\langle \Phi(x_i), \Phi(x_j) \right\rangle \tag{4.12}$$

where f_{κ} is the kernel function and $\Phi(x)$ is the principle defined by the kernel function.
For the purpose of transforming cost function (Eq. 4.12) into a dual problem, the Lagrangian multiplier is utilized as follows:

$$\begin{aligned} &\max_{\alpha} \sum \alpha_{ij} - \sum_{ij} \sum_{kl} \alpha_{kl}\alpha_{ij}(\Phi(x_k) - \Phi(x_l))\left(\Phi(x_i) - \Phi(x_j)\right) \\ &s.t.\ 0 \le \alpha_{ij} \le C \end{aligned} \tag{4.13}$$

During the computing process of the above quadratic programming issue, the calculation of the matrix $\left(\Phi(x_i) - \Phi(x_j)\right)(\Phi(x_k) - \Phi(x_l))$ takes up a lot of computing resources. Several researchers have proposed solutions to improve the calculating efficiency of it, like the 1-Relaxation structure method (Thorsten 2006), pair-problem reconstruction algorithm (Kuo et al. 2014), and theorem reorganization method (Hwanjo and Sungchul 2012). Based on the method proposed by *Hwanjo*, the Kernel Ranking SVM is improved. The kernel is applied to samples instead of sample couples and Eq. 4.14 is modified as below:

$$\begin{aligned} &\max_{\alpha} \sum \alpha_i - \sum_{i} \sum_{kl} \alpha_i f_{\kappa}(x_i, x_k) f_{\kappa}(x_i, x_l) \\ &s.t.\ 0 \le \alpha_i \le C \end{aligned} \tag{4.14}$$

Besides, the new ranking function is defined as follows:

$$f(x) = \sum_i \alpha_i f_\kappa(x_i, x) \tag{4.15}$$

(2) *Kernel selection*
Since the performance of an SVM model depends on its kernel function, it is crucial for nonlinear Ranking SVM to select a suitable kernel function. The most commonly used kernel is the RBF kernel, also called Gaussian kernel. But Gaussian kernel function is very sensitive to its hyper-parameter (Tanaka et al. 2007). Thus, many other kernel functions are studied by researchers from all over the world. Each kernel function has its own unique characteristic and applicable scene. The details of some popular kernel functions are introduced as below (Liu and Xie 2020):

(a) Gaussian Kernel
The Gaussian kernel is able to resist noise from the data. Its parameter determines the effective scope of the function. Besides, when the input data exceeds the scope, it would be ignored. However, by virtue of the sensitivity of the Gaussian kernel to its parameter, to select an optimal value of the hyperparameter is very difficult. Besides, the training time of the RBF kernel is also unsatisfactory. The Gaussian kernel can be expressed as below:

$$f_{RBF}(x_i, x_j) = \exp\left(-\frac{\|x_i - x_j\|^2}{2\sigma^2}\right) \tag{4.16}$$

(b) Linear Kernel
The linear kernel is the simplest kernel among all kernels. It is always used for comparison. The definition is introduced as follows:

$$f_L = x_i^T x_j \tag{4.17}$$

(c) Rational Quadratic Kernel
The rational quadratic kernel is proposed to solve the inefficiency of the Gaussian kernel. But it is also sensitive to the selection of its parameter. It is defined as below:

$$f_{RQ} = 1 - \frac{\|x_i - x_j\|^2}{\|x_i - x_j\|^2 + c} \tag{4.18}$$

(d) Generalized T-Student Kernel (TS kernel)
The TS kernel is also commonly used in machine learning. It belongs to the mercer kernel, which means that the TS kernel is also a semidefinite symmetric function. It is introduced as follows:

$$f_{TS} = \frac{1}{1 + \left\| x_i - x_j \right\|^d} \tag{4.19}$$

Aiming at finding an appropriate Ranking SVM model for smart device recognition, all of the kernels above will be tested on the REDD data.

4.2.5 Multi-label Classification Evaluation Indices

Due to the multiple labels of data, the multilabel classification evaluation indices are also different from the single-label classification. The evaluation measures of multilabel classification are usually divided into example-based evaluation indices and label-based evaluation indices (Pereira et al. 2018) according to the calculating method. The example-based measures indicate the performance of the multilabel classification model on sample classification while the label-based measures reflect the performance on label recognition. Besides, according to different prototypes of the metrics, the evaluation indices can also be grouped into ranking-based indices and bipartition indices. As shown in Fig. 4.7, there are several evaluation indices belonging to the two groups, and the detailed computing methods of some common metrics are given as follows:

(1) *Example-based evaluation indices*
(a) Example-based bipartition evaluation indices

(i) Hamming Loss
Hamming Loss (Sokolova and Lapalme 2009) is a popular multilabel classification metric that indicates the classification error of a multilabel classification model. As Eq. 4.20 expresses, it is obtained by calculating the proportion of missing error and prediction error to the number of all samples and labels.

$$Hamingloss = \frac{1}{N}\sum_{i=1}^{N}\frac{error_i}{N_L} \tag{4.20}$$

where N_L represents the number of labels on a signal, *error* stands for the quantity of labels which are missed or wrongly predicted.

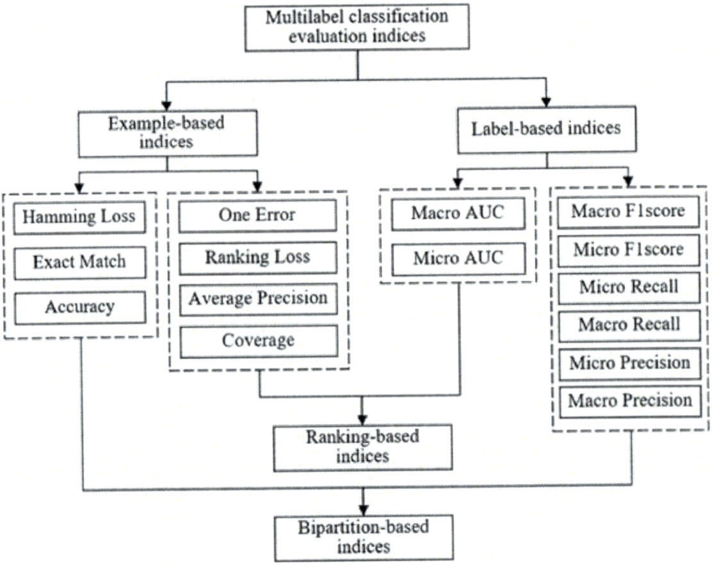

Fig. 4.7 Multilabel classification evaluation indices

(ii) Exact Match
The Exact Match index (Sokolova and Lapalme 2009) is defined by calculating the proportion of the samples whose label set is predicted totally correctly. The metric ignores the labels that are partially correct, thus it cannot distinguish the performance of classifiers that always give near exact predictions or wrong predictions.

$$Exactmatch = \frac{1}{N}\sum_{i=1}^{N}(P_i == Lm_i) \tag{4.21}$$

in which *P* and *G* are the predicted label set and true label set respectively.

(b) Example-based ranking evaluation indices

(i) One Error
The metric represents the proportion of samples where the label which is top-ranked is not in the relevant labels (Tsoumakas et al. 2010).

$$Oneerror = \frac{1}{N}\sum_{i=1}^{N}h(argmin\ Rank_i(L)) \tag{4.22}$$

where $Rank_i(L)$ is the position of label *L* in ranking order, *h* is a function that determines whether the top-ranked label is in the relevant labels.

(ii) Ranking Loss
Ranking Loss (Tsoumakas et al. 2010) indicates the ratio of the case that irrelevant labels are ranked higher than the relevant labels.

$$Rankingloss = \frac{1}{N}\sum_{i=1}^{N}\frac{1}{|Lm_i == 1||Lm_i == 0|}|\{(L_a, L_b) : Rank_i(L_a) > Rank_i(L_b)\}| \quad (4.23)$$

where |*A*| represents the size of vector *A*.

(iii) Coverage
Coverage (Pereira et al. 2018) reflects the sample mean of the distance required to cover all relevant labels and it is defined as below:

$$Coverage = \frac{1}{N}\sum_{i=1}^{N}\max(Rank_i(L)) - 1 \quad (4.24)$$

(2) *Label-based evaluation indices*

(a) Label-based bipartition evaluation indices
There are two groups of label-based bipartition metrics derived from the metrics (recall, precision, F1score) of single-label classification. The first group of metrics is generated by applying the macro approach to the traditional single classification evaluation measures. The macro approach is conducted by computing the single classification metrics for all labels independently, averaging the values, and getting macro-recall, macro-precision, and macro-F1score (Gibaja and Ventura 2014). The other label-based bipartition metrics are obtained by applying the micro method, which is conducted by taking into account predicted labels of all samples together, calculating the final metrics across all of the labels, and obtaining micro-recall, micro-precision, and micro-F1score (Gibaja and Ventura 2014). Table 4.3 gives the relationship of the multi-label classification evaluation indices and traditional single-label classification evaluation indices.

Table 4.3 The multilabel classification metrics and their prototypes

Single-label classification metrics	Label-based bipartition multilabel classification metrics	
	Macro method modified	Micro method modified
Recall	Macro-Recall	Micro-Recall
Precision	Macro-Precision	Micro-Precision
F1score	Macro-F1score	Micro-F1score

(b) Label-based ranking evaluation indices

The Area Under Curve (AUC) classification metric refers to the proportion of the area under the receiver operating characteristic curve (ROC) curve. By applying macro and micro methods, the macro-AUC metric and micro-AUC metric are obtained.

(3) *Indices utilized in the section*

In this chapter, One Error, Hamming Loss, Ranking Loss, macro-F1score, micro-AUC, and micro-F1score are adopted to validate the performance of multilabel classification models. These five metrics are the most commonly used evaluation indexes in the field of multilabel classification, and they can make a comprehensive evaluation of the models.

4.2.6 Evaluation of Ranking SVM in Terms of Multi-label Device Recognition

Aiming at evaluating the multilabel SVM model, all the experiments are conducted on the REDD dataset. Unusually, in the chapter, only the main circuit signals are used for training and testing, which is different from the traditional device recognition. In this chapter, by using main circuit signal data and multilabel learning, the intrusive data of all devices are not in need and the non-intrusive device recognition is realized in a real sense. For the same reason, some inevitable issues take up. The label with large sample size is associated with other labels, thus the data imbalance problem is hard to be solved. The performance of multilabel models would be limited. In order to evaluate the performance of the Ranking SVM model with different kernels, some comparison experiments are conducted in the section.

After applying the feature extraction method and data labeling method mentioned in Sects. 4.2.2 and 4.2.3, the multilabel data from House 3 are obtained. In the experiment, 80% of the data are used for training, and the rest 20% for testing. The Ranking SVM models with all of the four kernels are trained and evaluated on the data. Table 4.4 gives the labels of devices. The experiment results and comparative analysis are also clarified in the section.

(1) *Data training and testing*

Considering that there is a data imbalance problem in the REDD dataset, a modified cross-validation method is designed to solve the problem. After data analysis, we find that the sample sizes of lights, electronics, and refrigerators are extremely large. Thus, the samples which only have labels of the above three devices are divided into five parts. And then, each part is mixed with the samples that also contain labels of other devices as shown in Fig. 4.8. Finally, apply cross-validation on the five parts, select the model with the best results, and test the selected model on the test data.

Table 4.4 The labels of devices in the experiments

Labels of the devices	Names of the devices
1	Lighting
2	Electronics
3	Refrigerator
4	Dishwasher
5	Furnace
6	Washer dryer
7	Microwave
8	Bathroom gfi
9	Kitchen outlets

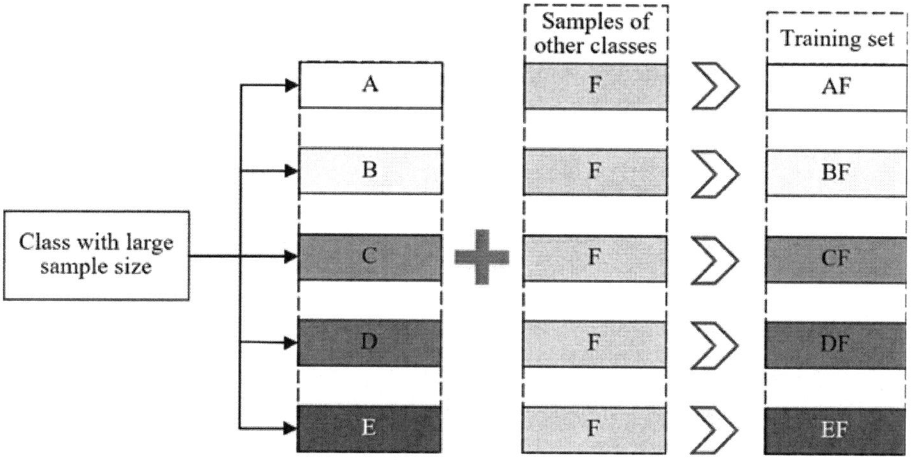

Fig. 4.8 Data dividing method in the modified cross-validation procedure

(2) *Multilabel confusion matrix*
Considering that the confusion matrix is an effective method to evaluate the performance of classification, the results of multilabel classification will be more visualized if the confusion matrix can be applied to the field. Since that in terms of the multilabel classification, one sample has several labels, the confusion matrix cannot be used in the field directly. By dealing with the labels, a multilabel confusion matrix is proposed in the section.
The multilabel confusion matrix is generated by transforming the multilabel data into single-label data. Firstly, the labels $Lm = \{L_1, L_2, \ldots, L_i, \ldots\}$ are transformed as below:

$$L_i = \begin{cases} 0, & if\ L_i = 0 \\ i, & if\ L_i = 1 \end{cases} \tag{4.25}$$

Then the sample with N labels is changed into N same samples that each one has a single label. In the procedure, 0 is also treated as a label. Finally, the samples with multiple labels are transformed into single-label samples, and the multilabel confusion matrix can be calculated. In the multilabel confusion matrix, the '0' label represents the off state of devices in the sample. The first row represents the devices predicted on but actually off, and the first column represents the devices predicted off but actually on. The labels on the diagonal line are the number of devices whose states are correctly predicted.

Figures 4.9, 4.10, 4.11 and 4.12 are the multilabel confusion matrixes of experiment results of the Ranking SVM model with four different kernels. From the pictures, it can be concluded that: (1) the Gaussian kernel performs better and more stable than the other three kernels on the multilabel device classification issues; (2) the classification results of devices whose sample size is too small is not satisfying enough, which proves that the performance of kernel methods is affected by sample size; (3) the multilabel Ranking SVM model is able to solve non-intrusive device recognition when the appropriate kernel and hyper-parameters are selected.

(3) *ROC curve and multilabel metrics of the experiments*

Figure 4.13 shows the ROC curve of the Ranking SVM models based on different kernels. It can be found in the figure that: (1) the classification results of kitchen

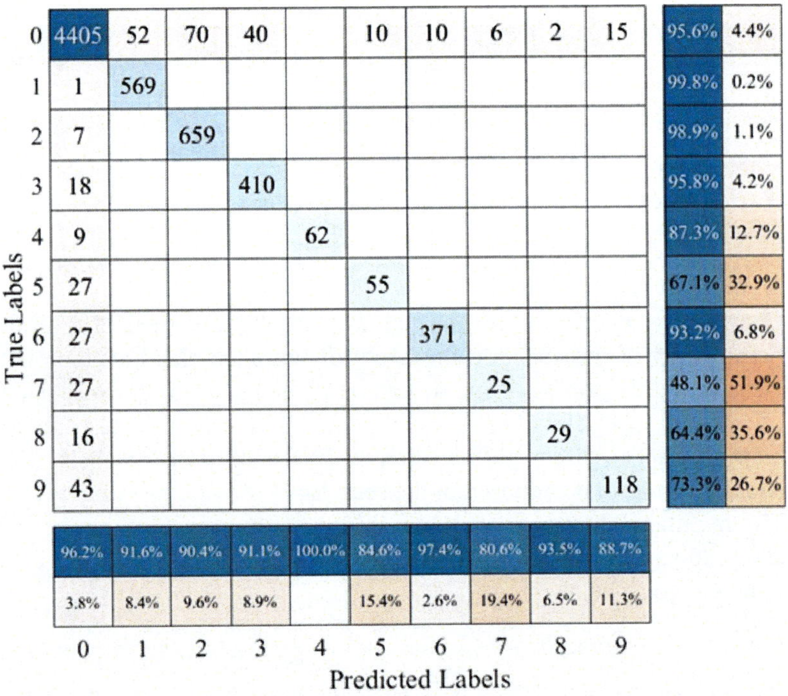

Fig. 4.9 Confusion matrix of ranking SVM based on Gaussian Kernel

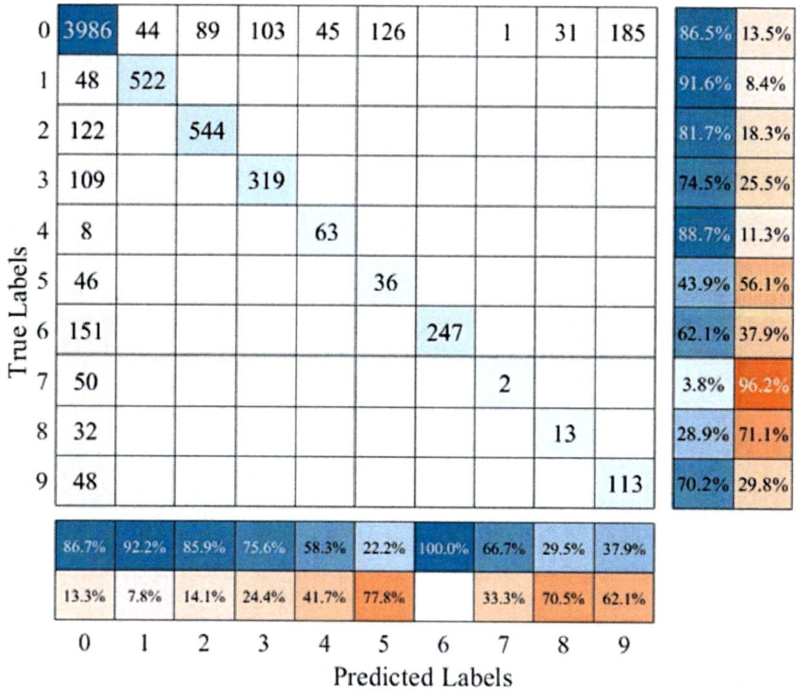

Fig. 4.10 Confusion matrix of ranking SVM based on TS Kernel

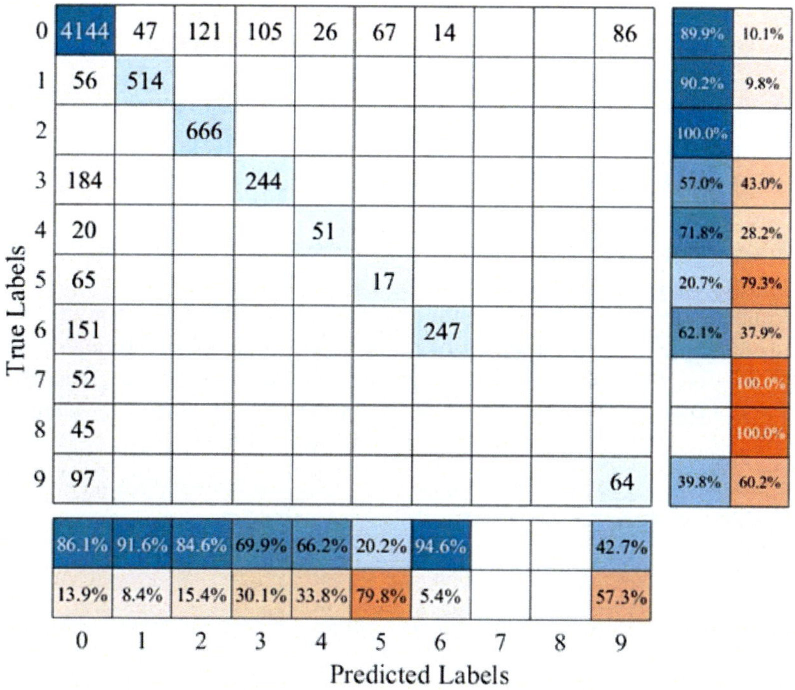

Fig. 4.11 Confusion matrix of ranking SVM based on Linear Kernel

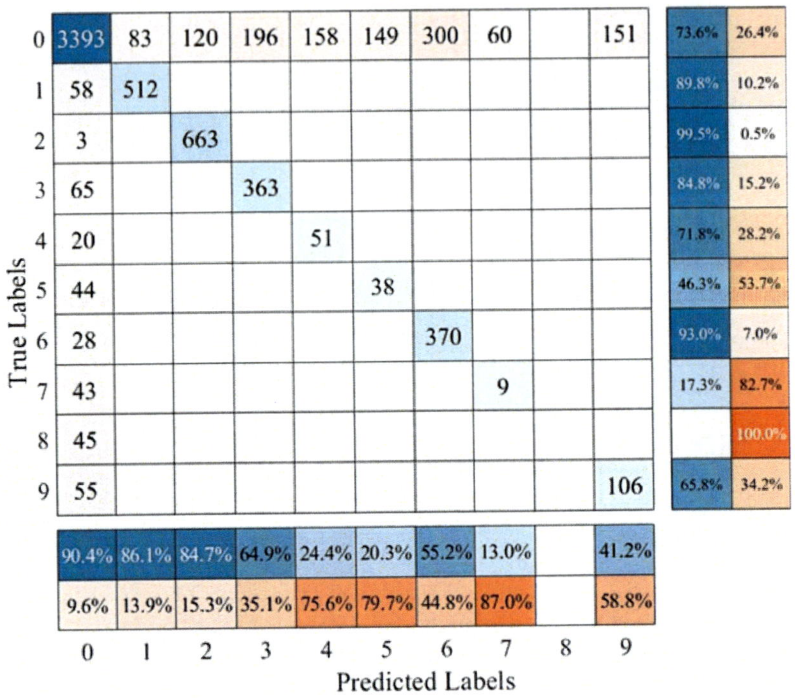

Fig. 4.12 Confusion matrix of ranking SVM based on rational quadratic kernel

outlets are always the worst as a result of the unstable features of kitchen outlet; (2) the classification results of the dishwasher are the best due to its distinctive characteristics; (3) the results of kernels except for Gaussian Kernel are all unsatisfying and the dishwasher labels are all correctly predicted by the Ranking SVM model with Gaussian Kernel.

The multilabel metric values of the experiments are given in Table 4.5 and Fig. 4.14 and some conclusions are drawn as follows: (1) the macroF1score values are always smaller than the microF1score values in all experiments, which indicates that some labels cannot be predicted accurately enough; (2) among all of the kernels, the metrics of the Ranking SVM model based on Gaussian Kernel are the best, which also indicates the effectiveness of Gaussian kernel; (3) the Ranking loss, One Error, and Hamming loss of the Gaussian kernel is very low, which reflects that the classification error is extremely small and the Ranking SVM model with Gaussian kernel is stable enough for non-intrusive device recognition.

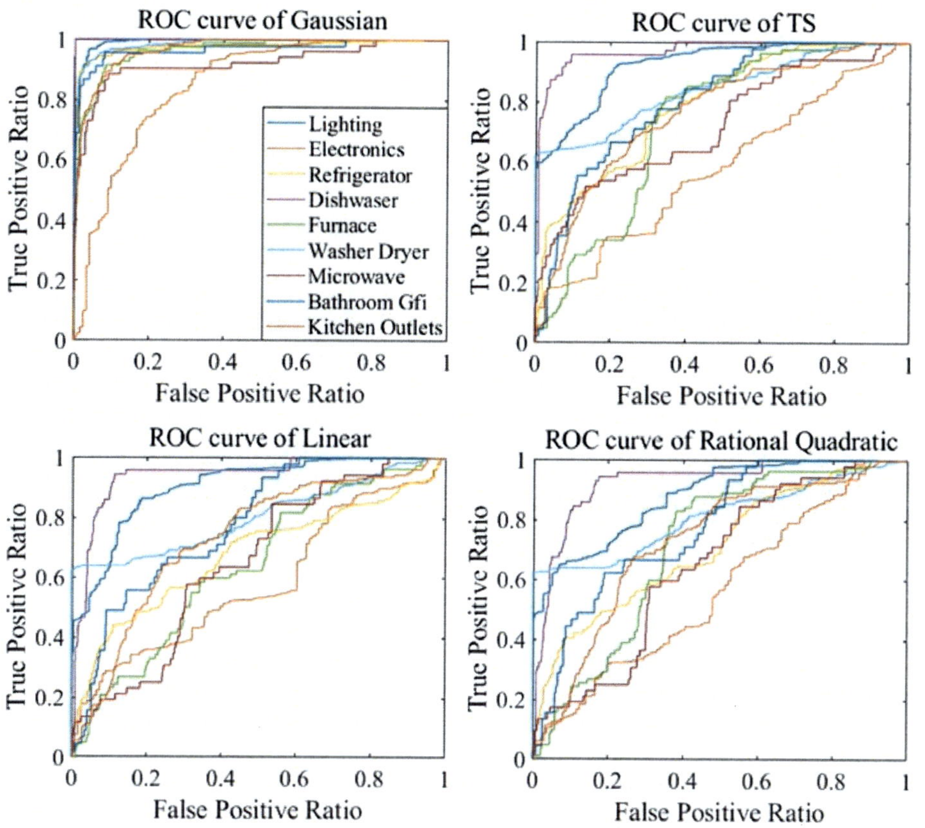

Fig. 4.13 The ROC curve of ranking SVM models based on different kernels

Table 4.5 The multilabel metrics of experiment results

Kernel	Micro-F1score	Macro-F1score	One error	Ranking loss	Hamming loss	Micro-AUC
Gaussian Kernel	0.9236	0.8484	0.0013	0.0990	0.0536	0.9581
Generalized T-Student Kernel	0.7502	0.5699	0.0267	0.3246	0.1748	0.7939
Linear Kernel	0.7504	0.5010	0.0089	0.3240	0.1604	0.7458
Rational Quadratic Kernel	0.7280	0.5028	0.0076	0.3615	0.2228	0.7513

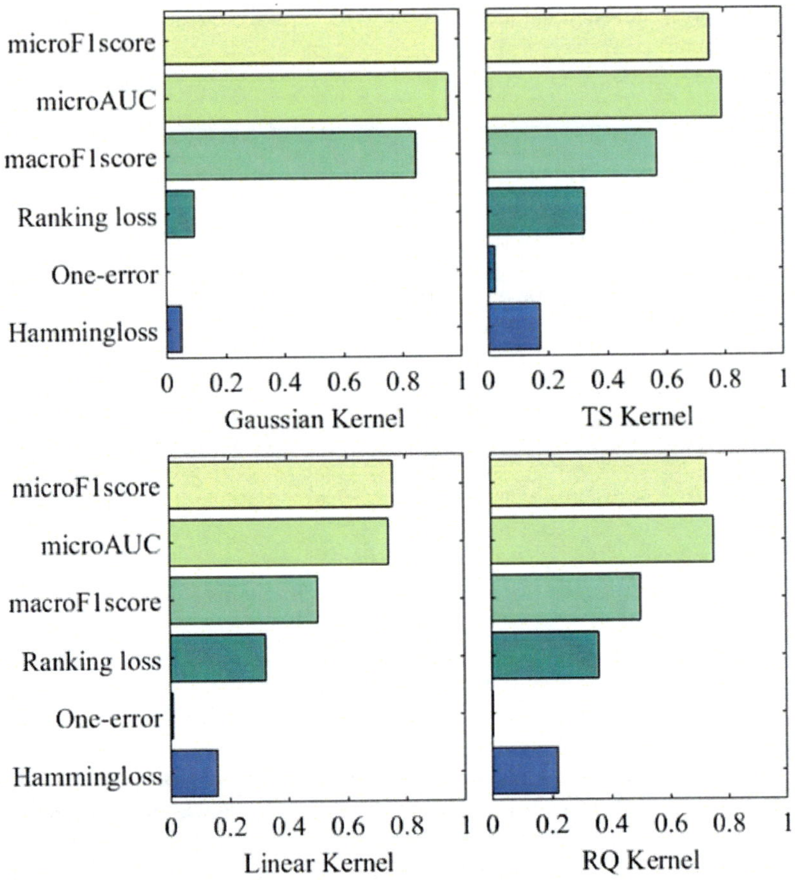

Fig. 4.14 The evaluation indexes of Ranking SVM based on different kernels

4.3 Device Recognition Method Based on Multi-label K-Nearest Neighbors Algorithm

In the field of classification, the K-Nearest Neighbors (KNN) method is a very commonly used method owing to its efficiency and stability. By adapting the KNN method to the multilabel classification field, the Multilabel K-Nearest Neighbors (MLKNN) algorithm is generated. Unlike other multilabel classification methods whose training time increases obviously when the method is modified for multilabel classification, the MLKNN model still requires little computing resources. Besides, for real-time data analysis problems such as non-intrusive device recognition, the efficiency of MLKNN is a great advantage in practical application. In the section, the basic theory of MLKNN is clarified in detail and the application procedure of the MLKNN-based non-intrusive device model is also illustrated.

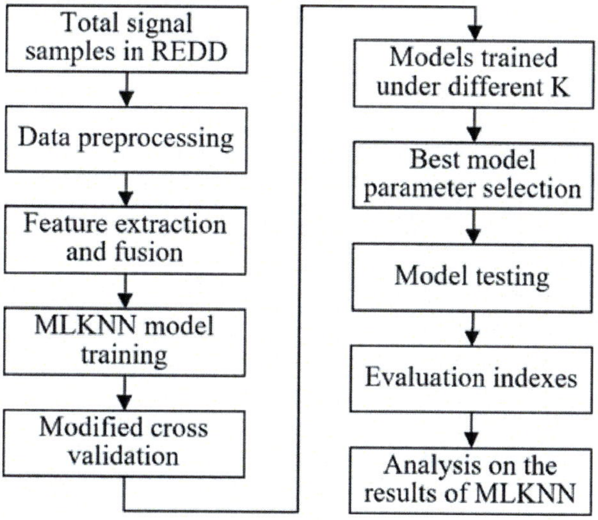

Fig. 4.15 The flow chart of the multilabel K-Nearest neighbors algorithm

4.3.1 Model Framework

Figure 4.15 briefly shows the procedure of the MLKNN model applicated in the non-intrusive device recognition. The detailed steps of the MLKNN model applied to the non-intrusive device recognition is given as follows:

(a) Label the main circuit data with the states of devices, calculate the features of the data, and then build a feature database.
(b) Reconstruct the features of main circuit data.
(c) Randomly select 80% of the samples as a training set, and the rest samples are used as the test set. Apply cross-validation during the training process and then select the model with the best results among all models trained by cross-validation.
(d) Calculate evaluation metrics of the results of models performed on the test set and analyze the efficiency and performance of the MLKNN model.

4.3.2 Data Preprocessing

The procedure of determining the state of devices and labeling the main circuit data is identical to the method clarified in Sect. 4.2.2. Besides, the features shown in Table 4.2 are also extracted by the method introduced in Sect. 4.2.3, and by applying the PCA data dimension reduction method, the fused feature library is constructed. The processed data are finally utilized to evaluate the MLKNN model in the field of non-intrusive device recognition.

4.3.3 The Basic Theory of Multi-label K-Nearest Neighbors

The Multilabel K-Nearest Neighbors (MLKNN) model derived from the k-nearest neighbors (KNN) algorithm is first proposed by *Zhihua Zhou* and *Minling Zhang* (Zhang and Zhou 2007). The MLKNN model is a multilabel classification algorithm. It inherits the high efficiency and stability of the KNN algorithm, and it is suitable for multi-label classification problems with chaotic data distribution or overlapping tags of data (Zhang and Zhou 2014).

Recently, the MLKNN model was applied in terms of multilabel non-intrusive device recognition. In the experiments conducted by *Seyed Mostafa* and his team, it is proved that the MLKNN model has a better performance than the Rank k-Label method based on time-domain features (Basu et al. 2015). The MLKNN model becomes popular as a result of its high efficiency and stable performance. In the segment, the basic theory of MLKNN is introduced thoroughly and the steps of applying MLKNN are also contained.

(1) *Traditional k-nearest neighbors algorithm*
The core idea of the traditional KNN algorithm applied to single-label classification is to find the k nearest neighbors in the training samples of the test samples. And then by voting on each label of the k neighboring samples, we can obtain the label of the test samples. The KNN algorithm does not need a model training process, thus, it saves a lot of training time compared with other classification models. As the label of the sample is only affected by its k nearest neighbors, KNN performs better for data sets with chaotic data distribution. However, there is no exact method that is always effective to select the best value of k. The trial and error method and cross-validation are usually adopted.

(2) *The MLKNN model*
By using Bayesian conditional probability to calculate the tag value probability, the MLKNN model for multilabel classification is proposed. MLKNN still needs to find k nearest neighbors of the test samples in the training samples. But different from KNN, it uses the Bayesian method to calculate the probability of 0 or 1 for each label of the sample in the process of determining the label of the test samples, and the value with higher probability is the final value of the label. The detailed implementation steps of MLKNN is given as below:
Step 1. Calculate and find out the k nearest neighbors of each sample in the dataset.
Step 2. Calculate the prior probability $P(h_i)$ and compute the frequency array κ_i of each label whose value is 1. Then, calculate the prior probability $P(\neg h_i)$ and compute the frequency array $\tilde{\kappa}_i$ of each label whose value is 0. The calculating method is introduced as the following formulas (Zhang and Zhou 2007):

$$P(h_i) = \frac{1 + \sum_{j=1}^{M} Count\{L_{ij} = 1\}}{2 + M} \tag{4.26}$$

$$P(\neg h_i) = 1 - P(h_i) \tag{4.27}$$

$$\kappa_i(g) = \sum\nolimits_{j=1}^{m} Count\{O(L_{ij}) = g\} * Count\{L_{ij} = 1\}, \quad 0 \le g \le k \tag{4.28}$$

$$\tilde{\kappa}_i(g) = \sum\nolimits_{j=1}^{m} Count\{O(L_{ij}) = g\} * Count\{L_{ij} = 0\}, \quad 0 \le g \le k \tag{4.29}$$

in which $\kappa_i(g)$ represents the quantity of samples whose i-th label values are all equal to 1 and the number of neighbors the i-th label value being 1 is equal to g; $\tilde{\kappa}_i(g)$ represents the quantity of samples whose i-th label values are all equal to 0 and the number of neighbors the i-th label value being 0 is equal to g; $O(L_{ij})$ is the number of the current sample's neighbors whose i-th label is 1; M is the size of samples.

Step 3. Applying step 1 and step 2 for all labels of all training samples.

Step 4. Calculate the k nearest neighbors of each test sample and get the number of the sample's neighbors whose i-th label is 1, expressed as o_i.

Step 5. Calculate the value of each label of the test sample:

Firstly, compute the contingent probabilities of 1 and 0 value of the i-th label:

$$P(o_i|h_i) = \frac{1 + \kappa_i(o_i)}{k + 1 + \sum_{g=0}^{k} \kappa_i(g)} \tag{4.30}$$

$$P(o_i|\neg h_i) = \frac{1 + \tilde{\kappa}_i(o_i)}{k + 1 + \sum_{g=0}^{k} \tilde{\kappa}_i(g)} \tag{4.31}$$

Then, get the label of the test sample by the following equation:

$$\frac{P(h_i|o_i)}{P(\neg h_i|o_i)} = \frac{P(h_i)P(o_i|h_i)}{P(\neg h_i)P(o_i|\neg h_i)} \tag{4.32}$$

$$L_{ij} = \begin{cases} 1, & if\ \dfrac{P(h_i|o_i)}{P(\neg h_i|o_i)} > 1 \\ 0, & if\ \dfrac{P(h_i|o_i)}{P(\neg h_i|o_i)} < 1 \end{cases} \tag{4.33}$$

where $P(h_i|o_i)$ is the probability of the i-th label value of the test sample being 1 on the condition that the number of neighbors whose i-th label values are equal to 1 is o_i; $P(o_i|h_i)$ is the probability of that the number of neighbors whose i-th label values are equal to 1 is o_i on the condition that the i-th label value of the test sample is 1.

In order to validate the effectiveness of the MLKNN model on non-intrusive device recognition, the experiments will be conducted on the REDD dataset.

4.3.4 Evaluation of MLKNN in Terms of Multi-label Device Recognition

(1) *Evaluation indexes*
To evaluate the performance of MLKNN applicated in terms of the non-intrusive device recognition, six multilabel classification measures are selected, namely One Error, Hamming Loss, Ranking Loss, macro-F1score, micro-AUC, and micro-F1score.

(2) *Models performance analysis under different values of K*
As shown in Fig. 4.16, the six metrics values are changing along with the parameter k of MLKNN. It can be found that the Hamming Loss, Ranking Loss, macro-F1score, and micro-F1score metrics are getting worse as the value of k increases, but the range of variation is not large. It indicates that the MLKNN

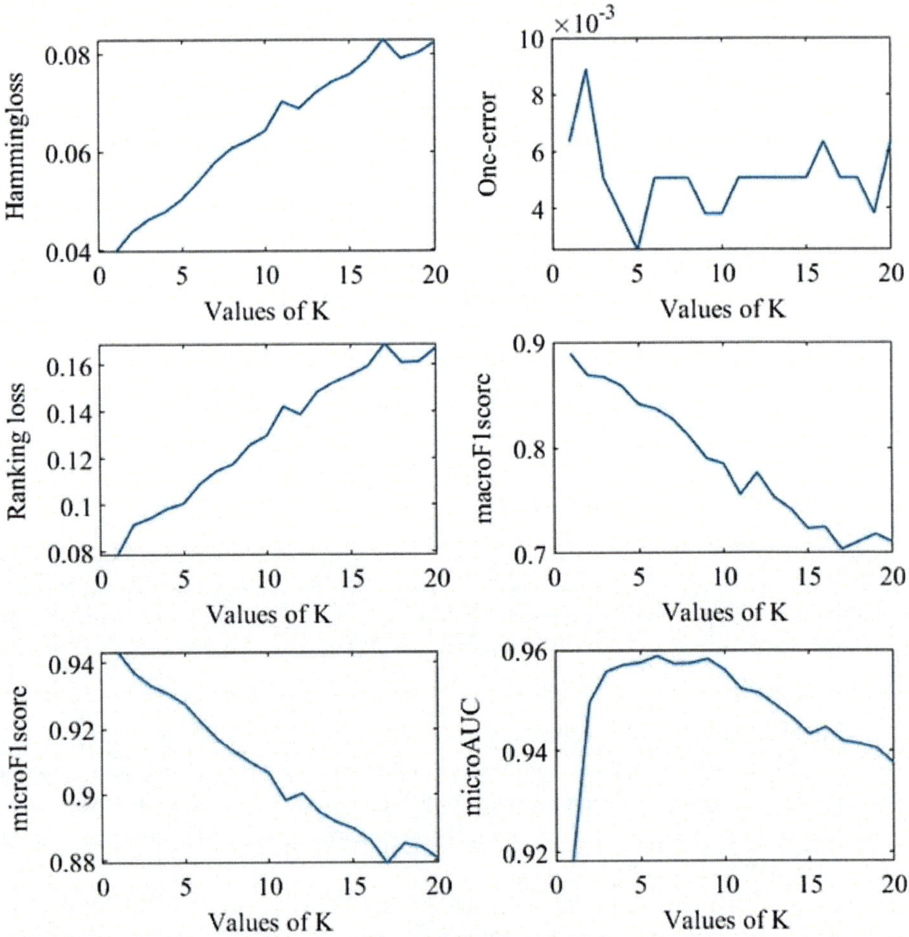

Fig. 4.16 Evaluation indexes of MLKNN model with different values of k

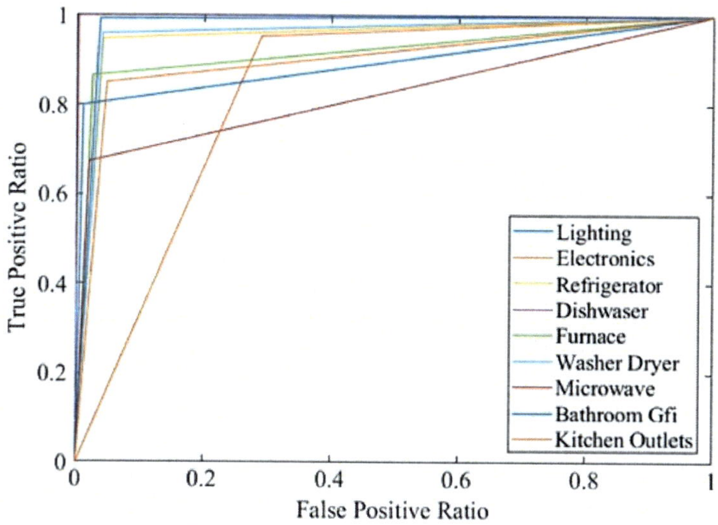

Fig. 4.17 The ROC curve of MLKNN

model performs a little worse when the value of k gets larger but it still has satisfying performance. The micro-AUC index has a peak value when k is equal to 6, which indicates that the accuracy of predictions of some devices is enhanced a lot at the point.

Finally, the k value is set to 1 to make the MLKNN model have its best performance on the problem of multilabel non-intrusive device recognition.

(3) *Detailed performance analysis of MLKNN with the optimal K value*
The optimal value of k is 1 in the experiment. From Figs. 4.17, 4.18, and Table 4.6, we can find that accuracy of the dishwasher is the highest that all of the samples in the test set are correctly predicted. Besides, the prediction results of the Bathroom GFI and microwave are worse than other devices, which is due to the small sample size of these two devices. On the whole, the results of MLKNN is accurate enough, which is able to meet the requirements for application.

(4) *Efficiency comparison analysis of Ranking SVM model and MLKNN model*
The training time and testing time of Ranking SVM and MLKNN are proposed in Table 4.7. It proves that compared to the Ranking SVM model, the MLKNN model requires fewer computing resources and the training time is reduced a lot.

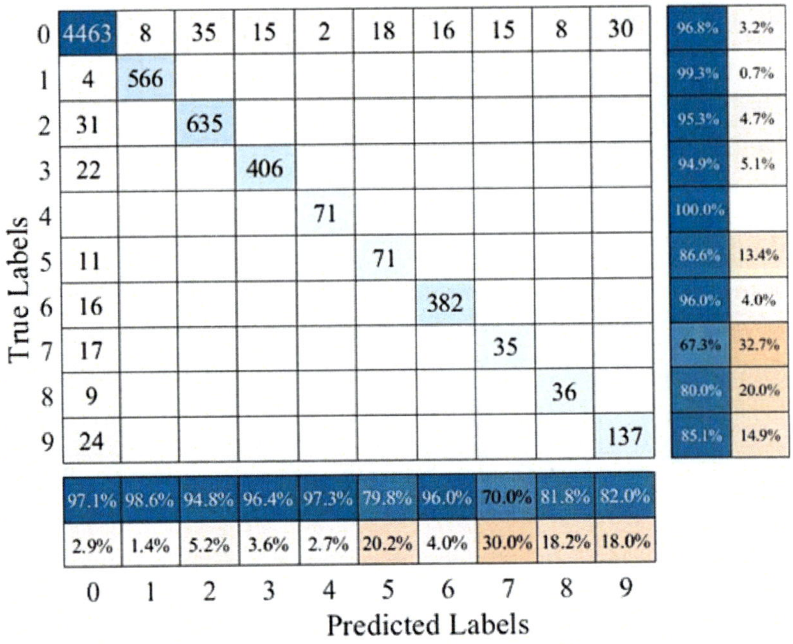

Fig. 4.18 The confusion matrix of MLKNN

Table 4.6 Evaluation indexes of MLKNN

Model	Micro-F1score	Macro-F1score	One error	Ranking loss	Hamming loss	Micro-AUC
MLKNN ($k = 1$)	0.9439	0.8879	0.0064	0.0780	0.0389	0.9168

Table 4.7 The training time and testing time of different models

Model	Training time	Testing time
Ranking SVM	859.58 s	17.922 s
MLKNN	26.85 s	12.53 s

4.4 Device Recognition Method Based on Multi-label Neural Networks

The application of the BP Neural Network in classification and data prediction has been quite mature. Many scholars have devoted themselves to improving BP neural network to apply it to different engineering problems. *Zhou* first applied the BP neural network to multi-label classification problems by defining a global error function and proposed Backpropagation for Multilabel Learning (BPMLL) (Zhang and Zhou 2006). At present, the BPMLL model has been applied to text recognition, biological information feature recognition, and many other aspects, and has been proved to be

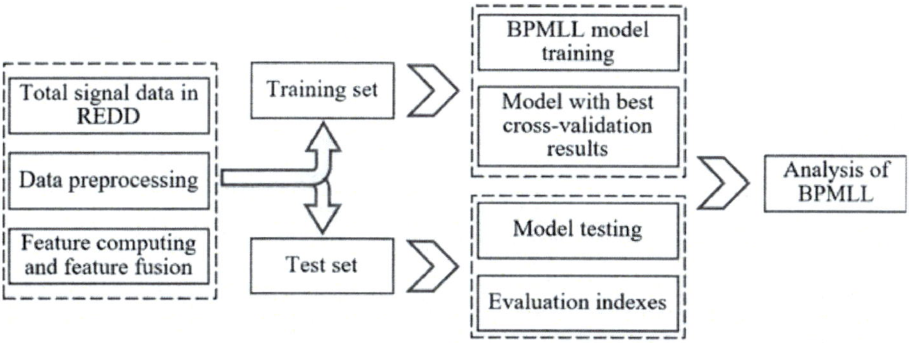

Fig. 4.19 The flow chart of the Backpropagation Multilabel learning model

effective in multilabel classification. This chapter studies its application in the field of multilabel non-intrusive device identification, proposes a non-intrusive device recognition method based on BPMLL, and designs experiments to test the influence of parameter selection on its performance.

4.4.1 Model Framework

The application procedure of BPMLL in terms of non-intrusive device recognition is briefly introduced in Fig. 4.19 and the thorough introduction is given as follows:

(a) Data processing. The raw data from the REDD dataset needs to be labeled and the features are desired to be extracted and fused from the raw current and voltage data.
(b) Dividing all of the samples into two sets, one is for training and the other for testing. 80% of the samples are utilized to train the BPMLL model and the rest is used for testing. the best hyper-parameters of BPMLL are obtained by performing cross-validation.
(c) Six multilabel evaluation measures are selected to assess the validation of the BPMLL model.

4.4.2 Preprocessing of the Raw Data

According to data labeling and feature extraction methods introduced in Sects. 4.2.2 and 4.2.3, the raw current and voltage data are processed into samples each of which has several labels and features. Besides, the PCA method is adopted to reduce the dimension of the characteristics. Finally, the samples with fused features are utilized in training the BPMLL model and evaluating the performance of the model.

4.4.3 The Basic Theory of Backpropagation Multi-label Learning

The BPMLL model proposed by *Zhou Zhihua* (Zhang and Zhou 2006) is derived from BP neural networks. Actually, the BP neural networks model is able to be applied to multilabel classification directly, but due to its unstable performance, the method hasn't been promoted a lot. However, by using a new global error function and the stochastic gradient descent algorithm, the BPMLL method has more stable and satisfactory performance in multilabel learning. The introduction of BPMLL is illustrated thoroughly as below (Zhang and Zhou 2006):

Step 1. Determine the quantity of neurons in each layer. Besides, the quantity of input neurons is $(M + 1)$ and the quantity of output neurons is N. The quantity of the hidden layer neurons is decided by cross-validation.

Step 2. Determine the initial values of the parameters. Determine the adjustment step size, the upper limit of iteration, and the error threshold of BPMLL, initialize the weights between the input and hidden layers and between the output and hidden layers, define the global error function. The global error of sample j is calculated as follows:

$$E_j = \frac{1}{\left|Lm_j\right| \times \left|\overline{Lm}_j\right|} \sum_{(L_a, L_b) \in Lm_j \times \overline{Lm}_j} \exp\left(-\left(\sigma_{ja} - \sigma_{jb}\right)\right) \tag{4.34}$$

where σ_{ja} represents the confidence coefficient of label a of sample j.

Step 3. For each sample, firstly, the forward transmission is performed according to the neural network. Besides, the global error is also calculated. And then the backward propagation is performed, the weights between the output and hidden layers and between the input hidden layers are updated; finally, repeat step 3 until the maximum number of iterations is reached or the global error exceeds the threshold.

Step 4. Calculate the confidence coefficient of the test sample by utilizing the final weights obtained in Step 3 and get the labels of the test sample according to the confident coefficient values.

4.4.4 Evaluation of BPMLL in Terms of Multi-label Device Recognition

(1) *Evaluation indexes*

By performing the BPMLL model on the REDD dataset, the accuracy of BPMLL on multilabel device classification can be evaluated. Six multilabel classification metrics, namely One Error, Hamming Loss, Ranking Loss, macro-F1score, micro-AUC, and micro-F1score, are calculated in the experiments to make the evaluation more credible and visualizable.

Table 4.8 Evaluation indexes of BPMLL

Model	Micro-F1score	Macro-F1score	One error	Ranking loss	Hamming loss	Micro-AUC
BPMLL	0.9129	0.8107	0.0051	0.1178	0.0608	0.9287

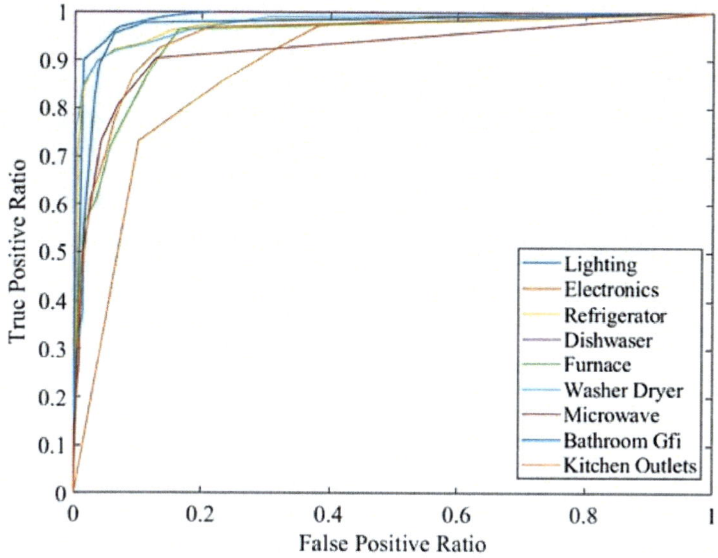

Fig. 4.20 The ROC curve of BPMLL

(2) *Experiments*
From Table 4.8, Figs. 4.20, and 4.21, it can be concluded that the BPMLL model also has acceptable performance on the problem of non-intrusive device recognition. The predictions of the dishwasher samples in the test set are all correct and the model performance on kitchen outlets and Bathroom Gfi is a little worse but still acceptable.

4.5 Experiment Analysis

In the section, the performance of Ranking SVM, MLKNN, and BPMLL are analyzed together. Table 4.9 gives the values of multilabel classification evaluation metrics of the three models. The performances of the three models are similar. According to the One Error and micro-AUC metrics, the Ranking SVM model has the best performance, which demonstrates that Ranking SVM has similar performance for all of the samples. In terms of micro-F1score and macro-F1score, MLKNN performs

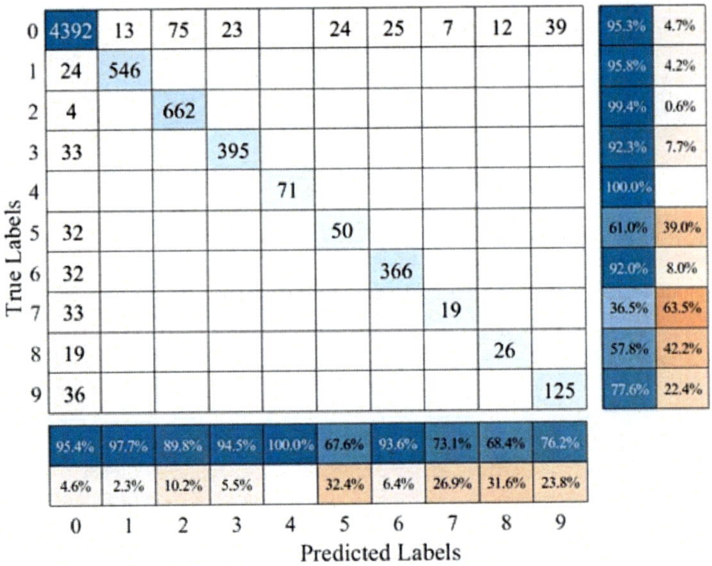

Fig. 4.21 The confusion matrix of BPMLL

Table 4.9 Evaluation indexes of Ranking SVM, MLKNN, BPMLL

Model	Micro-F1score	Macro-F1score	One error	Ranking loss	Hamming loss	Micro-AUC
Ranking SVM	0.9236	0.8484	0.0013	0.0990	0.0536	0.9581
MLKNN	0.9439	0.8879	0.0064	0.0780	0.0389	0.9168
BPMLL	0.9129	0.8107	0.0051	0.1178	0.0608	0.9287

better than the other two models, which indicates MLKNN has a good comprehensive performance. The performance of BPMLL falls somewhere in between the former two models. All of the three models are able to meet the requirements of practical application. However, for some devices, the recognition accuracy still needs to be enhanced. Thus, further researches in improving the classification precision of some special devices or some devices with few samples are still in need.

References

Abe, S. (2015). Fuzzy support vector machines for multilabel classification. *Pattern Recognition, 48*(6), 2110–2117.

Basu, K., Debusschere, V., Bacha, S., Maulik, U., & Bandyopadhyay, S. (2015). Non intrusive load monitoring: A temporal multi-label classification approach. *IEEE Transactions on Industrial Informatics, 11*(1), 262–270.

Bouhouras, A. S., Gkaidatzis, P. A., Panagiotou, E., Poulakis, N., & Christoforidis, G. C. (2019). A NILM algorithm with enhanced disaggregation scheme under harmonic current vectors. *Energy and Buildings, 183,* 392–407. https://doi.org/10.1016/j.enbuild.2018.11.013.

Buddhahai, B., Wongseree, W., & Rakkwamsuk, P. (2018). A non-intrusive load monitoring system using multi-label classification approach. *Sustainable Cities and Society, 39,* 621–630.

Gibaja, E., & Ventura, S. (2014). Multilabel learning: A review of the state of the art and ongoing research. *Wiley Interdisciplinary Reviews: Data Mining and Knowledge Discovery, 4*(6), 411–444.

He, D., Liang, D., Yi, Y., Harley, R. G., & Habetler, T. G. (2012). Front-end electronic circuit topology analysis for model-driven classification and monitoring of appliance loads in smart buildings. *IEEE Transactions on Smart Grid, 3*(4), 2286–2293.

Hwanjo, Y., & Sungchul, K. (2012). SVM tutorial: Classification, regression, and ranking. In *Handbook of Natural Computing* (pp. 479–506). Springer. https://doi.org/10.1007/978-3-540-92910-9_15.

Jayadeva, P. H., Sharma, M., & Soman, S. (2019). Twin neural networks for the classification of large unbalanced datasets. *Neurocomputing, 343,* 34–49. https://doi.org/10.1016/j.neucom.2018.07.089.

Kolter, J. Z., & Johnson, M. J. (2011). REDD: A public data set for energy disaggregation research. Paper presented at the Proceedings of the SustKDD Workshop on Data Mining Applications in Sustainability (pp. 59–62).

Kouziokas, G. N. (2020). SVM kernel based on particle swarm optimized vector and Bayesian optimized SVM in atmospheric particulate matter forecasting. *Applied Soft Computing, 93,* 106410. https://doi.org/10.1016/j.asoc.2020.106410.

Kuo, T. -M., Lee, C. -P., & Lin, C. -J. (2014). Large-scale Kernel RankSVM. Paper presented at the Proceedings of the 2014 SIAM International Conference on Data Mining.

Liu, Y., & Xie, M. (2020). Rebooting data-driven soft-sensors in process industries: A review of kernel methods. *Journal of Process Control, 89,* 58–73. https://doi.org/10.1016/j.jprocont.2020.03.012.

Liu, Y., Xue, W., Lin, Z., Liu, Y. J. E., & Buildings. (2018). Admittance-based load signature construction for non-intrusive appliance load monitoring. *Energy and Buildings, 171,* 209–219.

Machlev, R., Tolkachov, D., Levron, Y., & Beck, Y. (2020). Dimension reduction for NILM classification based on principal component analysis. *Electric Power Systems Research, 187,* 106459. https://doi.org/10.1016/j.epsr.2020.106459.

Mahmoudi, M. R., Heydari, M. H., Qasem, S. N., Mosavi, A., & Band, S. S. (2020). Principal component analysis to study the relations between the spread rates of COVID-19 in high risks countries. *Alexandria Engineering Journal,* 2–8. https://doi.org/10.1016/j.aej.2020.09.013.

Moyano, J. M., Gibaja, E. L., Cios, K. J., & Ventura, S. (2018). Review of ensembles of multi-label classifiers: Models, experimental study and prospects. *Information Fusion, 44,* 33–45. https://doi.org/10.1016/j.inffus.2017.12.001.

Pereira, R. B., Plastino, A., Zadrozny, B., & Merschmann, L. H. C. (2018). Correlation analysis of performance measures for multi-label classification. *Information Processing and Management, 54*(3), 359–369. https://doi.org/10.1016/j.ipm.2018.01.002.

Sivasathya, M., & Joans, S. M. (2012). Image feature extraction using non linear principle component analysis. *Procedia Engineering, 38,* 911–917. https://doi.org/10.1016/j.proeng.2012.06.114.

Sokolova, M., & Lapalme, G. (2009). A systematic analysis of performance measures for classification tasks. *Information Processing and Management, 45*(4), 427–437. https://doi.org/10.1016/j.ipm.2009.03.002.

Tabatabaei, S. M., Dick, S., & Xu, W. (2017). Towards non-intrusive load monitoring via multi-label classification. *IEEE Transactions on Smart Grid, 8,* 26–40.

Tanaka, A., Imai, H., Kudo, M., & Miyakoshi, M. (2007). Integrated kernels and their properties. *Pattern Recognition, 40*(11), 2930–2938. https://doi.org/10.1016/j.patcog.2007.02.014.

Thorsten, J. (2006). Training linear SVMs in linear time. Paper presented at the Proceedings of the ACM SIGKDD International Conference on Knowledge Discovery and Data Mining, 217–226.

Tsoumakas, G., Katakis, I., & Vlahavas, I. (2010). Mining multi-label data. In *Data Mining and Knowledge Discovery Handbook* (pp. 667–685). Springer. https://doi.org/10.1007/978-0-387-09823-4_34.

Zhang, H., Cheng, S., Li, H., Fu, K., & Xu, Y. (2020). Groundwater pollution source identification and apportionment using PMF and PCA-APCA-MLR receptor models in a typical mixed land-use area in Southwestern China. *Science of the Total Environment, 741,* 140383. https://doi.org/10.1016/j.scitotenv.2020.140383.

Zhang, M.-L., & Zhou, Z.-H. (2007). ML-KNN: A lazy learning approach to multi-label learning. *Pattern Recognition, 40*(7), 2038–2048. https://doi.org/10.1016/j.patcog.2006.12.019.

Zhang, M.-L., & Zhou, Z.-H. (2006). Multilabel neural networks with applications to functional genomics and text categorization. *IEEE Transactions on Knowledge and Data Engineering, 18*(10), 1338–1351.

Zhang, M.-L., & Zhou, Z.-H. (2014). A review on multi-label learning algorithms. *IEEE Transactions on Knowledge and Data Engineering, 26*(8), 1819–1837.

Chapter 5
Smart Non-intrusive Device Recognition Based on Intelligent Clustering Methods

5.1 Introduction

5.1.1 Background

With the rise of UEIOT and the continuous construction of the smart grid, more and more electrical metering devices like the smart meters have been applicated and a great deal of load data has been acquired. To mine and classify the relationship and composition of loads in different types and regions for the sake of a better understanding of users and stronger theoretical support for Demand-Side Management (DSM), load clustering analysis is frequently performed in a deep-going way.

Clustering is a typical unsupervised algorithm (Jain et al. 1999), the basic function of which is partitioning a given dataset into homogeneous groups i.e. clusters. There are broadly five categories of clustering methods, including segmentation methods, grid-based methods, hierarchical methods, model-based methods, and density-based methods. This chapter is restricted to the Fast Global K-Means (FGKM) method for segmentation clustering and the DBSCAN method for density-based clustering.

As early as when *Hart* first put forward the conception of NILM, the clustering method was taken to cluster the power points on the P-Q plane to identify the electrical appliances, which initiated the research of NILM (Hart 1992). Currently, the researches of load clustering basically put more emphasis on the issue of clustering methods, including the ascertainment of the cluster number, the analysis and evaluation of clustering performance, selection of the initial value of centroid, and the effective convergence of the algorithm. Factors that affect load clustering, such as the distance calculation method, and the influence of data dimension reduction, have also been studied. Generally, there are two types of clustering methods, one is in a direct way and the other in an indirect way, as shown in Fig. 5.1. The direct method performs clustering algorithms based on the load curve of power users. However, due to the high dimension and large quantity of the original load data, the calculation and storage requirements are relatively high in the clustering process. The indirect one is to first process the original load data to reduce its complexity. There are generally two

H. Liu et al., *Smart Device Recognition*,
https://doi.org/10.1007/978-981-33-4925-4_5

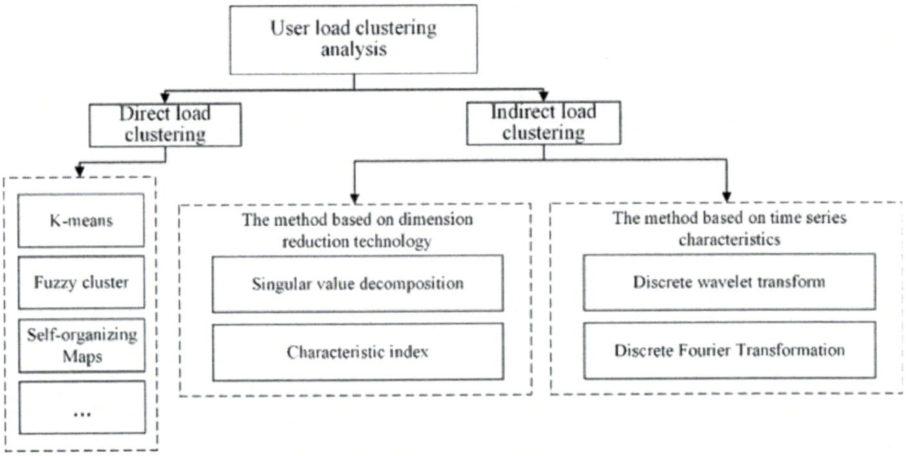

Fig. 5.1 Methods of load clustering

processing methods, one of which is to simplify the original load data by dimension reduction techniques, such as Principle Component Analysis (PCA) (Zakaria and Lo 2009), Singular Value Decomposition (SVD), Auto Encoder (AE) (Varga et al. 2015). Besides, another way is to process the load in the time domain or frequency domain based on time series (Teeraratkul et al. 2018).

Aiming at the defects of determining the cluster number in traditional clustering algorithms, Chicco et al. improved the k-means algorithm combined with adaptive learning theory (Chicco et al. 2006). Kwac et al. improved the FCM algorithm, according to the validity function to carry out dynamic clustering, which can make it choose the optimal number of clusters independently to implement load analysis (Kwac et al. 2014). Kong et al. proposed an extensible non-intrusive load decomposition method (Kong et al. 2018), in which an iterative k-means clustering algorithm is performed to establish a hidden Markov model, which is solved by an integer programming method. The method is validated on the REDD dataset and AMPds dataset respectively. Lin et al. proposed an improved spectral clustering method based on correlation measurement to carry out multi-scale clustering analysis of load morphological characteristics (Lin et al. 2019). The validity of the algorithm is evaluated by 100 different types of commercial loads. Li et al. proposed a new multi-resolution clustering algorithm for the massive data of smart meters (Li et al. 2016). Firstly, a Gaussian mixture model was established for massive data. Secondly, the load data were decomposed into different levels by different characteristics, such as peak value and overall trend. Finally, the dimension reduction algorithm based on spectrum analysis was used k-means clustering, the algorithm can be directly from the massive data load clustering. Teeraratkul et al. carried out the k-means clustering and used the dynamic time warping algorithm for distance measurement (Teeraratkul et al. 2018). Markov model is also established on the basis of classification for load forecasting by category. Panapakidis et al. proposed a method based on the combination of k-means++ and Self-Organizing Map (SOM) neural network (Panapakidis et al. 2014). SOM determines the optimal number of clusters, and k-means++ is used for stable load clustering.

5.1.2 Cluster Validity Index

Throughout the research and application of clustering, there are usually two essential problems to be solved (Giancarlo and Utro 2012). One is how to divide a given dataset so that the classification result is optimized, the other is how many classes the dataset is divided into. Among them, the former can be solved by the usage of clustering algorithms. However, the latter often referred to as the issue of cluster validation, is much tougher. To address this, an index function i.e. the validity index is established to evaluate the clustering algorithms under different circumstances. The optimal index function corresponds to the optimal clustering partition (Halkidi et al. 2001).

The cluster validity index can be divided into three categories (Salem and Nandi 2009): internal validity index, external validity index, and relative validity index. The internal validity index evaluates the geometry structure of the clusters, such as compactness, discreteness, and overlap. The external validity index compares the matching degree between partition results and the external criteria, which are usually the known class labels (Fukui and Numao 2012). In contrast, the relative index establishes a set of criteria before clustering to obtain the optimal partition under different parameter sets (Bolshakova et al. 2005).

In the chapter, the actual category of devices is given beforehand in the benchmarking dataset. Therefore, external indexes are employed to validate whether the selected devices are well segregated into distinct clusters and whether the unknown device is properly detected. Four adopted evaluation indexes are given below:

1. Purity
   Purity shows the proportion of the members that are properly categorized to the total members. For each cluster C_i, purity $P_i = \max\{P_{ij}\}$, where P_{ij} refers to the probability that members in cluster i belong to class j:

$$P_{ij} = \frac{m_{ij}}{m_i} \tag{5.1}$$

   where m_i is the size of cluster i, m_{ij} is the number of members in cluster i belong to class j. The Purity of the whole cluster partition can be calculated as follows:

$$Purity = \sum_{i=1}^{k} \frac{m_i}{m} P_i \tag{5.2}$$

   where k and m denote the number of clusters and the number of objects involved in the clustering, respectively.
   Purity takes value from 0 to 1. 0 for complete error, 1 for complete correctness.

2. Entropy
   Similar to purity, the entropy is defined as follows:

$$Entropy = \sum_{i=1}^{k} \frac{m_i}{m} E_i \tag{5.3}$$

$$E_i = -\sum_{j}^{L} P_{ij} \log P_{ij} \tag{5.4}$$

   where L is the number of classes. Entropy reflects uncertainty. The lower the Entropy, the more stable the system. The higher the Entropy, the more chaotic the system.
3. *Normalized Mutual Information*
   Mutual information (MI) denotes the dependence between two variables. A higher value of MI indicates a closer relationship. The MI between two variables A and B is calculated as follows:

$$MI(A, B) = \sum_{a} \sum_{b} \log \frac{p(a, b)}{p(a)p(b)} p(a, b) \tag{5.5}$$

   where $p(a)$, $p(b)$, $p(a, b)$ are respectively the edge probability density of A, the edge probability density of B, and the joint probability density of (A, B).
   Normalized Mutual Information (NMI) limits the magnitude of MI between 0 and 1and is given by the following definition:

$$NMI(A, B) = 2\frac{I(A, B)}{H(A) + H(B)} \tag{5.6}$$

$$H(A) = \sum_{x} p(a)MI(a), \quad MI(a) = \log \frac{1}{p(a)} \tag{5.7}$$

   In this chapter, A and B represent the actual and clustering partitions respectively.
4. *Rand Index*
   Assume the clustering results of dataset X = {X_1, X_2, ⋯, X_n} is C = {C_1, C_2, ⋯, C_k} and the real partition is P = {P_1, P_2, ⋯, P_j}. Based on the matching degree between C and P, the evaluation of clustering quality can be carried out. For any pair of data points (x_e, x_f) in the dataset, calculate the following terms (Halkidi et al. 2001):

   (a) SS: both points are included in the same partition in P and the same cluster C.
   (b) SD: points are included in the same partition in P but the different clusters in C.

Table 5.1 Confusion matrix of RI

Real partition P	Clustering results C	
	Same cluster	Different clusters
Same class	SS	SD
Different classes	DS	DD

(c) DS: points are included in the different partitions in P but the same cluster in C.

(d) DD: both points are included in the different partitions in P and the different clusters in C.

Table 5.1 gives the confusion matrix of RI.

The value of the Rand Index (RI) is calculated as the following formula:

$$RI = \frac{SS + DD}{SS + SD + DS + DD} \tag{5.8}$$

The range of values for the RI is 0 to 1. The higher the RI value, the higher the quality of clustering.

5.1.3 *Data Preprocessing*

(1) *COOLL dataset*

In the chapter, device recognition experiments are carried out on the Controlled On/Off Loads Library (COOLL), a new public dataset whose measurements are sampled at 100 kHz (Picon et al. 2016). It contains 12 types of electrical appliances with 42 instances in total, including Hedge trimmer, Drill, Router, Saw, Hair-dryer, Lamp, Grinder, Vacuum cleaner, Paint Stripper, Fan, Planer, and Sander. The summary of the COOLL dataset is given in Table 5.2. Based on the

Table 5.2 Summary of COOLL dataset

Appliance type	Hedge trimmer	Drill	Router	Saw
# of instances	3	6	1	8
# of samples	60	120	20	160
Appliance type	Hair-dryer	Lamp	Grinder	Vacuum cleaner
# of instances	4	4	2	7
# of samples	80	80	40	140
Appliance type	Paint stripper	Fan	Planer	Sander
# of instances	1	2	1	3
# of samples	20	40	20	60

Table 5.3 Load feature space and corresponding variable representation

Load feature vectors	Variable
Active power	P
Reactive power	Q
Apparent power	S
Resistance	R
Impedance	Z
Reactance	X
Admittance	Y
Conductance	G
Electrical susceptance	B
The imaginary and real parts of the first 5 odd harmonics	H_i

mined-out original voltage and current measurements, 19 sets of feature vectors are extracted. The feature space and the corresponding variable representation are shown in Table 5.3.

Considering that there is a large disparity in the size of samples of different appliances, for instance, the size of samples of the Saw is 160, while that of the Paint stripper is only 20. In this case, the appliances with a small sample size may be identified as the other devices. Thus, it results in adversely affecting the performance of clustering and leads to the wrong classification. To this end, data reconciliation is done. Paint stripper, Planer, and Router are excluded, the sample sizes of Drill, Saw, Vacuum cleaner are reduced to 80. The information on appliance samples after reconciliation is shown in Table 5.4.

Table 5.4 Information on appliance samples after reconciliation

Appliance type	Hedge trimmer	Drill	Saw
# of instances	3	4	4
# of samples	60	80	80
Label	1	2	3
Appliance type	Hair-dryer (HD)	Lamp	Grinder
# of instances	4	4	2
# of samples	80	80	40
Label	4	5	6
Appliance type	Vacuum cleaner	Fan	Sander
# of instances	4	2	3
# of samples	80	40	60
Label	7	8	9

(2) *PCA based feature selection*
For the conventional low-dimensional samples with high computation and more feature quantities, the clustering algorithm can perform efficiently, calculating the distance between samples and making the right classification, which is not the case with high-dimensional load data. The problems of sparse data samples and distance calculation can be serious obstacles for all clustering algorithms. To this end, dimensionality reduction is in need. On the one hand, it can greatly reduce the storage space of datasets. On the other hand, it can shorten the time of calculating the distance between data points to increase the efficiency of the clustering algorithm. For this purpose, the famous dimensionality reduction method PCA described in Sect. 5.4 can be utilized. The flow chart of feature selection based on PCA is shown below in Fig. 5.2.
The eigenvalues underline the significance of the corresponding principal components. The eigenvalues and cumulative contribution rate (CCR) of principal components after PCA are shown in Table 5.5. The CCR of all principal components is shown in Fig. 5.3. As can be seen from Table 5.5 and Fig. 5.3, the CCR of the first five principal components already surpasses 95%, which indicates a comprehensive representation of the original data. Consequently, they are selected as features after dimension reduction.
The chapter introduces two fundamental clustering methods as well as their application in non-intrusive device recognition for twelve household appliances. After preprocessing the real data from the COOLL dataset, more than twenty features are extracted. Five principle features are further selected via the PCA method. Then the clustering algorithms are carried out to separate classes from load data. As a result, appliance samples are assigned to clusters or labeled as 'Noise'. In the end, several evaluation indexes are established to assess the validation of different clustering algorithms. The model framework for smart non-intrusive device recognition with different clustering methods is shown in Fig. 5.4.

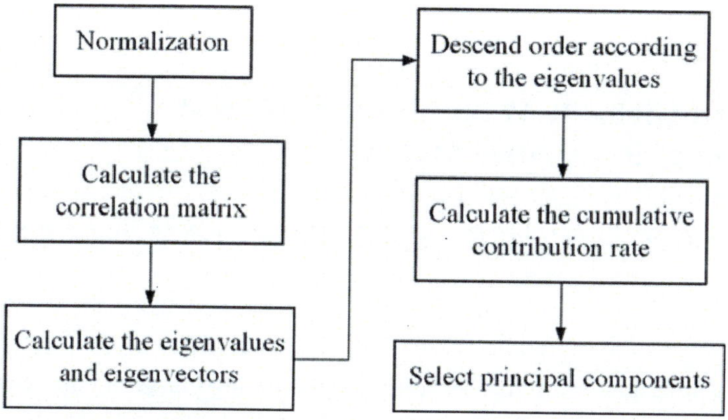

Fig. 5.2 The flow chart of feature selection based on PCA

Table 5.5 Eigenvalues and the CCR of components

Principle component	Eigenvalues	CCR (%)
1	2.1703	49.38
2	1.1489	75.51
3	0.4409	85.54
4	0.3253	92.94
5	0.1129	95.51
6	0.0548	96.76
7	0.0471	97.83
8	0.0376	98.68
9	0.0226	98.90
10	0.0164	99.20
11	0.0097	99.57
12	0.0053	99.79
13	0.0033	99.91
14	0.0020	99.99
15	0.0003	100

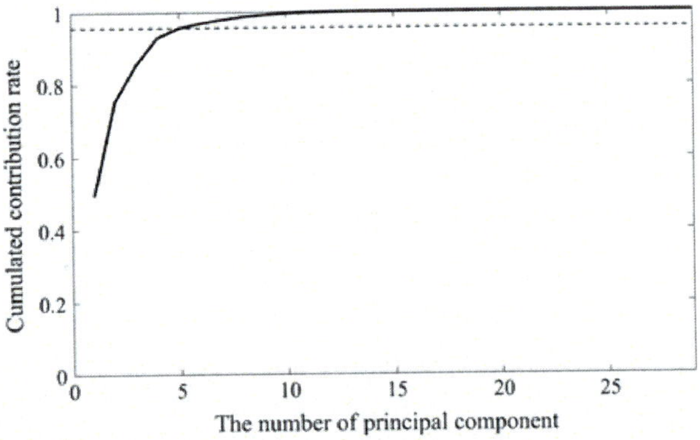

Fig. 5.3 The CCR of all principal components

5.2 Fast Global K-Means Clustering-Based Device Recognition Method

5.2.1 The Theoretical Basis of K-Means, GKM and FGKM

(1) *K-means*

Originally proposed by *Macqueen* in 1967 (Macqueen 1965), the k-means (KM) algorithm now is commonly used in data mining clustering analysis due to its simplicity and fast convergence speed. It realizes clustering through continuous iterations.

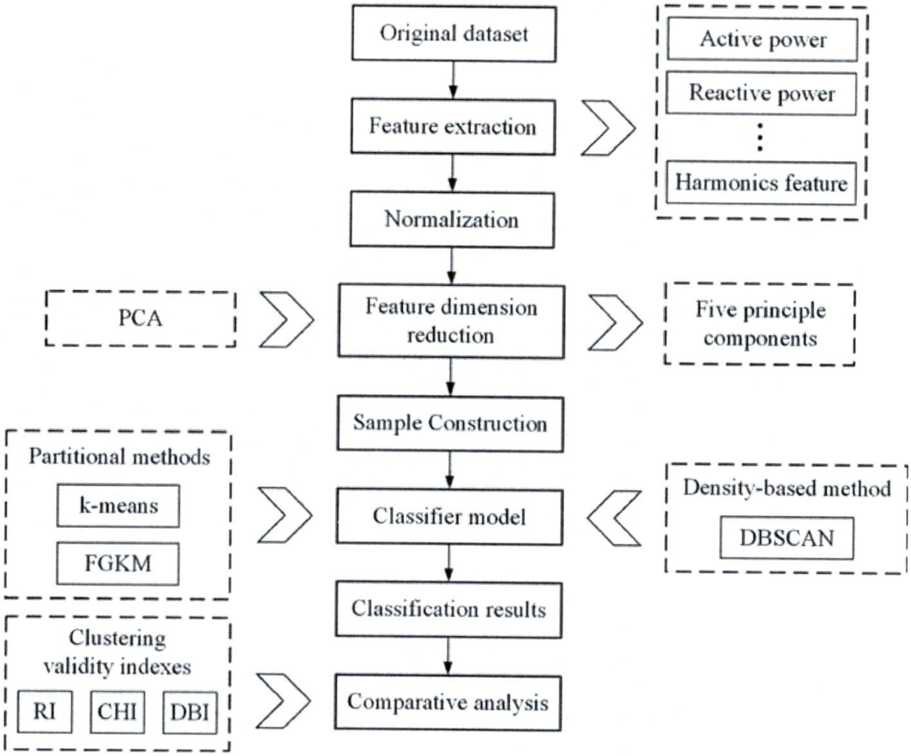

Fig. 5.4 The model framework of the device recognition with clustering methods

The KM algorithm aims to divide data points into k clusters and define a cluster centroid for each cluster (Zhao et al. 2015). The steps of the KM algorithm can be described as follows:

Step 1. For a given dataset containing N data points $x_i(1 \leq i \leq N)$, select randomly k points as the initial cluster centers $c_j(1 \leq j \leq N)$.

Step 2. Calculate the distance (generally Euclidean distance) between the rest of the objects and the selected centers. Assigned each object to the nearest cluster center.

Step 3. Once all the assignments are done, recalculate the cluster centers as follows:

$$c_j = \frac{1}{b_j} \sum\nolimits_{x \in C_j} x (1 \leq j \leq k) \tag{5.9}$$

where b_j is the quantity of data points in the cluster C_j.

Step 4. Define a criterion as the cases that objects are on longer reassigned to other clusters, the centroids no longer alter, or the sum of the squared errors (SSE) is locally minimized. When the criterion is satisfied, the iteration will be terminated and the clustering results will be obtained.

The SSE is defined as follows:

$$SSE = \sum_{j=1}^{k} \sum_{x \in C_j} d(c_j, x)^2 \tag{5.10}$$

The distance $d(x_k, y_k)$ is denoted as the Euclidean distance. It can be obtained as follows:

$$d(x_k, y_k) = \sqrt{\sum_{k=1}^{N} (x_k - y_k)^2} \tag{5.11}$$

However, the KM clustering method also has some shortcomings. The most important one is its strong reliance on the initial choice of the cluster centroids. If the given initial cluster centroid is not well selected, it will only guarantee a local minimum rather than a global minimum, which leads to the poor performance of clustering.

(2) *GKM*
To tackle the problem of the initial positions in the KM algorithm, various types of clustering algorithms have been modified via global search methods. Among them, the global k-means (GKM) clustering model proposed by *Likas* et al. in 2003 (Likas et al. 2003) is a representative model that effectively solves the local minimum problem in an incremental approach without being subject to any initial parameters. The basic idea of GKM is to determine a new optimal initial cluster center in each iteration through a global search, which is realized by N executions of the KM algorithm. More specifically, the process of the GKM algorithm is described as follows (Bai et al. 2013):
Step 1. The mean value of all data points is computed and taken as the initial cluster centroid $c_1 = \sum_{i=1}^{N} x_i / N$ of the first cluster C_1. Set $g = 1$.
Step 2. Add 1 to g each loop. If $g > k$, then stop.
Step 3. ($c_1, c_2, \cdots, c_{g-1}$) is taken as the centers of (g-1)-clusters and each object x_i is taken as the initial center of the g-th cluster. The KM clustering algorithm is executed for a local search and new centers of each cluster (c_1*, c_2*, ⋯, c_N*) are obtained.
Step 4. Set c_i= c_i* and perform step 2.

(3) *FGKM*
The GKM clustering algorithm is computationally expensive and is not appropriate for massive data clustering. Hence, *Likas* et al. proposed a modified method named the fast global k-means (FGKM) algorithm to shorten the computation time (Likas et al. 2003).
The modification lies in introducing a parameter f_i for the selection of the initial center of the next cluster. The value of parameter f_i at each point in the dataset is calculated. The initial cluster center of the next cluster is the point where the f_i value reaches the maximum. f_i is defined as follows:

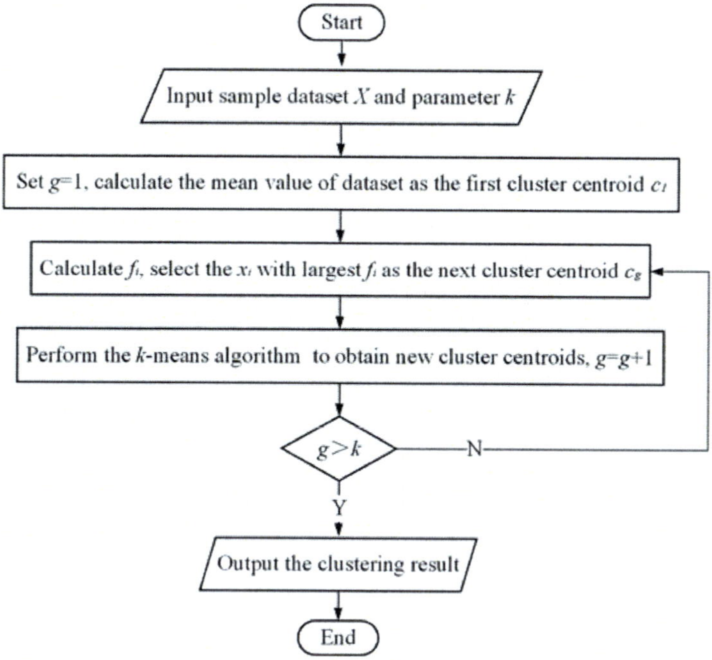

Fig. 5.5 The flow chart of the FGKM algorithm

$$f_i = \sum_{j=1}^{N} \max(d_{k-1}^j - \|x_i - x_j\|^2, 0) \tag{5.12}$$

where d_{k-1}^j denotes the squared Euclidean distance between the nearest center and x_j. f_i represents the reduction of the SSE when an object x_i in the dataset is taken as the initial cluster center of the next cluster. In other words, the larger the reduction, the smaller the corresponding value of the SSE. Figure 5.5 shows the flow chart of the FGKM algorithm, which can be described as follows (Bai et al. 2013):

Step 1. The mean value of all data points is computed and taken as the initial cluster centroid $c_1 = \sum_{i=1}^{N} x_i / N$ of the first cluster C_1. Set $g = 1$.

Step 2. Add 1 to g each loop. If $g > k$, then stop.

Step 3. (c_1, c_2, ···, $c_{g\text{-}1}$) is taken as the centers of (g-1)-clusters. The value of parameter f_i corresponding to each object x_i is calculated respectively so that the object x_i where f_i gets the maximum value is the best initial center of the g-th cluster. Then the KM algorithm is executed again to optimize the initial centers and new centers of each cluster (c_1*, c_2*, ···, c_N*) are obtained.

Step 4. Set c_i= c_i * and carry out step 2.

5.2.2 Steps of Modeling

(a) Preprocess the COOLL dataset. Twelve typical devices such as Drill, Fan, etc., are selected. Over twenty load feature vectors are extracted and further normalized.
(b) Reduce the dimension of features obtained in *step(a)*. To this end, the PCA method is applied and five principal features are selected.
(c) Perform the FGKM method with different k on the feature space and the clustering results are obtained.
(d) Count the number of the identified appliances in the clusters, and calculate the four evaluation index values including Purity, Entropy, NMI, and RI to analyze the performance of the classifier.

5.2.3 Clustering Results

This section compares the performance of the fast global k-means algorithm under different k values. First, set $k = 6,9,12$ respectively. After the clustering work is executed, the unknown appliances are labeled. The visualization results are illustrated in Figs. 5.9, 5.10 and 5.11. Secondly, the types and quantities of appliances detected in each cluster are counted and displayed in the form of a histogram in Figs. 5.6, 5.7 and 5.8. In each subfigure, the 1st to 9th column corresponds to the number of HTs, Drills, Saws, HDs, Lamps, Grinders, VCs, Fans, and Sanders. Finally, the cluster validity indexes under different k values are calculated. The comparison results are shown in Table 5.6 and Fig. 5.12.

According to Figs. 5.6, 5.7, 5.8, 5.9, 5.10, 5.11, 5.12 and Table 5.6, it can be concluded that:

(a) Appliances are well identified by FGKM. Among all the nine types of appliances, Fans have the highest recognition accuracy, followed by HDs and Sanders, whose distribution barely changes with the cluster refinement. The features of HTs, Drills, and Saws seem to share a high similarity, which makes it difficult to distinguish them from each other.
(b) As k takes a higher value, appliances in clusters with a large number of samples are more easily separated to form a new cluster. However, it is also likely that a certain type of appliances will be divided into two or more clusters. For example, Cluster#6 in Fig. 5.6 contains up to six appliances, while the largest cluster in Fig. 5.8 contains three. Appliances like HTs, Drills, Saws appear in Cluster#6 of Fig. 5.6 but scatter to multiple clusters in Fig. 5.8, such as Cluster#8, Cluster#11, and Cluster#12.
(c) The optimal cluster number of the model is about 7. It is apparent that in Fig. 5.12, the Purity gradually decreases, while the Entropy, NMI, and RI show an overall upward trend. When $k = 7$, RI, NMI, and Purity are relatively high, and the entropy value is also controlled below 2. As k continues to increase,

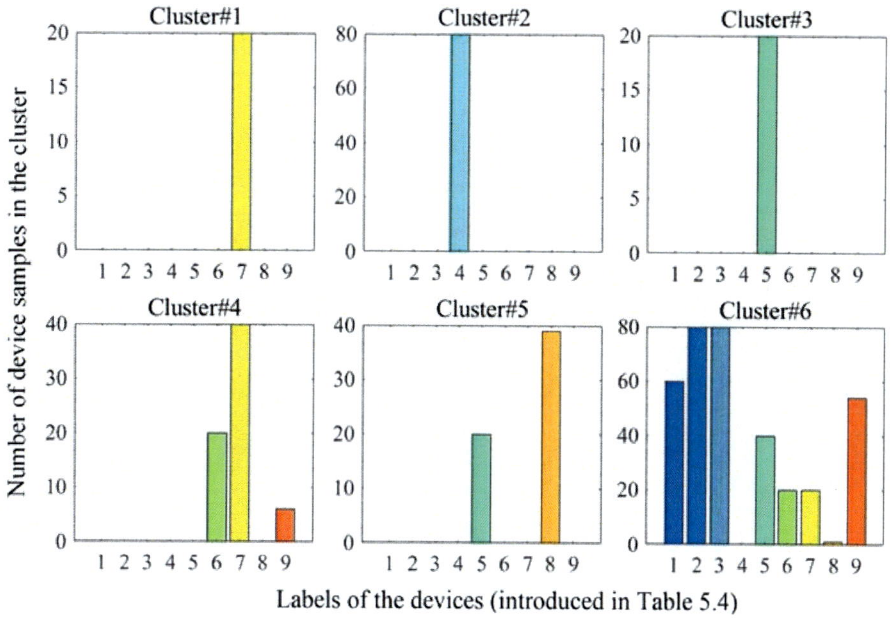

Fig. 5.6 Appliances in Cluster#1 to Cluster#6 ($k = 6$)

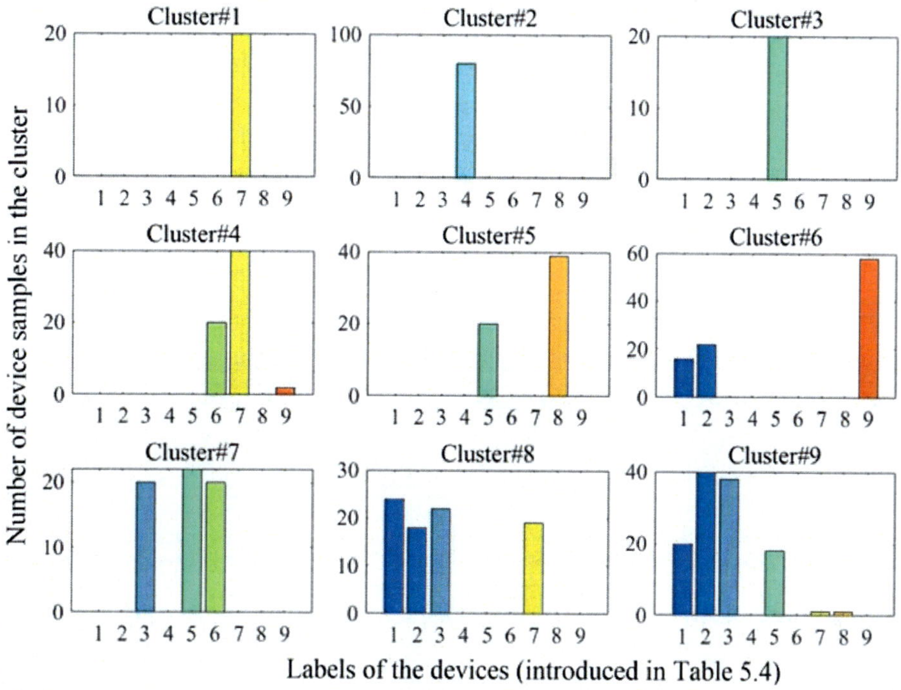

Fig. 5.7 Appliances in Cluster#1 to Cluster#9 ($k = 9$)

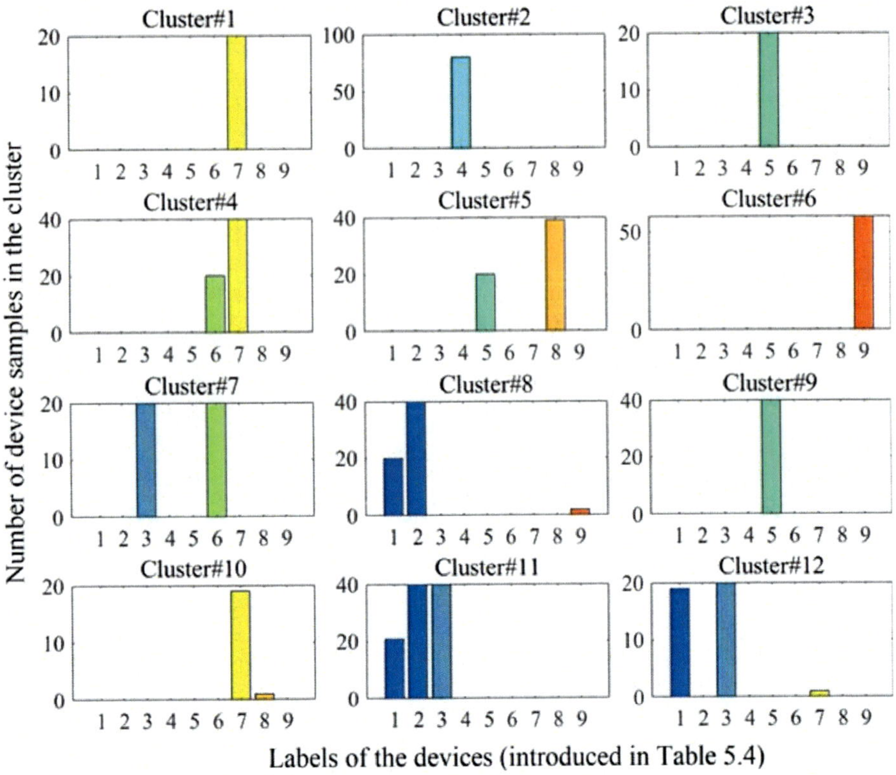

Fig. 5.8 Appliances in Cluster#1 to Cluster#12 ($k = 12$)

Table 5.6 The clustering validity indexes of FGKM under different k

Model	Value of k	Purity	Entropy	NMI	RI
FGKM	5	0.8567	0.9942	0.4444	0.5653
	6	0.8217	1.2768	0.5306	0.6725
	7	0.6750	1.6750	0.5582	0.8003
	8	0.6917	1.9016	0.6466	0.8575
	9	0.6017	2.1649	0.6095	0.8699
	10	0.6500	2.1625	0.7025	0.8894
	11	0.6367	2.2929	0.7205	0.9049
	12	0.6300	2.3691	0.7392	0.9110
	13	0.6017	2.4774	0.7215	0.9115
	14	0.5583	2.5328	0.7035	0.9067
	15	0.5550	2.5754	0.7001	0.9070

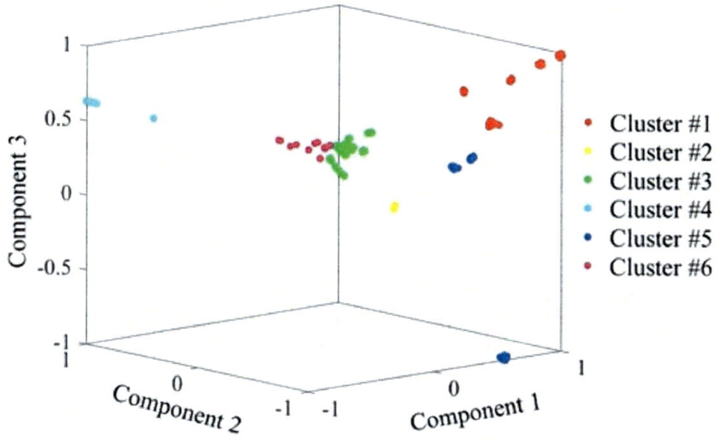

Fig. 5.9 The clusters found by FGKM ($k = 6$)

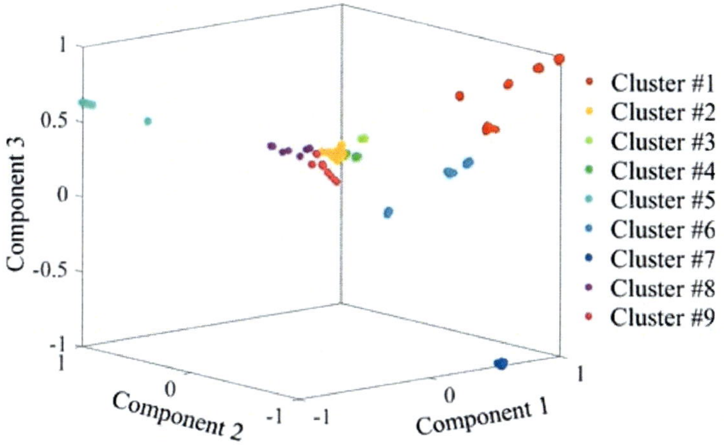

Fig. 5.10 The clusters found by FGKM ($k = 9$)

the growth trend in RI, NMI, and Purity gradually slows down, but the Entropy has already mounted over 2.5. As demonstrated in Sect. 5.1.2, a good clustering performance requires both high values of Purity, NMI, RI, and low value of Entropy. Thus, the optimal value of k can be determined near 7.

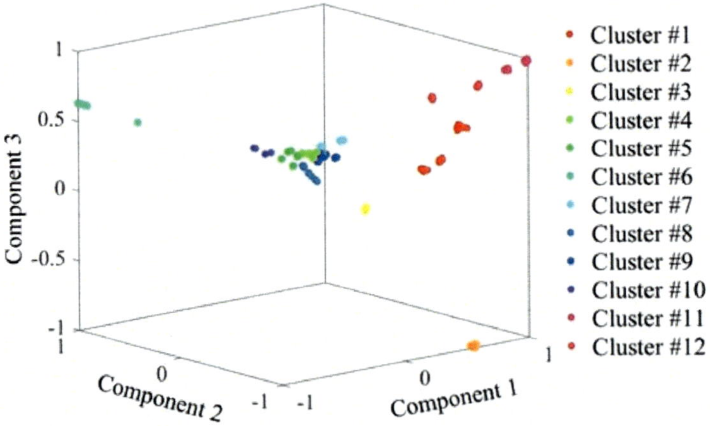

Fig. 5.11 The clusters found by FGKM ($k = 12$)

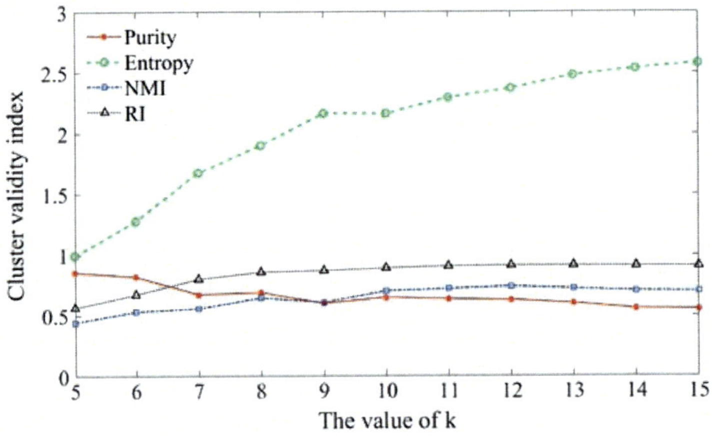

Fig. 5.12 Comparison of cluster validity indexes under different k

5.3 DBSCAN Based Device Recognition Method

5.3.1 The Theoretical Basis of DBSCAN

The density-based spatial clustering of applications with noise (DBSCAN) was first proposed by Ester et al. in 1996). In DBSCAN, there is no need to determine the cluster number beforehand as is the case with the KM method. The clusters are grouped through the data points closely arranged with high density. For each object in a certain cluster, the number of objects in its given radius domain cannot be less than a given minimum number. The DBSCAN method is capable of recognizing clusters with arbitrary shapes as well as correctly isolating the noise from the rest of

Fig. 5.13 The flow chart of the DBSCAN algorithm

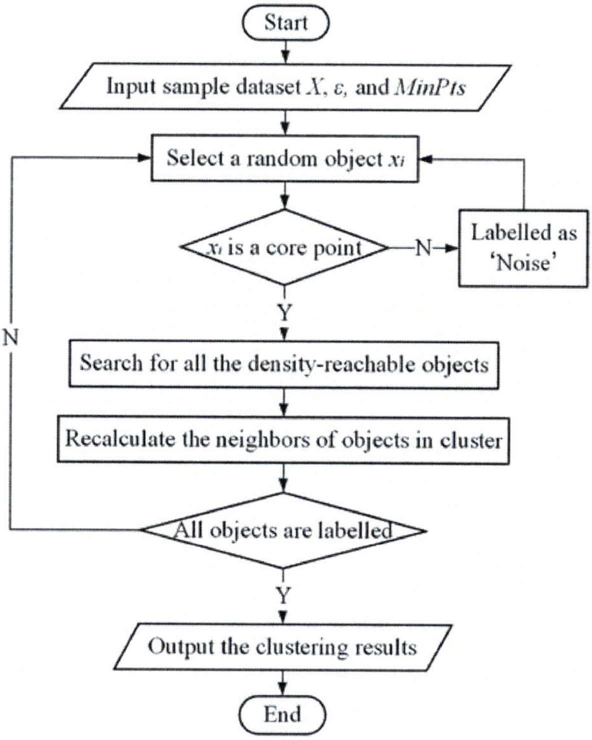

the data. For a dataset *D* having *n* objects, several basic concepts of the DBSCAN are given as follows (Galán 2019):

(a) ε-neighborhood: For object $x_i \in D$, the ε-neighborhood is denoted as $N_\varepsilon(x_i)$.
(b) *MinPts*: Neighborhood density threshold.
(c) Core points: For object $x_i \in D$, if the number of points in its neighborhood exceeds *MinPts*, then x_i is defined as a core point.
(d) Directly density-reachable: For objects $(x_e, x_f) \in D$, if x_f is in the ε-neighborhood of x_e (which is a core point), then x_f is directly density-reachable from x_e.
(e) Density-reachable: Given a sequence of *m* objects $(x_1, x_2, \cdots, x_m)$ where $2 \leq m \leq n$ and $1 \leq g \leq m\text{-}1$, if x_{g+1} is directly density-reachable from x_g, then x_m is density-reachable from x_1.
(f) Density-connected: For object $x_k \in D$, if not only x_i is density-reachable from x_k but also x_j is density-reachable from x_k, then x_i is density-connected to x_j.

Figure 5.13 shows the flow chart of DBSCAN, which can be described as follows:

Step 1. An unprocessed object x_i is randomly selected from dataset *D* and its ε-neighborhood satisfies the *MinPts* requirement, that is, x_i must be a core point.

Step 2. Traverse the entire dataset to search for all the objects that are density-reachable from to x_i so that a new cluster is obtained.

Step 3. For the objects obtained in the cluster in Step 2, each of their neighbors is recalculated until the cluster cannot be expanded further.

Step 4. Repeat Step 2 and Step 3 until all objects in the dataset are labeled. The object not included in any cluster is labeled as outliers (noise).

5.3.2 *Steps of Modeling*

(a) Data preprocessing. Mine the original current and voltage samples. Compute electrical and harmonics feature quantities.
(b) Feature dimensionality reduction. Input normalized feature vectors, utilize the PCA method to select several principal components that can comprehensively represent the whole features.
(c) Clustering. Perform the DBSCAN algorithm with different parameters to realize the ideal classification results.
(d) Clustering performance evaluation.

5.3.3 *Clustering Results*

In this section, the transformed input samples are clustered with the DBSCAN method. The parameter *MinPts* is not trained but heuristically set to 5. By altering the value of ε, several clustering configurations are set. The classification results are visualized in Figs. 5.14, 5.15 and 5.16. The identification of various appliances is shown in Figs. 5.17, 5.18, 5.19. In each subfigure, the 1st to 9th column respectively represents the number of HTs, Drills, Saws, HDs, Lamps, Grinders, VCs, Fans, and Sanders. The comparison of evaluation indexes of the DBSCAN algorithm under different parameter configurations is shown in Table 5.7 and Fig. 5.20.

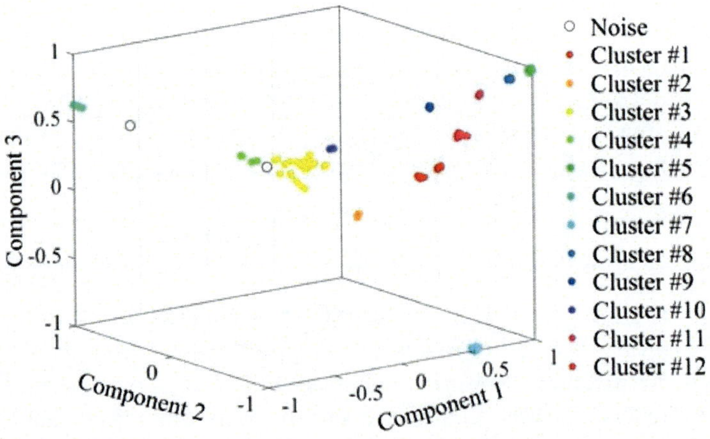

Fig. 5.14 The clusters found by DBSCAN ($\varepsilon = 0.10$, *MinPts* = 5)

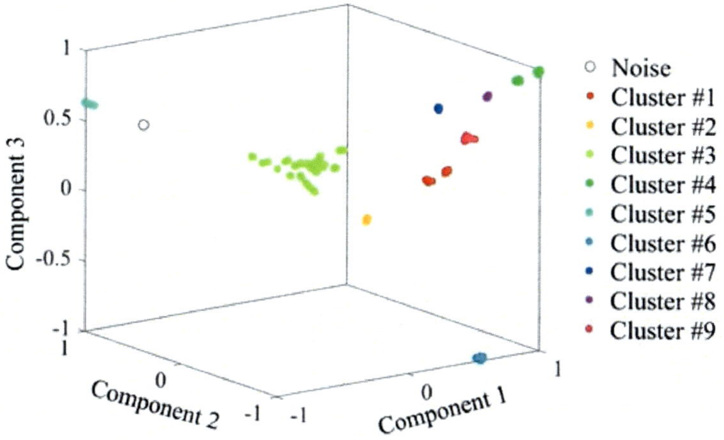

Fig. 5.15 The clusters found by DBSCAN ($\varepsilon = 0.20$, $MinPts = 5$)

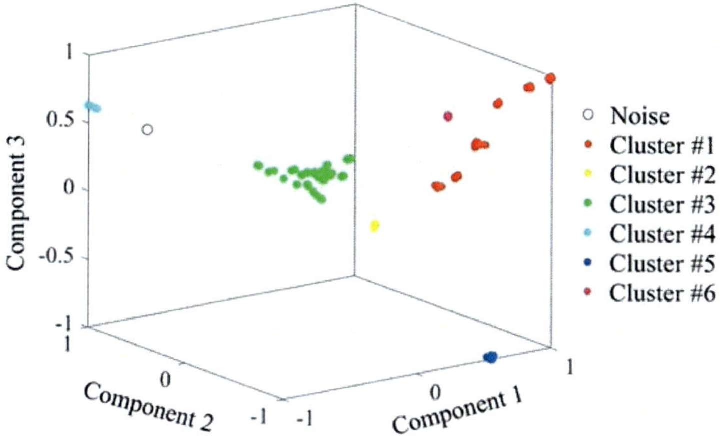

Fig. 5.16 The clusters found by DBSCAN ($\varepsilon = 0.30$, $MinPts = 5$)

According to Figs. 5.14, 5.15, 5.16, 5.17, 5.18, 5.19, 5.20 and Table 5.7, it can be concluded that:

(a) A massive cluster can be found in each clustering result in Figs. 5.14, 5.15 and 5.16, which suggests that many appliance samples overlap in this region of feature space. Besides, with the increase ε, fewer clusters are obtained and fewer data points are labeled as 'Noise'.

(b) Appliances are not well identified by DBSCAN. According to Figs. 5.17, 5.18 and 5.19, too many appliances, such as HTs, Drills, Saws, Lamps, and Sanders, are concentrated in one cluster, while appliances like HDs are included in multiple clusters. This may be due to the fact that each appliance has several operating states and brands, which makes the samples of each appliance quite different from each other. In the feature space, the distance between the features

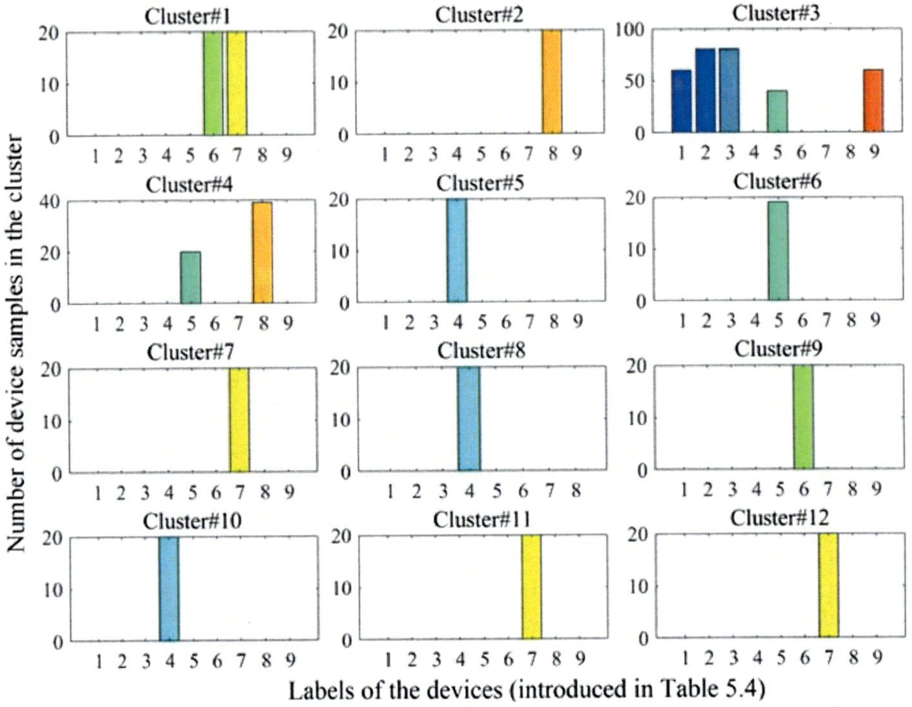

Fig. 5.17 Appliances in Cluster#1 to Cluster#12

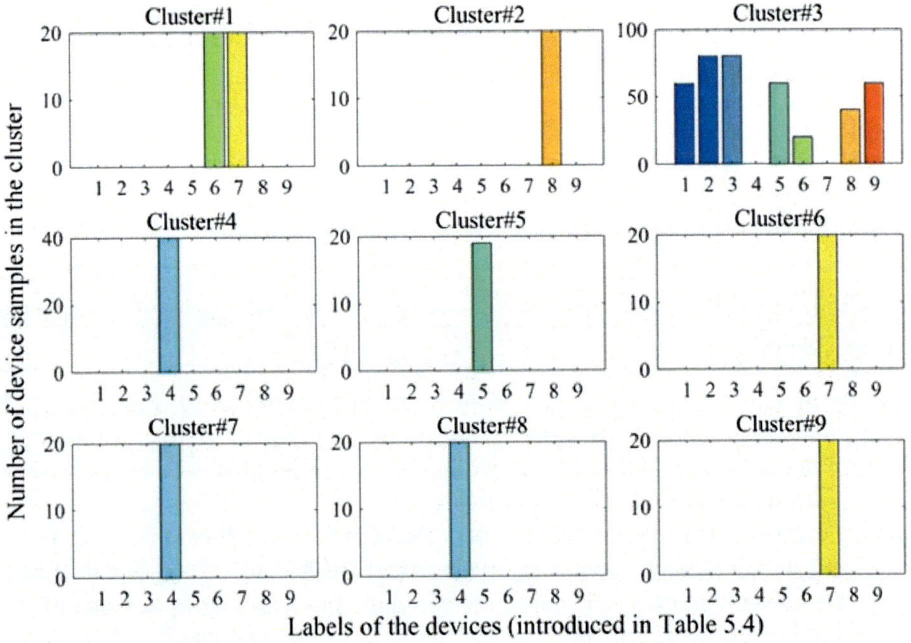

Fig. 5.18 Appliances in Cluster#1 to Cluster#9

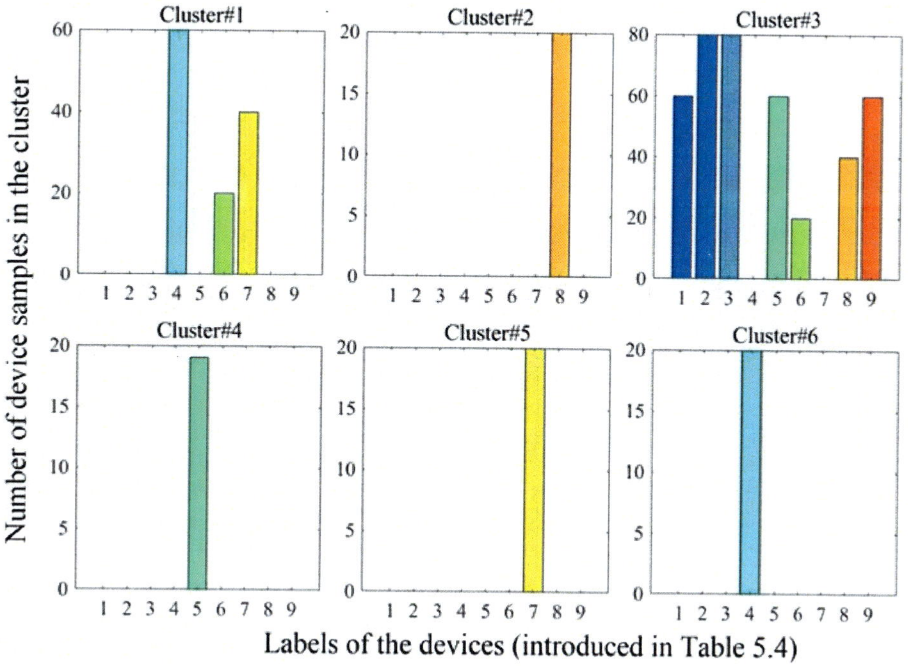

Fig. 5.19 Appliances in Cluster#1 to Cluster#6

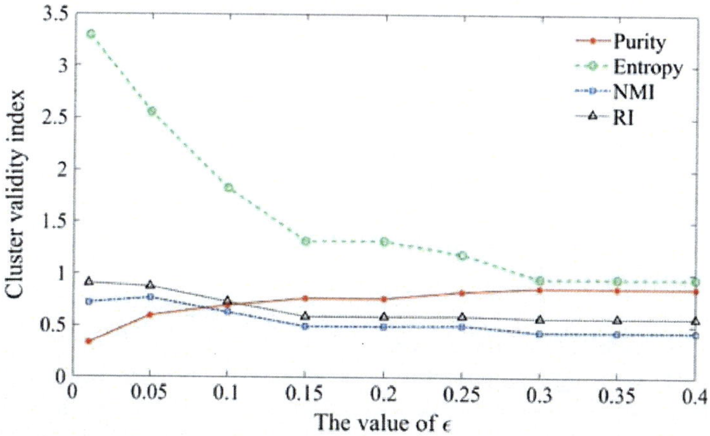

Fig. 5.20 Comparison of cluster validity indexes under different ε

of the same appliance is even larger than that of the features of different appliances. As a consequence, the identification and classification of other appliances are seriously affected.

(c) 5 for *MinPts* and 0.15 for ε is the optimal parameter configuration. In Fig. 5.20, as ε the value increases, the Entropy goes through a drastic decline from 3.2973

Table 5.7 The cluster validity indexes of DBSCAN under different ε

Model	*MinPts*	ε	Clusters	Purity	Entropy	NMI	RI
DBSCAN	5	0.01	28	0.3333	3.2973	0.7227	0.9074
		0.05	19	0.5983	2.5557	0.7682	0.8773
		0.10	13	0.6983	1.8277	0.6306	0.7305
		0.15	9	0.7667	1.3182	0.5063	0.5948
		0.20	9	0.7667	1.3182	0.4978	0.5903
		0.25	7	0.8333	1.1909	0.4978	0.5903
		0.30	5	0.8667	0.9566	0.4422	0.5747
		0.35	5	0.8667	0.9566	0.4422	0.5747
		0.40	5	0.8667	0.9566	0.4422	0.5747

to 0.9566, while the Purity slightly rises from 0.3333 to 0.8667. NMI shares a similar downward trend with RI. Both NMI and RI take the maximal value at $\varepsilon = 0.01$, where, however, the Entropy is too high to meet the requirements. It is worth mentioning that when $\varepsilon = 0.15$, the Entropy drops the fastest and suddenly reaches the elbow of the broken line. The Purity and RI are relatively high with 0.7667 and 0.5948 respectively at this point. All the indexes remain unchanged after $\varepsilon = 0.4$. Put all of it together, 0.15 can be determined as the optimal value of ε.

5.4 Experiment Analysis

This chapter mainly analyzes the performance of the fast global k-means algorithm and the DBSCAN algorithm in non-intrusive device recognition. Based on the COOLL dataset, 19 feature quantities of 9 appliances are extracted and further fused by the PCA method. Then, two clustering methods are performed to identify the appliance samples. Finally, the clustering performance evaluation is realized.

In order to substantiate the superiority of FGKM and DBSCAN in load identification, k-means, as the most basic clustering method, is also tested. Table 5.8 and Fig. 5.21 give the comparison of cluster validity indexes of k-means, FGKM, DBSCAN under the optimal parameter configuration. In k-means clustering, k is set to 9. In FGKM clustering, k is set to 7. In DBSCAN clustering, *MinPts* is 5 and ε is 0.15.

According to Table 5.8 and Fig. 5.21, the following conclusions can be drawn:

(1) It is hard to conclude which model is absolutely superior when taking all the evaluation indexes into account. FGKM has the highest NMI and RI values, but its Purity and Entropy are not so satisfactory. DBSCAN has the highest Purity and the lowest Entropy, but its NMI and RI values are relatively low. But in general, both FGKM and DBSCAN perform better than k-means clustering.

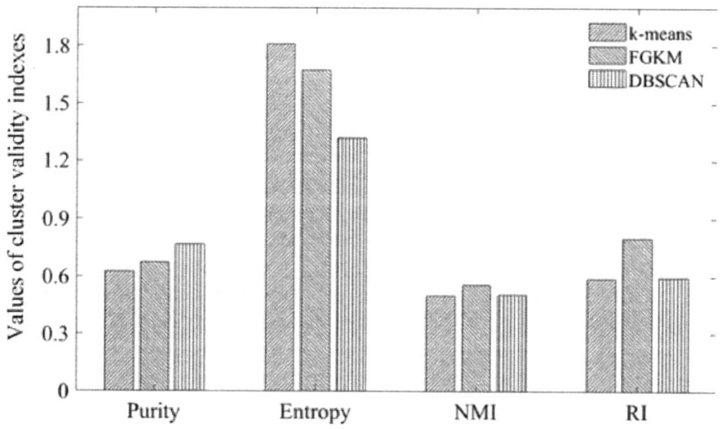

Fig. 5.21 Comparison of cluster validity indexes of k-means, FGKM, DBSCAN

Table 5.8 The cluster validity indexes of k-means, FGKM, DBSCAN

Model	Purity	Entropy	NMI	RI
k-means	0.7667	1.3116	0.4988	0.5904
FGKM	0.6750	1.6750	0.5582	0.8003
DBSCAN	0.7667	1.3182	0.5063	0.5948

Apparently, one cluster validity index is far from enough to evaluate the clustering performance. Thus, it is necessary to introduce two or more indexes to conduct cross-evaluation so as to determine the optimal parameter configuration.

(2) Despite its high Entropy, the FGKM method is still more suitable in recognizing unknown devices, especially Fans, HDs, and Sanders. In contrast, the DBSCAN method performs relatively poorly. Due to the overlap in the feature space, some appliances with similar features, such as HTs, Drills, and Saws, cannot be effectively identified and separated. Further studies on the modifications towards DBSCAN must go on.

(3) Cluster refinement will not only lead to the improvement of the clustering performance but also be accompanied by the possibility that appliance samples end up in different clusters, making it even worse from the perspective of order and certainty. On one hand, as the number of clusters expands, the appliances that were originally mixed with other types of appliances can be identified and grouped into another cluster. On the other hand, after the appliance samples are classified, they may continue to be divided into two clusters. As a result, one type of appliances may be identified as two, which is not worth the gain. Therefore, how to reach a compromise between these two aspects is still a subject to be further explored.

References

Bai, L., Liang, J. Y., Sui, C., & Dang, C. Y. (2013). Fast global k-means clustering based on local geometrical information. *Information Sciences, 245,* 168–180. https://doi.org/10.1016/j.ins.2013.05.023.

Bolshakova, N., Azuaje, F., & Cunningham, P. (2005). A knowledge-driven approach to cluster validity assessment. *Bioinformatics, 21*(10), 2546–2547. https://doi.org/10.1093/bioinformatics/bti317.

Chicco G, Napoli R, Piglione F (2006) Comparisons Among Clustering Techniques for Electricity Customer Classification. IEEE Transactions on Power Systems 21 (2):p. 933–940

Ester M, Kriegel H-P, Sander J, Xu X (1996) A density-based algorithm for discovering clusters in large spatial databases with noise. Paper presented at the Proceedings of the Second International Conference on Knowledge Discovery and Data Mining, Portland, Oregon,

Fukui K-i, Numao M (2012) Neighborhood-Based smoothing of external cluster validity measures. Paper presented at the Proceedings of the 16th Pacific-Asia conference on Advances in Knowledge Discovery and Data Mining - Volume Part I, Kuala Lumpur, Malaysia,

Galán, S. F. (2019). Comparative evaluation of region query strategies for DBSCAN clustering. *Information Sciences, 502,* 76–90. https://doi.org/10.1016/j.ins.2019.06.036.

Giancarlo, R., & Utro, F. (2012). Algorithmic paradigms for stability-based cluster validity and model selection statistical methods, with applications to microarray data analysis. *Theoretical Computer Science, 428,* 58–79. https://doi.org/10.1016/j.tcs.2012.01.024.

Halkidi, M., Batistakis, Y., & Vazirgiannis, M. (2001). On clustering validation techniques. *J Intell Inf Syst, 17*(2–3), 107–145. https://doi.org/10.1023/A:1012801612483.

Hart, G. W. (1992). Nonintrusive appliance load monitoring. *Proceedings of the IEEE, 80*(12), 1870–1891.

Jain, A. K., Murty, M. N., & Flynn, P. J. (1999). Data clustering: A review. *ACM Computing Surveys, 31*(3), 264–323. https://doi.org/10.1145/331499.331504.

Kong, W. C., Dong, Z. Y., Ma, J., Hill, D. J., Zhao, J. H., & Luo, F. J. (2018). An Extensible Approach for Non-Intrusive Load Disaggregation With Smart Meter Data. *Ieee Transactions on Smart Grid, 9*(4), 3362–3372. https://doi.org/10.1109/Tsg.2016.2631238.

Kwac, J., Flora, J., & Rajagopal, R. (2014). Household Energy Consumption Segmentation Using Hourly Data. *IEEE Transactions on Smart Grid, 5*(1), 420–430.

Li, R., Li, F. R., & Smith, N. D. (2016). Multi-Resolution Load Profile Clustering for Smart Metering Data. *IEEE Transactions on Power Systems, 31*(6), 4473–4482. https://doi.org/10.1109/Tpwrs.2016.2536781.

Likas, A., Vlassis, N., & Verbeek, J. J. (2003). The global k-means clustering algorithm. *Pattern Recognition, 36*(2), 451–461.

Lin, S. F., Li, F. X., Tian, E. W., Fu, Y., & Li, D. D. (2019). Clustering Load Profiles for Demand Response Applications. *Ieee Transactions on Smart Grid, 10*(2), 1599–1607. https://doi.org/10.1109/Tsg.2017.2773573.

Macqueen J (1965) Some Methods for Classification and Analysis of MultiVariate Observations. In: Proc of Berkeley Symposium on Mathematical Statistics & Probability, 1965.

Panapakidis, I. P., Papadopoulos, T. A., Christoforidis, G. C., & Papagiannis, G. K. (2014). Pattern recognition algorithms for electricity load curve analysis of buildings. *Energy and Buildings, 73,* 137–145. https://doi.org/10.1016/j.enbuild.2014.01.002.

Picon T, Meziane MN, Ravier P, Lamarque G, Novello C, Bunetel JCL, Raingeaud Y (2016) COOLL: Controlled On/Off Loads Library, a Public Dataset of High-Sampled Electrical Signals for Appliance Identification.

Salem, S. A., & Nandi, A. K. (2009). Development of assessment criteria for clustering algorithms. *Pattern Analysis and Applications, 12*(1), 79–98. https://doi.org/10.1007/s10044-007-0099-1.

Teeraratkul, T., O'Neill, D., & Lall, S. (2018). Shape-Based Approach to Household Electric Load Curve Clustering and Prediction. *Ieee Transactions on Smart Grid, 9*(5), 5196–5206. https://doi.org/10.1109/Tsg.2017.2683461.

Varga, E. D., Beretka, S. F., Noce, C., & Sapienza, G. (2015). Robust Real-Time Load Profile Encoding and Classification Framework for Efficient Power Systems Operation. *IEEE Transactions on Power Systems, 30*(4), 1897–1904. https://doi.org/10.1109/Tpwrs.2014.2354552.

Zakaria Z, Lo KL (2009) Two-stage Fuzzy Clustering Approach for Load Profiling. Upec: 2009 44th International Universities Power Engineering Conference:976-+

Zhao, C. H., Li, X. C., & Cang, Y. (2015). Bisecting k-means clustering based face recognition using block-based bag of words model. *Optik, 126*(19), 1761–1766. https://doi.org/10.1016/j.ijleo.2015.04.068.

Chapter 6
Smart Non-intrusive Device Recognition Based on Intelligent Optimization Methods

6.1 Introduction

6.1.1 Background

Ubiquitous Electric Internet of Things (UEIOT) is an intelligent system that integrates mobile Internet, big data, artificial intelligence, and other advanced technologies of communication and information in the whole process of a power system to realize network connectivity, data continuity, comprehensive state perception of human-computer interaction, efficient information disposal, simple and flexible application, and other characteristics. As a key technology in UEIOT, device recognition is about real-time monitoring of the equipment on each socket, and the obtained data is transmitted to the server. And the server runs a process including identification algorithms to identify the type of equipment and various other parameters, thereby issuing necessary optimization control instructions. And optimization is one of the essential steps for device recognition.

In terms of optimization approaches used in device recognition, machine learning is studied and adopted widely to perform the optimization. The supervised methods are most popular in recent years, appealing numbers of researchers to explore and improve. Lei et al. adopts a Multiple-class Support Vector Machine (M-SVM) (Jiang et al. 2012) to monitor each device in the house with the power consumption for electrical consumption managing. Peng et al. adopts K-Nearest Neighbor (KNN) (Tao et al. 2019) to perform nonintrusive load identification and disaggregation. Though the accuracy calculated by these models is high and satisfying enough, there is a fatal flaw when it comes to unknown appliances. Hidden Markov Model (HMM) (Heracleous et al. 2014) is the most popular method in the unsupervised models. There are many variants generated by HMM such as Factorial Hidden Markov Model (FHMM). Gautam A. Raiker et al. used FHMM (Raiker et al. 2018) to monitor the unknown loads in many places. However, these algorithms contain a complicated training part that needs plenty of computing resources and more data.

H. Liu et al., *Smart Device Recognition*,
https://doi.org/10.1007/978-981-33-4925-4_6

As another kind of optimization approaches that are different from the machine learning algorithms, mathematical optimization methods are used to find the optimal combination of appliances. According to the number and complexity of objective functions, the mathematical optimization methods contain single-objective optimization as well as multi-objective optimization. For single-objective optimizations, Liang et al. proposed an enhanced gesture recognition method which is multi-touched based on optimization algorithms to save storage (Shen et al. 2017). In NILM, Hamed Ahmadi set a feature library by adopting an optimization method to recognize the data from smart meters (Ahmadi and Martí 2015). As for multi-objective optimizations, Wu et al. build an improved NSGA-II model based on simulation to explore the multi-objective optimization operation principles between the generation of power and the consumption of energy (Wu et al. 2019). Feng et al. designed a multi-objective particle swarm optimization algorithm in order to check out the anticipated task scheduler on resources considering more than 1 factors including total task execution time, reserving resource, and QOS of each task (Feng et al. 2012). Nevertheless, the multi-objective optimization has been seldom adopted in NILM. In this chapter, three multi-objective optimization methods are introduced known as Non-dominated Sorting Genetic Algorithm-II (NSGA-II), Multi-object Particle Swarm Optimization (MOPSO), and Multi-object Grey Wolf Optimization (MOGWO).

6.1.2 Steady-State Current Decomposition

The steady-state current decomposition in NILM is to obtain important information such as the content, state, phase, and power of each load by analyzing the electrical data collected at the general end of the system based on the steady-state process. Since the power frequency voltage is relatively constant so that the power can be computed directly from the voltage and current, the steady-state current signal is decomposed in the chapter.

According to the Fourier analysis theory, the current of a single load in a certain measurement can be decomposed as:

$$i_t = a_1 \sin(\omega t + \varphi_1) + a_2 \sin(2\omega t + \varphi_2) + \cdots + a_k \sin(k\omega t + \varphi_k) \tag{6.1}$$

where a is the amplitude of each harmonic, ω is the fundamental angle frequency, and φ is the initial phase angle of each harmonic in the measurement.

For the same load, the current is measured multiple times, and the amplitude values of the fundamental wave and each harmonic are constant. The frequency is reflected by the harmonic order and is also constant. But the phase angle is not constant, because the waveform position recorded at the beginning of the measurement may be different. For a steady-state current containing multiple harmonics, it can have a fixed waveform, not only because the fundamental wave and each harmonic have a constant amplitude, but also because the fundamental wave and each harmonic have a fixed relative position. This relative position information is hidden in the phase angle of each measurement. By finding the "phase difference" between the fundamental wave and each harmonic, the original waveform can be reformed according to the amplitude.

The multi-load steady-state current is generally decomposed horizontally and vertically. The horizontal decomposition is defined as the decomposition of the total current waveform into the form of the fundamental wave and the addition of the harmonics, while the vertical decomposition is defined as the decomposition of the total current waveform into the form of the superposition of single load current waveforms. The horizontal decomposition can be directly obtained by harmonic analysis.

According to the superposition theorem, the total load current composed of n-type load current can be defined as:

$$I_t = \beta_1 i_{t1} + \beta_2 i_{t2} + \cdots + \beta_n i_{tn} \tag{6.2}$$

where i is the single load current, and β are the weights of various loads, which is the number of corresponding loads. I is expressed to be a matrix form as:

$$\begin{bmatrix} a_{I1}\angle\theta_{I1} \\ a_{I2}\angle\theta_{I2} \\ \vdots \\ a_{Ik}\angle\theta_{Ik} \end{bmatrix} = \begin{bmatrix} a_{11}\angle\theta_{C1} + \Delta\theta_{11} & a_{12}\angle\theta_{C2} + \Delta\theta_{12} & \cdots & a_{1n}\angle\theta_{Cn} + \Delta\theta_{1n} \\ a_{21}\angle 2\theta_{C1} + \Delta\theta_{21} & a_{22}\angle 2\theta_{C2} + \Delta\theta_{22} & \cdots & a_{2n}\angle 2\theta_{Cn} + \Delta\theta_{2n} \\ \vdots & \vdots & \cdots & \vdots \\ a_{k1}\angle k\theta_{C1} + \Delta\theta_{k1} & a_{k2}\angle k\theta_{C2} + \Delta\theta_{k2} & \cdots & a_{kn}\angle k\theta_{Cn} + \Delta\theta_{kn} \end{bmatrix} \begin{bmatrix} \beta_1 \\ \beta_2 \\ \vdots \\ \beta_n \end{bmatrix} \tag{6.3}$$

where θ_I and a_I are the phase angle and amplitude of the lateral decomposition of the multi-load current I, a is the amplitudes of harmonics of each load, $\Delta\theta$ and θ_C are the harmonic position angle and the fundamental phase angle of each load, and β is the weight coefficient.

The formula above can be simplified as:

$$I_t = A * B \tag{6.4}$$

where $I_t = (a_{I1}\angle\theta_{I1}, a_{I2}\angle\theta_{I2}, \cdots, a_{Ik}\angle\theta_{Ik},)^T$, which is the matrix form of the total load current obtained by the lateral decomposition. A is the characteristic parameter matrix, which contains the shape information of a single load current waveform and the position information of different load current waveforms. B is the weight coefficient matrix, which contains the quantity of each load in the total current waveform.

The steady-state current decomposition of NILM is to calculate matrix B on the basis of the measured total load current matrix I and matrix A.

6.1.3 Data Description

In the chapter, the WHITED dataset (Kahl et al. 2016) is used to performs the experiments. It includes the first 5 s of various appliances' start-ups, covering 47 different appliance types. Table 6.1 shows the information about appliances selected from the WHITED dataset. And Fig. 6.1 provides information about the power supply and appliances distribution.

Since the data in the WHITED dataset is collected from different zones, some distinctions exist for the voltage frequency of various appliances. These distinctions may result in actual erroneousness in the aggregated current generated by synthesis. To solve this problem, an algorithm is put forward to standardize the frequency by using the larger periods of voltage to shrink the appliance currents, which is defined as:

Table 6.1 The information of appliances selected from the WHITED dataset

Types of appliance	Abbreviations	Number of samples	Number of appliances
Air conditioning	Ai	10	1
Charger	Ch	50	5
Drilling machine	Dr	20	2
Fan heater	Fa	10	1
Game console	Ga	40	4
Hairdryer	Ha	60	6
Iron	Ir	30	3
Kettle	Ke	60	6
LED light	Le	90	9
Light bulb	Li	60	6
Massager	Ma	30	3
Microwave	Mi	20	2
Mixer	Mx	40	4
Fridge	Fr	10	1
Stove	St	10	1
Toaster	To	40	4
TV	Tv	20	2
Vacuum cleaner	Va	40	4
Washing machine	Wm	10	1
Water heater	Wh	40	4
Water pump	Wa	10	1

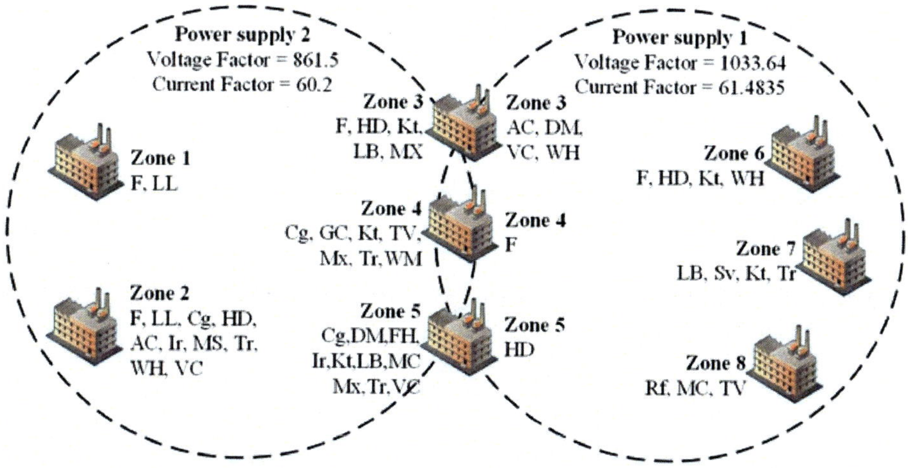

Fig. 6.1 The information about the power supply and appliances distribution

$$M_i = \frac{L_i}{L_i - \min(L_1 \cdots L_m)} \tag{6.5}$$

where M_i is the data points that will be deleted to shorten the voltage periods, L_i is the average length of voltage period for each appliance.

Moreover, the current data in WHITED dataset is only the record of the first 5 s of each appliance, which is not suitable for scientific research. To acquire a longer running time that is enough for experiments, the origin data is duplicated according to two rules: firstly, the repeated part is only when the appliance works in a steady state, which does not include the start-up part. Besides, the picked points to be connected should be from the identical position in the voltage phases, and there should be an upper boundary of the differences between their voltage values. Thus, the zero-crossing point is obviously proved to be the best junction in which the slope is the biggest so that it has the lowest false drop rate.

In terms of the test dataset, there are three distinct types of aggregated loads selected in this experiment, which means three combinations of appliances, and every combination contains three different appliances. As displayed in Table 6.2, instance #1 contains a drilling machine, vacuum cleaner, and water heater, which are all large appliances with high power consumption. Instance #2 contains fridge, microwave,

Table 6.2 The information about aggregated loads

Number of load instances	Appliances in the load	Operating status of each appliance	Total time
#1	Dr + Va + Wh	36.2 + 48.9 + 20.3	60.7
#2	Fr + Mi + Tv	31.2 + 70.6 + 69.8	108.8
#3	Ir + Mx + Ga	65.2 + 84.7 + 111.4	162.8

and TV, which are all large appliances with low power consumption, while instance #3 contains iron, mixer, and game console, which are all small appliances with low power consumption.

6.1.4 Feature Extraction

In the former Research, since the features extracted are unilateral and cannot reflect the whole information in the data, the load information is not completely utilized. Thus, a novel feature *weighted harmonic vectors* (Liu et al. 2019) are adopted in the chapter which can fully utilize the information of the sampling data.

As for a certain appliance, it is found that σ_p and σ_a which refer to the deviation of phase angle and amplitude of the whole samples can reflect the dependability of the harmonic well. Furthermore, the deviation of vectors for current harmonics can be calculated as below:

$$\begin{aligned}\sigma &= \sqrt{\left[(A+\sigma_a)\cos(\theta+\sigma_p) - A\cos\theta\right]^2 + \left[(A+\sigma_a)\sin(\theta+\sigma_p) - A\sin\theta\right]^2} \\ &\approx 2\sqrt{A^2 + A\sigma_a}\sin\frac{\sigma_p}{2}\end{aligned} \tag{6.6}$$

where θ is the phase angle and A is the amplitude of the harmonic.

Considering the distinct influence of each order of harmonics, the weight is introduced. For each harmonic order, the weight is detailed as below:

$$w_i = \gamma_i\left(1 - \frac{\sigma_i}{A_i}\right) \tag{6.7}$$

$$w = \frac{1}{n}\sum_{i=1}^{n} w_i \tag{6.8}$$

where w_i is described to reflect the weights of harmonics for device i, and w is the total weights of the harmonic vector of the aggregated current in a combination; n is the certain number of devices that a combination contains; γ is introduced as a coefficient vector concerned with the amplitudes of harmonics. Adopting the weight vector, it comes to more precise and robust for the decomposition of the current.

6.1.5 Objective Function

In the chapter, since the steady-state current decomposition is based on weighted harmonic vectors, calculation of the summation and variance follows the principle of the vector calculating method. Thus, there is information about the vector extracted from the signal left out by adopting the vector calculating method. In the chapter, the first objective function minimizes the sum of errors of harmonics with different orders that are weighted, which is defined as:

$$\min f_1 = \sum w_j \bullet \frac{\left| \dot{I}_{kj} - \sum_{i=1}^{n} \dot{I}_{ij} \right|}{\left| \dot{I}_{kj} \right|}, \quad j\ is\ the\ first\ five\ odd\ orders \tag{6.9}$$

where $\dot{I}_{kj}$ is the j-th harmonic vector, $\sum_{i=1}^{n} \dot{I}_{ij}$ is accumulated by all the appliance in a combination with the j-th harmonic vectors, and w_j is the weight defined in Sect. 6.1.4 of the harmonic order j.

Additionally, the second objective function minimizes the standard deviation formed by the errors to show the global proximity, which defined as:

$$\min f_2 = \sqrt{\sum_{j=1,3,5,7,9} \left(\frac{\left| \dot{I}_{ij} - \sum_{i=1}^{n} \dot{I}_{ij} \right|}{\left| \dot{I}_{ij} \right|} - \frac{1}{5} \sum_{j=1,3,5,7,9} \frac{\left| \dot{I}_{ij} - \sum_{i=1}^{n} \dot{I}_{ij} \right|}{\left| \dot{I}_{ij} \right|} \right)^2} \tag{6.10}$$

6.1.6 Evaluation Indexes

To evaluate the validation of models adopted in this chapter methodically, F1score, precision, and recall are adopted in this chapter, which is the most widely used evaluation indexes (Zhang et al. 2020). These indexes are defined as:

$$\begin{cases} precision = \dfrac{TP}{TP + FP} \\ recall = \dfrac{TP}{TP + FN} \\ F_1 = \dfrac{(p^2 + 1) \cdot precision \cdot recall}{p^2 \cdot precision + recall} \end{cases} \tag{6.11}$$

where TP is the true positives, which means the appliance is on and predicted on; FP is the results of false positives that means the appliance is on but predicted off. And TN is the true negative, which means it is predicated on but actually off. p is a parameter of harmonic, which is generally set as 1.

6.2 NSGA-II Based Device Recognition Method

6.2.1 The Theoretical Basis of NSGA-II

NSGA-II algorithm was designed by Deb et al. in 2000 (Deb et al. 2000). Based on the genetic algorithm, the NSGA-II model also includes crossover, mutation, and selection operators. However, to solve more complex multi-objective optimization problems, some new operators are added as the fast nondominant sorting algorithm, individual crowdedness, and elite conservation.

(1) *The fast nondominant sorting algorithm*
Compared with the nondominant sorting algorithm, the fast nondominant sorting algorithm improves the operating efficiency and decreases the computational complexity of the algorithm. The algorithm flow is as below.
In population P, there are two parameters s_p and n_p for each individual p. s_p is described as a set of individuals dominated by p, while n_p is a set of individuals dominating p. To do the fast non-dominated sorting, there are two steps: firstly, put the individual into F_1 if the n_p of it is 0, and the non-dominated sorting number is set i_{rank}; then check s_p of individual p in the set F_I, and reduce the n_m of each individual m in the s_p by 1. If $n_q - 1 = 0$, switch the individual into another set M, and rank M with its non-dominated sequence number; repeat the above process until the classification for all individuals is done.

(2) *Individual crowdedness*
Individual crowdedness is to calculate the density around individuals in the population. The smaller the crowdedness of the individual, the greater the density around the individual, and the probability of the individual being selected is lower. Generally, i_d is used to represent the degree of the crowdedness of an individual, which essentially refers to the maximum perimeter of the rectangle that surrounds the individual i but does not contain other individuals.
The steps to calculate the individual crowding degree is as follows: first, initialize the crowding degree so that $I_i = 0$; then start non-dominated sorting of the whole individuals of the same level under the constraint of the objective function, and the two individuals at the edge are set to infinity, while the individuals in the middle are set as:

$$I_i = I_i + \frac{f_{i+1} - f_{i-1}}{f_{max} + f_{min}} \tag{6.12}$$

where f_i is the cost of each individual i calculated by the objective function, f_{max} and f_{min} are defined respectively as the maximum and minimum cost of individuals at the same level in the population sorted according to the objective function. For each objective function, the crowding degree of each individual marked as $i_{distance}$ is obtained according to the above operation.

After nondominant sorting and crowdedness calculation, each individual i of the population obtains two attributes as i_{rank} and $i_{distance}$. There are two situations: for any two different individuals i and j selected from the population, if $i_{rank} > j_{rank}$, the level of individual i is higher than it of j; if $i_{rank} = j_{rank}$ but $i_{distance} > j_{distance}$, i and j are in the same level while the crowdedness of i is larger than it of j, which proves that i is better than j, and i will be put into the mating pool.

(3) *Elite conservation*

The elite conservation is to merge the parent population P and the generated offspring Q into a new population R ($R = P \cup Q$), the size of which is 2 N. The new population R is fast non-dominated sorted and crowdedness calculated, selecting the best N individuals as the new generation parent population. By this means, the new population generated by the competition between the parent and the offspring not only retains the excellent genes of the parent, but also prevents the loss of excellent genes and optimizes the accuracy of the results.

6.2.2 Model Framework

Figure 6.2 demonstrates the framework for smart non-intrusive device recognition based on NSGA-II models. The steps of modeling are as below:

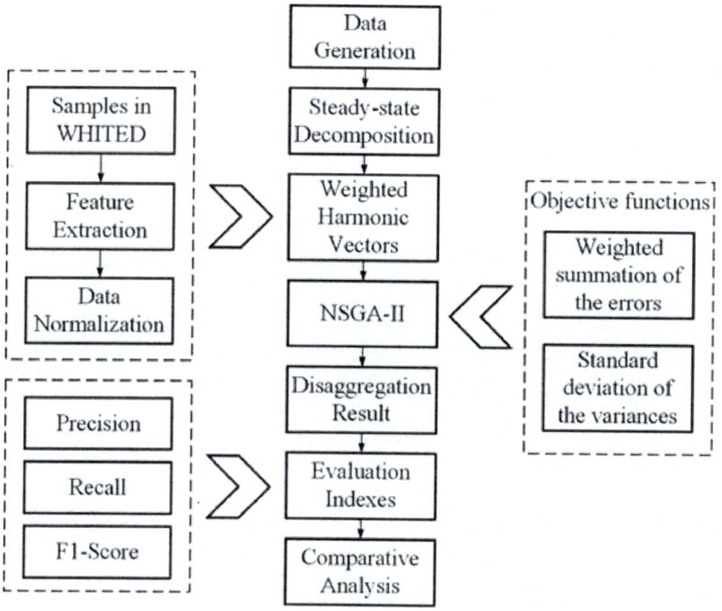

Fig. 6.2 Model framework of NSGA-II for smart non-intrusive device recognition

(a) Determining the combinations of appliances, and generating the aggregated current according to the combinations;
(b) Extracting the weighted harmonic vectors of single appliances as features, and decomposing the aggregated current;
(c) Confirming the objective functions and implementing the experiment by adopting NSGA-II;
(d) Calculating the values of the evaluation index, and comparatively analyzing the performance of the NSGA-II and Traversal search.

6.2.3 Evaluation of NSGA-II Model

In order to show the performance of the model better, a comparative model is set up to make the results of experiments more prominent. The traversal search method is adopted to perform the comparative model.

(1) *Performance of Traversal search*
The results of the three instances by traversal search are shown in Figs. 6.3, 6.4 and 6.5. As it is stated in Sect. 6.1.3, the three loads represent three different types of appliances or operating statues. Load #1 is a combination of three large appliances with high power consumption. Conversely, load #2 is a combination of three large appliances with low consumption, while load #3 is a combination of three small appliances with low consumption.
From the current curve in Fig. 6.3, it can be seen that the composed current calculated by traversal search method matches well with the actual current when the actual current is aggregated by large appliances with high power consumption, and the recall value is also of a good performance in this instance. However, the traversal search method performs terribly bad when it comes to low power consumption according to Figs. 6.4 and 6.5.

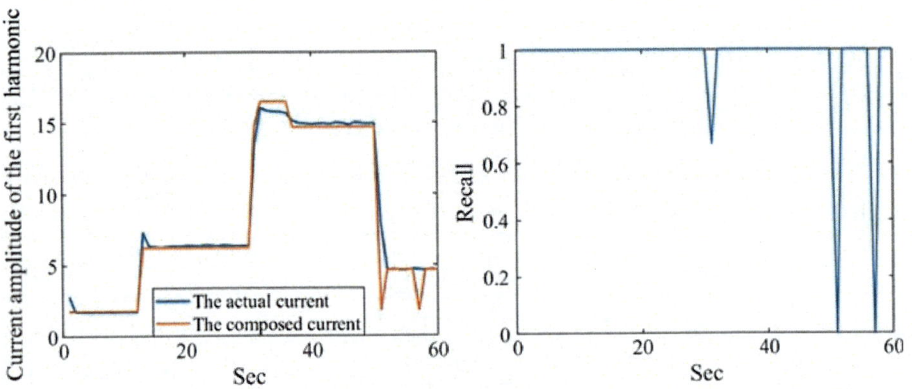

Fig. 6.3 The results of Load #1 calculated by traversal search

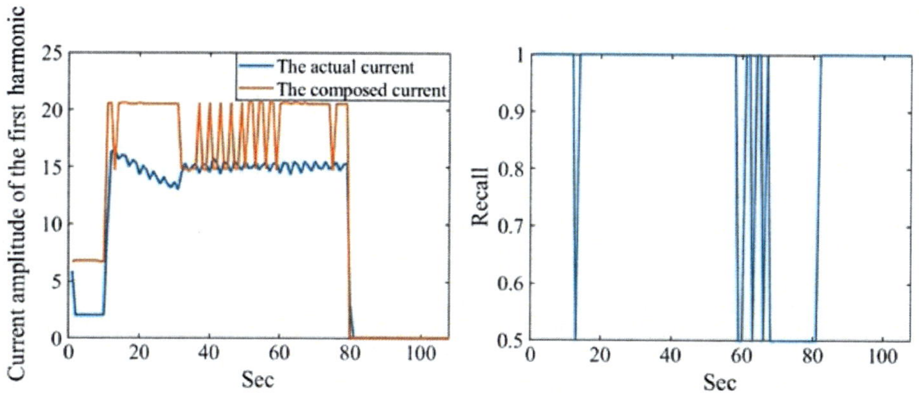

Fig. 6.4 The results of Load #2 calculated by traversal search

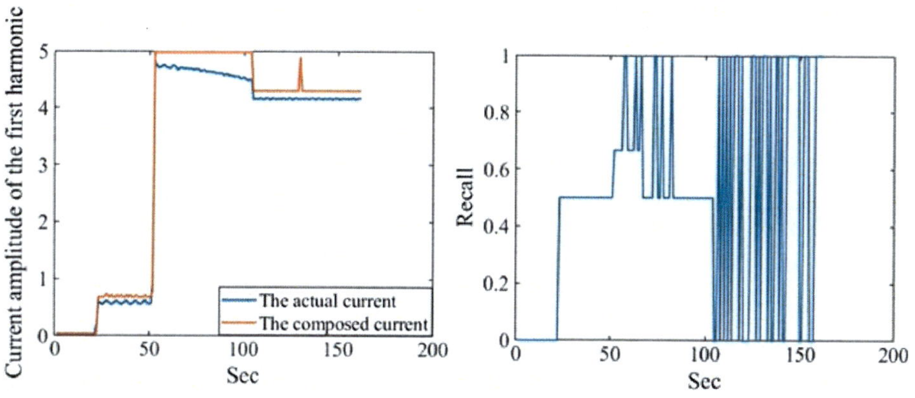

Fig. 6.5 The results of Load #3 calculated by traversal search

Considering the running period in every load, the predictions of a single appliance phase are more accurate than them of a multi appliances phase, but the results are still not so good. And with the number of appliances increase, the performance of the traversal search method is getting worse and worse.

(2) *Performance of NSGA-II*
The performance of the NSGA-II model can be observed in Figs. 6.6, 6.7 and 6.8 and Table 6.3. As can be seen in the three figures, the matching degrees of the composed currents with the actual currents are all pretty good in the three instances, especially in load #1 and #2. Thus, it is proved that the utilization of the NSGA-II model achieves better recognition in the instance that the loads are aggregated by large appliances. And in terms of the recall value, it is easily found that the recall values changing with time are quite stable in the variation and generally high in all three instances, which means the NSGA-II optimization method recognizes well for positive examples.

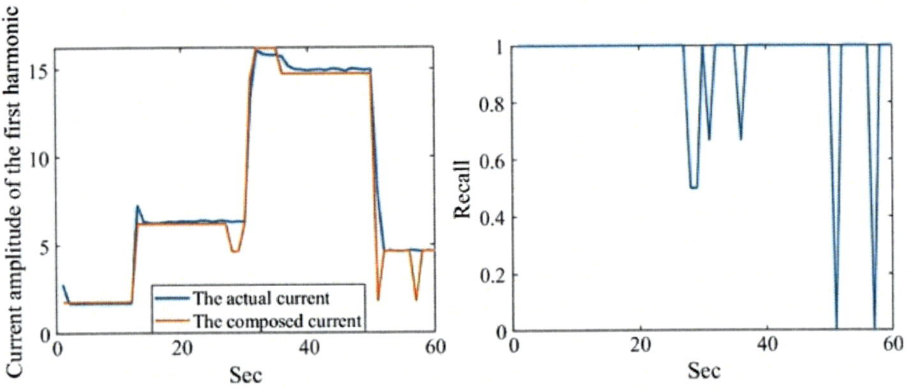

Fig. 6.6 The results of Load #1 calculated by NSGA-II

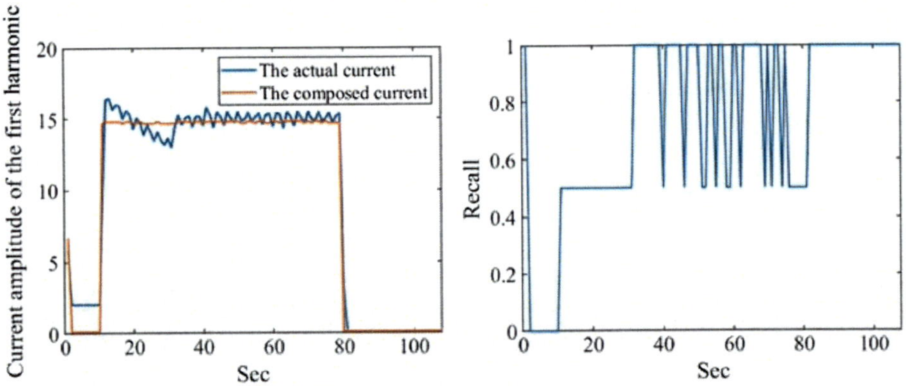

Fig. 6.7 The results of Load #2 calculated by NSGA-II

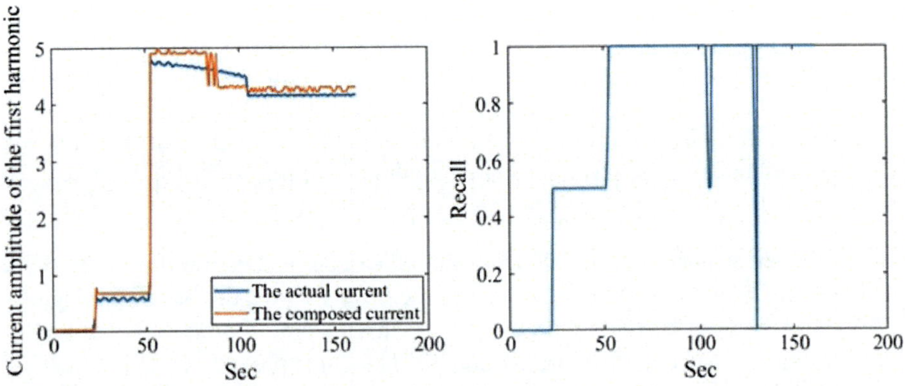

Fig. 6.8 The results of Load #3 calculated by NSGA-II

Table 6.3 The comparison between Traversal search and NSGA-II

Method		Traversal search	NSGA-II
#1	Precision	0.8994	0.9699
	Recall	0.8665	0.9156
	F1-Score	0.8990	0.9365
#2	Precision	0.8042	0.8793
	Recall	0.7654	0.8343
	F1-Score	0.7958	0.8652
#3	Precision	0.7896	0.8420
	Recall	0.6812	0.8142
	F1-Score	0.7463	0.8230

As for the running periods, it is obvious that the NSGA-II model predicts well both single appliance and multi appliances phases. However, it is seemly not so sensitive when the appliances switch on or off according to the current curve in the three figures. That may due to the fact that the convergence factor of NSGA-II is linear, which tends to fall into local optimum.

(3) *Comparative analysis of NSGA-II model and Traversal search*
Comparing the results of traversal search with NSGA-II in Table 6.3 and Figs. 6.6, 6.7 and 6.8, some conclusions are shown as below:

(a) It is evident that the composed current curve of the NSGA-II model matches better with the actual current curve than it of traversal search, especially in load #2 and load #3.
(b) For the result of three loads, the precision, recall, and F1-Score values of NSGA-II is much higher than it of Traversal search.
(c) In terms of the types of appliances scale and operating statue, NSGA-II performs much better in the instance that the appliances run with a low power consumption than traversal search. In the instance that the appliances run with high power consumption, both the traversal search and NSGA-II have similar performance, while NSGA-II performs much better when it comes to low power consumption, which indicates that NSGA-II is more effective in high-precision recognition.
(d) Compared with the traversal search, the NSGA-II model can recognize more accurately when there is more than one appliance on, while the recognition of multi appliances phases by adopting NSGA-II is not ideal.

6.3 Multi-object Particle Swarm Optimization Based Device Recognition Method

6.3.1 The Theoretical Basis of Multi-object Particle Swarm Optimization

On a normal occasion, most of the multi-objective optimization models are able to offer several Pareto optimal solutions for an optimization problem. The non-inferior solutions are evaluated through the "crowding distance", and then the optimal solutions in each generation of particles are stored to form a search set for the global optimal solution, that is, to introduce external archive set stores the optimal solutions. And the circle process is as Fig. 6.9 shows.

In MOPSO, the solution to handle a certain multi-objective optimization problem is called a particle, and it is generated by random. These particles form a group. By the means to adjust each particle's flight direction in a quasi-random manner, the particle group can search the space to obtain an approximate optimal solution for the objective function. The flight direction or trajectory created by the position vector of the particles and the movement mode of each particle is realized by a mixture of global search and local search in a random and definitive way.

The global optimal g-best is shared among all particles, and each particle also obtains its own optimal p-best. g-best and p-best guide each particle's search and point to its destination. At the same time, the element of randomness is also included in the movement of the particle to its destination. In the search process, the particle will update its own optimal position when it finds a better position than the last optimal position. At any point in the iterative process, all particles have a different single current optimal position and an unparalleled current global optimal position. Unless they saturate or the iteration is equal to the maximum number, the particles continue to find and improve the global best position.

The velocity iteration of the particles in MOPSO is as follows:

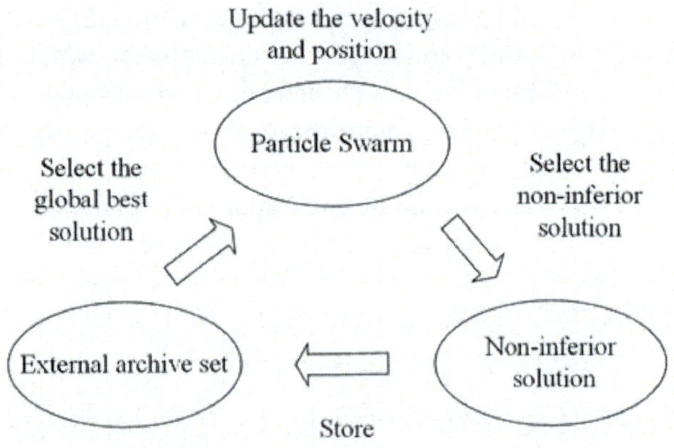

Fig. 6.9 The circle process of MOPSO

$$v_p(t) = \omega v_p(t-1) + c_1 r_1 (x_{pbest} - x_p(t)) + c_2 r_2 (x_{gbest}(t) - x_p(t)) \quad (6.13)$$

where v_p is the particle velocity, x_p is the current position of the particle, x_{pest} is the own optimal solution ever found of a particle, x_{pest} is the global optimal solution ever found, c_1 is cognitive constants while c_2 is social constants, ω is the constant of inertia, r_1 and r_2 are two randomly generated vectors whose values are between 0 and 1. In each iteration, each particle uses the above equation to update its own velocity, which is initialized to zero. The new position is updated by the equation follows:

$$x_p(t) = x_p(t-1) + v_p(t) \quad (6.14)$$

6.3.2 Model Framework

Figure 6.10 shows the model framework for Smart non-intrusive device recognition based on MOPSO models. As shown in Fig. 6.10, the detailed steps of modeling are shown below:

(a) Selecting the samples in WHITED dataset, and generating the aggregated current according to the samples;
(b) Performing single-load and multi-load decomposition;

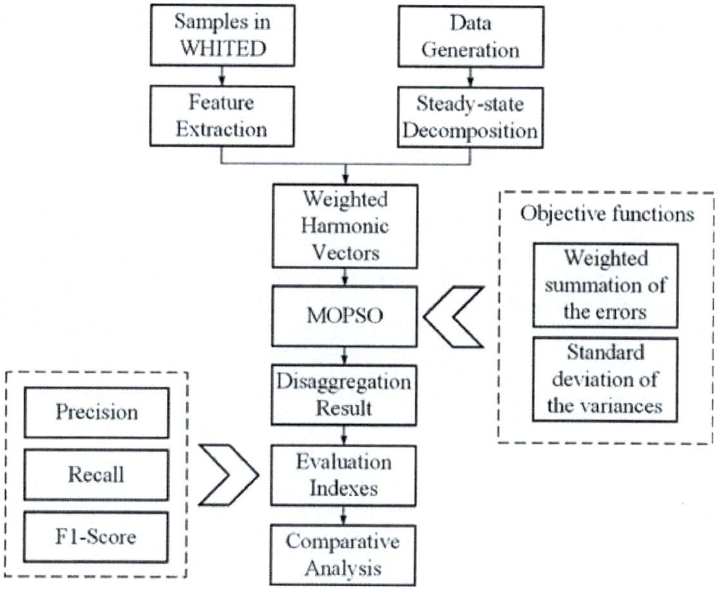

Fig. 6.10 The model framework of MOPSO for non-intrusive device recognition

(c) Confirming the objective functions and carrying out the experiment by adopting MOPSO;
(d) Calculating the values of the evaluation index, and comparatively analyzing the performance of the MOPSO and Traversal search.

6.3.3 Evaluation of MOPSO Model

(1) *Performance of MOPSO model*
Figures 6.11, 6.12 and 6.13 demonstrates the performance of the MOPSO model. It is evident that the composed currents have pretty good suitability with the actual currents in the whole three instances from the three figures. Therefore, it can be concluded that the NSGA-II model recognized correctly both on large and small appliances, whether the power consumption of appliances is high or

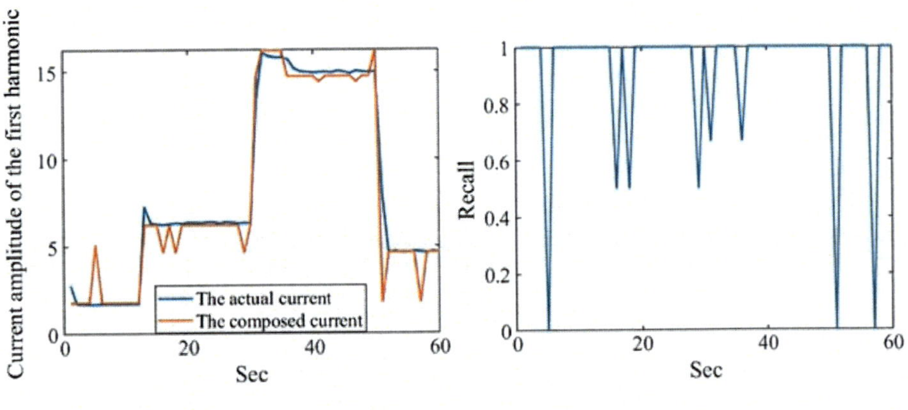

Fig. 6.11 Load #1 disaggregation results of the MOPSO

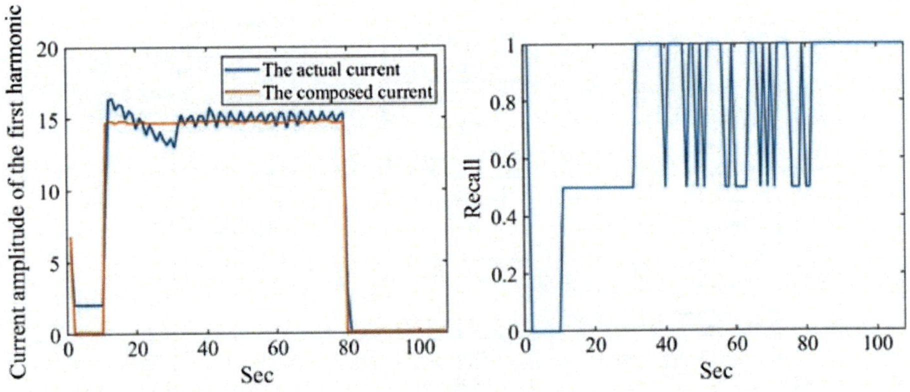

Fig. 6.12 Load #2 disaggregation results of the MOPSO

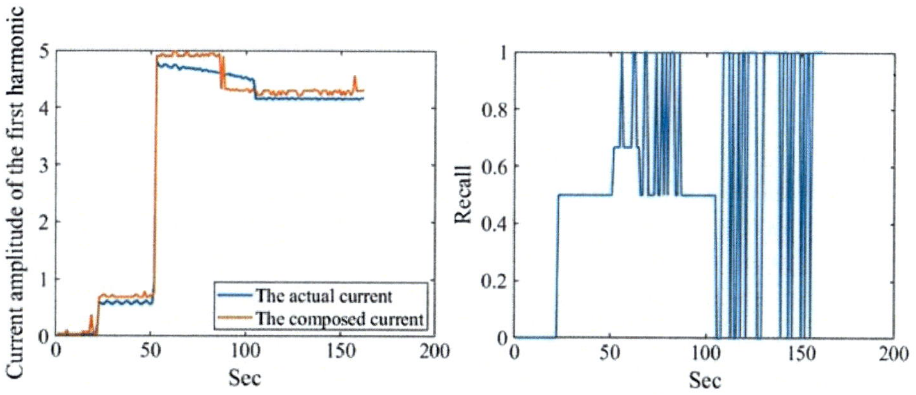

Fig. 6.13 Load #3 disaggregation results of the MOPSO

low. Additionally, the recall values of every second in every load are high, so that MOPSO is of an exact recognition of positive examples.
When it comes to the phase of load operating, MOPSO performs pretty well on the prediction of a single appliance phase, while the prediction of the multi appliances phase is not as expected, for it seems to be hard to recognize the switching of appliances in the load on time.

(2) *Comparative analysis of MOPSO model and Traversal search*
In a comparison of the evaluation indexes of traversal search and MOPSO in Table 6.4, some conclusions are shown as below:

(a) Obviously, the composed current calculated by the MOPSO model matches the actual current curve better than it calculated by the traversal search.
(b) Both the precision, recall, and F1-Score values scientifically calculated by MOPSO are much higher than it did by the traversal search method.
(c) Considering the types of appliances scale and operating statue, MOPSO performs much better in the instance that the appliances run with a low

Table 6.4 The comparison between Traversal search and MOPSO

Method		MOPSO	Traversal
#1	Precision	0.9688	0.8994
	Recall	0.9178	0.8665
	F1-Score	0.9452	0.8990
#2	Precision	0.8890	0.8042
	Recall	0.8283	0.7654
	F1-Score	0.8441	0.7958
#3	Precision	0.8644	0.7896
	Recall	0.8089	0.6812
	F1-Score	0.8114	0.7463

power consumption than traversal search. In the instance that the appliances run with high power consumption, both the traversal search and MOPSO have similar performance, while MOPSO performs much better when it comes to low power consumption, which indicates that MOSPO is more effective in high-precision recognition.

(d) On recognizing the multi appliances phases, the MOPSO model recognizes more accurately compared with the traversal search, though the recognition of the switching is not so accurate by adopting MOPSO.

6.4 Multi-object Grey Wolf Optimization Based Device Recognition Method

6.4.1 The Theoretical Basis of Multi-object Grey Wolf Optimization

Multi-objective Grey Wolf Optimizer (MOGWO) is proposed by Mirjalili in 2016 (Mirjalili et al. 2016). Based on Grey Wolf Optimizer (GWO), it contains two stages, named "Searching for prey" and "Attacking prey", which simulate the situation to explore as well as exploit the certain search space of real wolves.

(1) *Searching for prey*
The positions of a, β, and δ lead the wolf pack to find the place where the prey appears. The individuals in the pack spread out to better find the place where the prey appears. The wolves gather together to attack the prey. In order to imitate this process using mathematical theory, we stipulate that the vector A is a random value greater than 1 or less than -1 to achieve the effect of keeping the search agent away from the optimal solution. This part highlights the algorithm's ability to explore unknown spaces and helps the GWO algorithm to achieve a global search. Another factor that affects the global search capability of the GWO algorithm is the vector C, which is ranged in [0, 2] randomly. This vector can be regarded as a random weight reflecting the degree of prey influence, in order to randomly strengthen ($C > 1$) or weaken ($C < 1$) the influence degree of the optimal solution. This part highlights the randomness of the GWO algorithm in the optimization process to avoid the defect of falling into the local optimum appears. Here, compared with A, it can be found that C does not decrease linearly, but specifies that C is a random value at any time, in order not only to emphasize the search at the beginning of the iteration but also to reflect the global search ability at the end of the iteration. This vector helps the GWO algorithm avoid stagnation in the local optimum, especially in the final stage of the iteration. The vector C can be understood as an obstacle factor that simulates the grey wolf approaching its prey in nature. Generally speaking, it can be understood as certain obstacles in the hunting path of grey wolves under actual conditions. In fact, grey wolves cannot easily approach their prey. This is the role played

by the vector C in the equation. According to the grey wolf's position, the grey wolf is given a degree of difficulty in obtaining the prey, so that the distance between the grey wolf and the prey is getting farther or closer.

(2) *Attacking prey*
By reducing the value of a, the mathematical model can be used to simulate the process of wolves approaching their prey. A is at any value in the range of $[-a, a]$. As a decrease linearly from 2 to 0, the fluctuation range of A is also reduced. Once the random value of A is in the interval $[-1,1]$, the updated position of the wolf pack can appear anywhere from the current position to the prey's position. In the interval $|A| < 1$, the search agent successfully approaches the prey and launches an attack.
Based on GWO, MOGWO introduces two novel operators which are similar to them in MOPSO. The first one is the archive, defined to store the optimal solution ever found. The other component is the leader selection strategy, which helps select α, β and δ solutions from the archive as the leader of the search process.

6.4.2 Model Framework

Figure 6.14 shows the model framework for Smart non-intrusive device recognition based on MOGWO models. As shown in Fig. 6.14, the steps of the model are introduced as below:

(a) Standardizing the data of current and voltage, and performing feature extraction to establish the feature library of each appliance's weighted harmonic vectors;
(b) Decomposing the aggregated current and obtaining the superimposed weighted harmonic vectors.
(c) Conforming the objective functions and performing the MOGWO
(d) Calculating the values of the evaluation index, and comparatively analyzing the performance of the MOGWO and Traversal search.

6.4.3 Evaluation of MOGWO Model

(1) *Performance of MOGWO model*
The performance of the MOGWO model is shown in Figs. 6.15, 6.16 and 6.17. As we can see in the three figures, the matching degrees of the composed currents with the actual currents are all pretty good in the three instances, especially in load #1 and #2. Thus, it is proved that the utilization of the MOGWO model achieves better recognition in the instance that the loads are aggregated by large appliances. And in terms of the recall value, it is easily found that the recall values changing with time are quite stable in the variation and generally high in

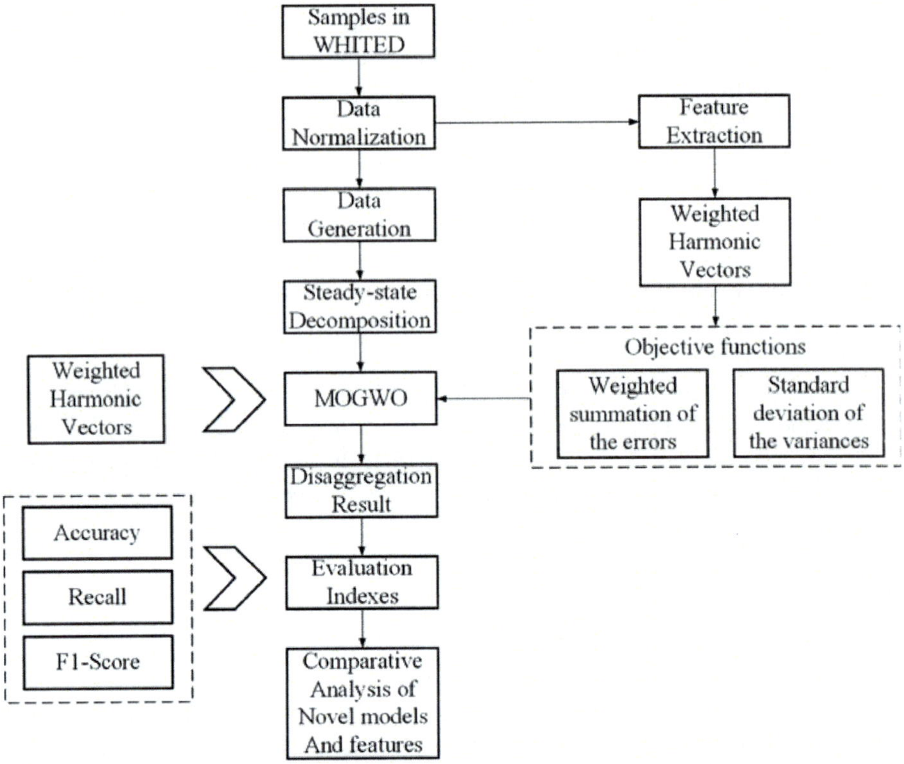

Fig. 6.14 The model framework of MOGWO for smart non-intrusive device recognition

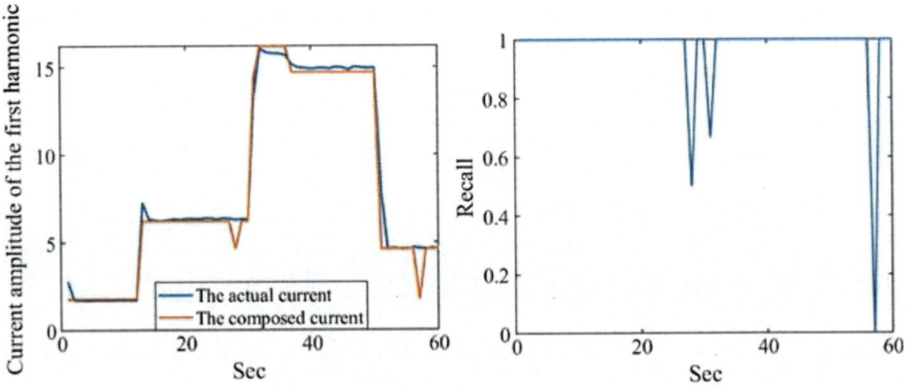

Fig. 6.15 Load #1 results of the MOGWO

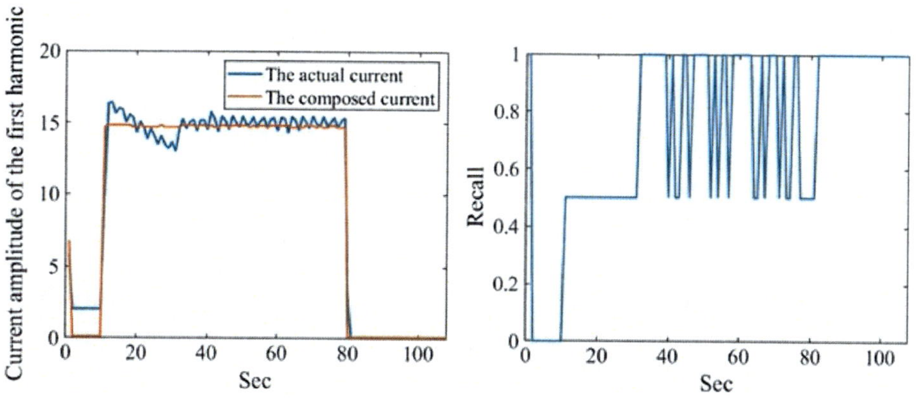

Fig. 6.16 Load #2 results of the MOGWO

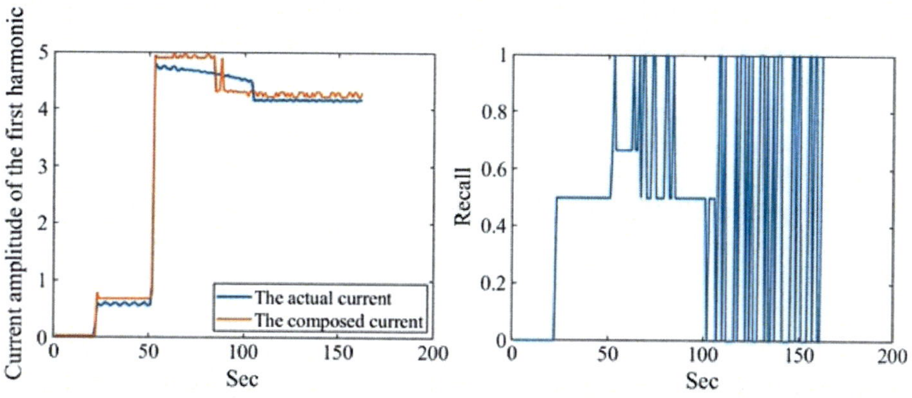

Fig. 6.17 Load #3 results of the MOGWO

all three instances, which means the MOGWO optimization method recognizes well for positive examples.

For the running periods, it is distinct that the prediction results of the multi appliances phase are unimaginably accurate by adopting the MOGWO model, not to mention the single appliance phase. It is highly sensitive to the switching off appliances and adapts to the changes as soon as possible when it recognizes that there are appliances switching on or off.

(2) *Comparative analysis of MOGWO model and Traversal search*
Comparing the evaluation indexes of traversal search and MOGWO in Table 6.5, some conclusions are shown as below:

(a) It is distinct that the composed current calculated by the MOGWO model matches the actual current curve better than it calculated by the traversal search.

Table 6.5 The comparison between Traversal search and MOGWO

Method		MOGWO	Traversal
#1	Precision	1.0000	0.8994
	Recall	0.9625	0.8665
	F1-Score	0.9873	0.8990
#2	Precision	0.8740	0.8042
	Recall	0.8544	0.7654
	F1-Score	0.8725	0.7958
#3	Precision	0.8364	0.7896
	Recall	0.8168	0.6812
	F1-Score	0.8246	0.7463

(b) All the values of evaluation indexes of NSGA-II are much better than them of the traversal search method, especially for F1-Score.
(c) In view of the scale of appliances and operating statue, MOGWO performs much better than traversal search in the instance that the appliances run with a low power consumption than traversal search, which indicates that MOGWO is more effective in high-precision recognition.
(d) MOPSO model recognizes much more accurately compared with the traversal search on recognizing the multi appliances phases, and the recognition of the switching is of high accuracy.

6.5 Experiment Analysis

As shown in Table 6.6 and Fig. 6.18, all the methods perform well in the experiments, and the F1score values calculated by the three multi-objective optimization methods are much better.

Table 6.6 The evaluation indexes of different methods

Method		Traversal	NSGA-II	MOPSO	MOGWO
#1	Precision	0.8994	0.9699	0.9688	1.0000
	Recall	0.8665	0.9156	0.9178	0.9625
	F1-Score	0.8990	0.9365	0.9452	0.9873
#2	Precision	0.8042	0.8793	0.8890	0.8740
	Recall	0.7654	0.8343	0.8283	0.8544
	F1-Score	0.7958	0.8652	0.8441	0.8725
#3	Precision	0.7896	0.8420	0.8644	0.8364
	Recall	0.6812	0.8142	0.8089	0.8168
	F1-Score	0.7463	0.8230	0.8114	0.8246

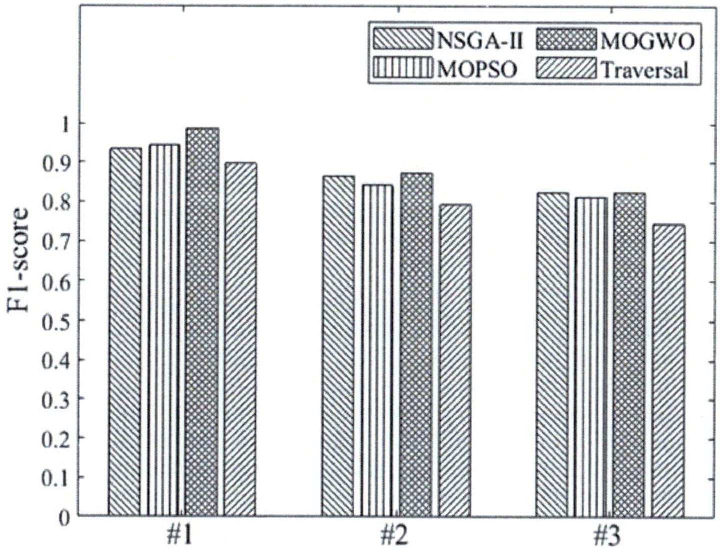

Fig. 6.18 The F1-Score of different models

Analyzing the results of the optimization in Table 6.6 and the figures of every model, some conclusions are drawn as below:

(a) The precision of device recognition is of great enhancement by adopting multi-objective optimization models compared with traversal search. And it is satisfying that the most values of F1-Score are over 0.8000.
(b) Among all the multi-objective optimization models adopted in this chapter, the MOGWO model performs best in the result, especially when the current has a large fluctuation. MOGWO is highly sensitive to the switching off appliances and adapts to the fluctuation as soon as possible when it recognizes that there are appliances switching on or off.
(c) As for different types of load, the models have accurate recognition when the loads are aggregated by the large appliances with high power consumption, while the recognition results are not ideal except for MOPSO when it comes to low power consumption.

In summary, the intelligent optimization methods NSGA-II, MOPSO, and MOGWO are all effective in smart non-intrusive device recognition, and the MOGWO model performs best.

References

Ahmadi, H., & Martí, J. R. (2015). Load decomposition at smart meters level using eigenloads approach. *IEEE Transactions on Power Systems, 30*(6), 3425–3436. https://doi.org/10.1109/TPWRS.2014.2388193.

Deb, K., Agrawal, S., Pratap, A., & Meyarivan, T. (2000). A fast elitist non-dominated sorting genetic algorithm for multi-objective optimization: NSGA-II. In M. Schoenauer, K. Deb, G. Rudolph, et al. (Eds.), *Parallel problem solving from nature PPSN VI, Berlin, Heidelberg, 2000* (pp. 849–858). Berlin Heidelberg: Springer.

Feng, M., Wang, X., Zhang, Y., & Li, J. (2012). Multi-objective particle swarm optimization for resource allocation in cloud computing. In *2012 IEEE 2nd International Conference on Cloud Computing and Intelligence Systems* (pp. 1161–1165). IEEE.

Heracleous, P., Angkititrakul, P., Kitaoka, N., & Takeda, K. (2014). Unsupervised energy disaggregation using conditional random fields. In *IEEE PES Innovative Smart Grid Technologies*, Europe, 12–15 Oct. 2014 pp. 1-5. https://doi.org/10.1109/isgteurope.2014.7028933.

Jiang, L., Luo, S., Li, J. (2012). An approach of household power appliance monitoring based on machine learning. In *2012 Fifth International Conference on Intelligent Computation Technology and Automation*, 12–14 January 2012. pp. 577-580. https://doi.org/10.1109/icicta.2012.151.

Kahl, M., Haq, A. U., Kriechbaumer, T., & Jacobsen, H. A. (2016). WHITED - A worldwide household and industry transient energy data set. *International Workshop on Non-intrusive Load Monitoring, 2016,* 1–5.

Liu, H., Yu, C., Wu, H., Chen, C., & Wang, Z. (2019) An improved non-intrusive load disaggregation algorithm and its application. *Sustainable Cities and Society*:101918. https://doi.org/10.1016/j.scs.2019.101918.

Mirjalili, S., Saremi, S., Mirjalili, S. M., & Coelho, Ld S. (2016). Multi-objective grey wolf optimizer: A novel algorithm for multi-criterion optimization. *Expert Systems with Applications, 47,* 106–119.

Piga, D., Cominola, A., Giuliani, M., Castelletti, A., & Rizzoli AEJIToCST, (2016). Sparse optimization for automated energy end use disaggregation. *IEEE Transactions on Control Systems Technology, 24*(3), 1044–1051.

Raiker, G. A., Reddy, S. B., Umanand, L., Yadav, A., Shaikh, M. M. (2018). Approach to non-intrusive load monitoring using factorial hidden markov model. In *2018 IEEE 13th International Conference on Industrial and Information Systems (ICIIS)*, 1–2 Dec. 2018 2018. pp. 381-386. https://doi.org/10.1109/iciinfs.2018.8721436.

Shen, L., Huo, S., Chen, M., Li, F., Yao, F., & Shen, H. (2017). Multi-touch gesture recognition algorithm of vehicle electronic devices-based on Bezier curve optimization strategy. In *2017 32nd Youth Academic Annual Conference of Chinese Association of Automation (YAC)*, 19–21 May 2017. pp. 720–723. https://doi.org/10.1109/yac.2017.7967503.

Zhang, Y., Yin, B., Cong, Y., & Du, Z. (2020). Multi-state household appliance identification based on convolutional neural networks and clustering. *Energies, 13,* 792. https://doi.org/10.3390/en13040792.

Tao, P., Liu, X., Zhang, Y., Li, C., & Ding, J. (2019). Multi-level non-intrusive load identification based on k-NN. In *2019 IEEE 3rd Conference on Energy Internet and Energy System Integration (EI2)*, 8–10 Nov. 2019. Pp. 1905-1910. https://doi.org/10.1109/ei247390.2019.9061896.

Wu, L., Bai, T., Huang, Q., Wei, J., & Liu, X. J. (2019). Multi-objective optimal operations based on improved NSGA-II for Hanjiang to Wei river water diversion project, China. *Water, 11*(6), 1159.

Chapter 7
Smart Non-intrusive Device Recognition Based on Ensemble Methods

7.1 Introduction

7.1.1 Background

Non-intrusive device recognition is an essential part of the grid (Nalmpantis and Vrakas 2018). It is essentially a multi-classification problem. The process of non-intrusive device recognition includes several parts: period extraction, feature extraction, and load classification (Aladesanmi and Folly 2015). Researchers have been dedicated to improving the classification performance of device recognition models.

In recent years, various load identification methods have been proposed. (Buddhahai et al. 2018) extracted electrical features including current, real power, power factor, and reactive power utilized RAndom k-labELsets (RAkEL) with Decision Tree to finish multi-label classification. The proposed model is proven to be effective for high power appliances, lightings, and plug-outlet utilities. (Yang et al. 2017) developed a dataset including 4 types of appliances and adopted Naive Bayes to achieve non-intrusive appliance load monitoring. The results demonstrate that satisfactory classification accuracies can be obtained by just adopting one feature (average power). (Chang et al. 2014) proposed a power spectrum of the wavelet transform coefficients obtained by Parseval's theorem. Then, the power spectrum was used to train back-propagation load identification model, which is proven to have high success rates. (Du et al. 2013) developed a hybrid supervised identifier for Plugged-in Electric Loads (PELs), which contains Self-Organizing Map (SOM) and Bayesian model. Its accuracy, robustness, and applicability are validated by real-world data.

As an important part of machine learning, ensemble learning has been attracting attention. Ensemble learning refers to the reasonable combination of several weak classifiers to form a much stronger classifier. In the process of ensemble learning, the necessity is to guarantee the diversity of individual parts. According to the sources of diversity, ensemble learning can be categorized as data diversity and learner diversity. Among them, data diversity methods are represented by Bagging (Breiman 1996)

H. Liu et al., *Smart Device Recognition*,
https://doi.org/10.1007/978-981-33-4925-4_7

and Boosting (Hastie et al. 2009). They obtain different datasets by re-sampling the data and other methods to train classification models. Learner diversity methods mainly adopt heterogeneous classifiers to generate diversity. Different classifiers have diverse performances when tackling with different datasets. However, the application of ensemble learning in load identification is relatively rare. For instance, (Sun et al. 2020) developed an framework for driven factors identifications and household load profiling based on ensemble clustering. (Himeur et al. 2020) proposed a non-intrusive appliance recognition system based on multiscale wavelet packet tree and ensemble bagging tree classifier, which showed average accuracies of 97.01% and 96.36% for two real datasets, named the GREEND and REDD.

The chapter analyzes two common ensemble methods and introduces their methodologies. For an optimized weighting strategy, several single-objective and multi-objective algorithms are tested. For boosting strategies, Adaptive Boosting (AdaBoost) and Linear Programming Boosting (LPBoost) are investigated. Finally, the classification performances of base and ensemble models are comprehensively compared by 1793 instances of actual dataset.

7.1.2 Data Description

In the chapter, 1793 instances of current data in PLAID are utilized. Each instance includes 9500 observations in total. Taking 11 different instances as an example, the current data are shown as Fig. 7.1.

It can be observed that the current time series of these appliances have a significant periodicity. Therefore, it is not significant to utilize the whole time series to achieve non-intrusive device recognition. Specify the period T as 500 observations, a simple method is proposed to extract the current data within one period as shown in Fig. 7.2.

As a result, the extracted current data of 11 different instances in Fig. 7.3 are shown as follows. The current time series are apparently more stationary and suitable for input and training of classification models. Although the lengths of time series are significantly reduced, the contained information still remains because the data within a single period contains most characteristics of the device.

To reasonably evaluate the performance of proposed classification models, the whole dataset is divided into a training set and a testing set.

When the number of instances is determined, the training set and the test set are randomly selected to guarantee a uniform division of samples. 80% instances are used as training set, and the rest 20% instances are used as testing set. The detailed description and division of PLAID are provided in Table 7.1. An illustration of data partition is shown in Fig. 7.4, where 200 instances are included. Among the 200 instances, 164 instances are categorized as training set and the rest 36 instances are categorized as testing set.

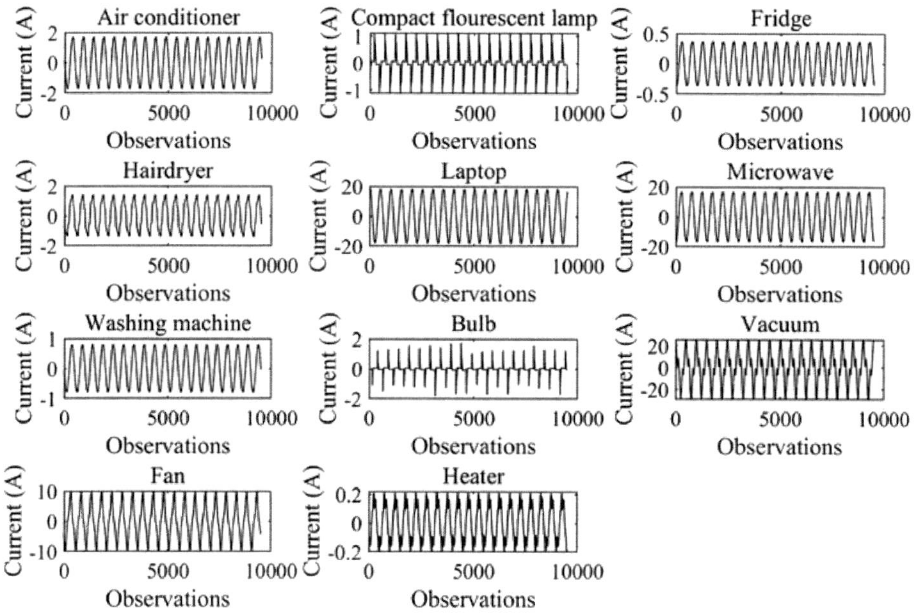

Fig. 7.1 The original current data of 11 different instances in the PLAID

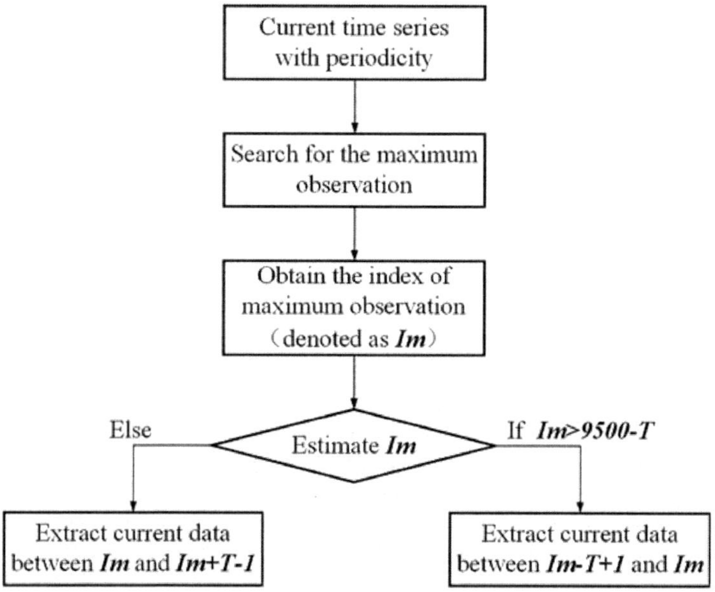

Fig. 7.2 Process of extracting the current data within one period

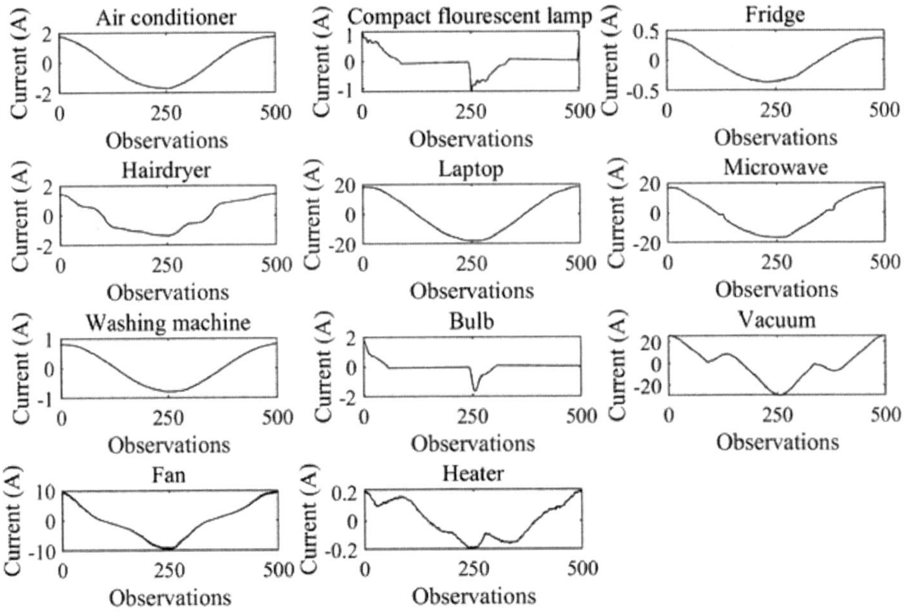

Fig. 7.3 Extracted current data of 11 different instances within one period

Table 7.1 The description and division of PLAID

Classification name	Appliance type	Number of instances in the training set	Number of instances in the testing set	Total
1	Air conditioner	169	39	208
2	Compact flourescent lamp	182	38	220
3	Fridge	168	42	210
4	Hairdryer	80	10	90
5	Laptop	185	63	248
6	Microwave	69	16	85
7	Washing machine	120	28	148
8	Bulb	173	34	207
9	Vacuum	168	61	229
10	Fan	62	11	73
11	Heater	58	17	75
Total		1434	359	1793

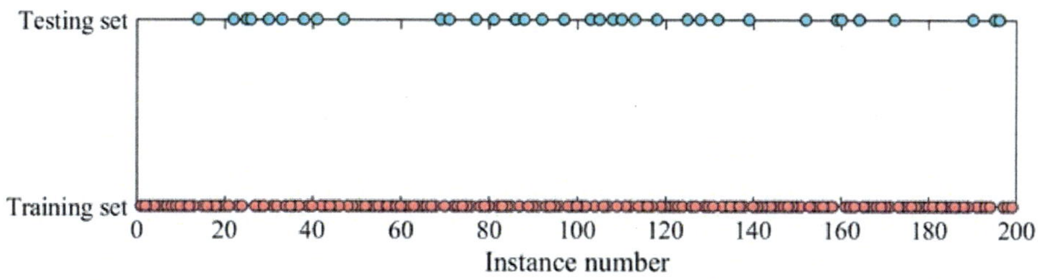

Fig. 7.4 An example of data partition. 200 instances are divided into two parts, where the training set and testing set include 164 and 36 instances, respectively

7.1.3 Feature Extraction

The extracted current time series contains 500 observations. To reduce redundancy, useful features should be extracted. The features should be able to represent the behavior pattern and characteristics of the appliance during the period. In the chapter, eight statistical features and one entropy feature are considered, including maximum, minimum, range, mean, median, variance, skewness, kurtosis, and sample entropy. Among them, maximum, minimum, range, mean, and median represents the value and the dispersion degree of current data. Skewness can quantify the degree of asymmetry. Kurtosis evaluates the normality of data distribution. Sample entropy measures the complexity of time series. Therefore, integrating the above features together can provide an overall description of the original current series. Taking the first instance as an example, its current time series is shown in Fig. 7.5. And the corresponding features are listed in Table 7.2.

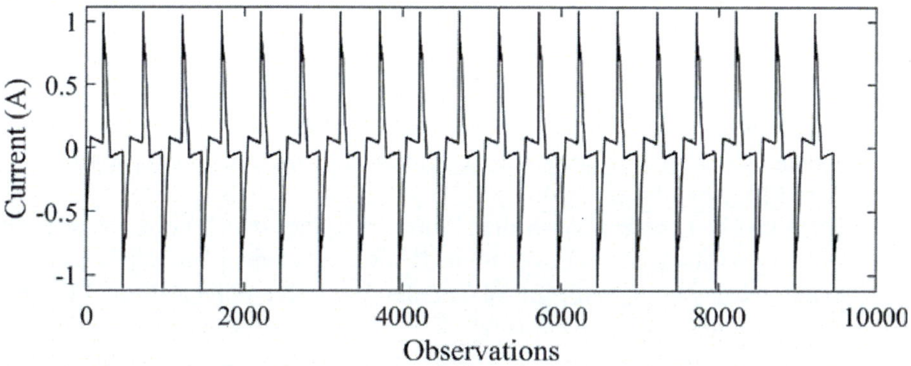

Fig. 7.5 Original current time series of the first instance

Table 7.2 The extracted features of the first instance in PLAID

Maximum	Minimum	Range	Mean	Median
1.1100	−1.1100	2.2200	0.0032	0.0050
Variance	Skewness	Kurtosis	Sample entropy	
0.1223	0.0045	4.8111	0.02718	

7.2 Ensemble Device Recognition Method Based on Optimized Weighting Strategy

7.2.1 Theoretical Basis of Base Classifiers

An optimized weighting strategy is implemented by base classifiers and optimization algorithms. Optimization algorithms calculate the weights of classifiers and combine their results to form ensemble models. In the chapter, the adopted base classifiers include four popular classifiers. They are Decision Tree (DT) (Gohari and Eydi 2020), K-Nearest Neighbor (KNN) (Gohari and Eydi 2020), Support Vector Machine (SVM) (Thorsten 2006), and Naive Bayesian Model (NBM) (Yang et al. 2017). Their theoretical basis of these base classifiers is shown as follows.

(1) *Decision Tree (DT)*
The DT is a common classification model which belongs to supervised learning. It utilizes certain rules to search for a mapping relationship between object attributes and object values. For classification problems in machine learning, the sample label is obtained through layer-by-layer decision-making after inputting the sample feature into the DT model.

(2) *K-Nearest Neighbor (KNN)*
The KNN is a simple classification model that does not need to be trained. It has no learning process. When k value, distance measure standard, and classification rule are defined, KNN finds k samples which are the nearest to the testing sample, and then classify the testing sample according to specific classification rule.

(3) *Support Vector Machine (SVM)*
The SVM is a class of generalized linear classifier which utilizes supervised learning to achieve dual classification. It introduces hyperplane and kernel function to make it competitive in solving classification problems. Kernel function can project the input data into a high-dimensional space. Then, SVM searches for a hyperplane to divide the data into two categories and maximizes the distance between the hyperplane and the nearest data. It solves the following problem:

$$\min_{w,b} \frac{1}{2}\|w\|^2 + C\sum_{i=1}^{N} \zeta_i$$
$$\text{s.t } y_i\left[w^T\phi(X_i) + b\right] \geq 1 - \zeta_i,\ \zeta_i \geq 0 \tag{7.1}$$

where C denotes the penalty parameter which can balance the size of soft margin against the classification error. ξ_i is a lack variable.

(4) *Naive Bayesian Model (NBM)*
The NBM is developed based on the Bayes theorem assuming that the characteristic conditions are independent of each other. The Bayesian method is characterized by avoiding the empirical error brought by prior probability, while preventing the over-fitting phenomenon of independent sample analysis. Assume sample set as $D = \{d_1, d_2, d_3, \ldots, d_n\}$, samples need to be classified as $Y = \{y_1, y_2, y_3, \ldots, y_m\}$, the feature set of sample data as $X = \{x_1, x_2, x_3, \ldots, x_n\}$, where the features are independent of each other. Naive Bayes is to determine the probability of D belonging to y_m. Then the prior probability of Y is $P_{prior}= P(Y)$, and the posterior probability of Y is $P_{post}= P(Y \mid X)$. According to the Bayesian algorithm, the posterior probability calculated as follows:

$$P(Y|X) = \frac{P(Y)P(X|Y)}{P(X)} \tag{7.2}$$

Because the feature set of naive Bayes is independent of each other, when the classification category y is determined, the equation can be written as follows:

$$P_{post} = P(Y|X) = \frac{P(Y)\prod_{i=1}^{d} P(x_i|Y)}{P(X)} \tag{7.3}$$

Since $P(X)$ is a fixed value, the posterior probability of samples belonging to y_i is:

$$P(y_i|x_1, x_2, x_3, \ldots, x_d) = \frac{P(y_i)\prod_{j=1}^{d} P\left(x_j|y_i\right)}{\prod_{j=1}^{d} P\left(x_j\right)} \tag{7.4}$$

7.2.2 *Theoretical Basis of Optimized Weighting Strategy*

Learner diversity is a frequently-used method to achieve ensemble learning. The simplest way is the average method, which assigns the same weights to different classifiers and calculates the mean value of their results. Another way to determine the weights of classifiers is regarding it as an optimization problem and applying optimization algorithms. In this case, the weak classifiers are combined through an optimized weighting strategy. According to the specific objective functions, the optimization methods can be categorized as single-objective optimization and multi-objective optimization.

The process of single-objective optimization is implemented as the following steps:

(a) The classification results and probabilities of weak classifiers are input into optimization algorithm. Assume the number of classes as N_c. The probability that the *i*-th weak classifier judges the sample as the *j*-th class can be denoted p_i^j $(j = 1, 2, \ldots, N_c)$.
(b) The algorithm adaptively determines the combination weights of these weak classifiers. Denote the combination weights of the *i*-th weak classifier as w_i.
(c) Obtain results of the ensemble model as follows:

$$pe^j = \sum_{i=1}^{n} w_i p_i^j \ (j = 1, 2, \ldots, N_c) \tag{7.5}$$

where pe^j represents the probability that the sample belongs to the j-th class.
(d) Obtain the final classification results of ensemble model by selecting the class with the highest probability. Denote the final classification results as c_f.
(e) Evaluate classification performance as the fitness of this solution. In the chapter, the objective function is set as the overall accuracy of classification results on all training samples:

$$Objective = accuracy\_all(c_F, c_A) \tag{7.6}$$

(f) According to the values of the objective function, the algorithm would learn to search for combination weights that can maximize classification accuracy. After several iterations, the optimal combination weights would be obtained.

Multi-objective optimization is different from single-objective optimization because it has multiple objectives and the optimal results cannot be easily obtained. Besides, multi-objective optimization and single-objective optimization differ in their solutions. The solution of single-objective optimization is unique, while multi-objective optimization gets a set of solutions because of the conflict of objectives. It is usually impossible to obtain the best solution for all objectives. If this happens, it is meaningless to solve this problem by multi-objective optimization because it can be directly transformed into single-objective optimization. With that in consideration, the objective functions of multi-objective optimization include not only accuracy, but also F1-macro. F1-macro is an evaluation index especially for multi-classification.

Due to the characteristics of accuracy and F1-macro, multi-objective optimization is desired to maximize these three objectives. The results of multi-objective optimization are called non-dominated solutions or Pareto set. And the corresponding objective values of these solutions can form a Pareto front.

In the chapter, the adopted single-objective optimization algorithms include Particle Swarm Optimization (PSO) (Chang et al. 2013), Genetic Algorithm (GA) (Michael and Jiirgen 2004), Differential Evolution (DE), Firefly Algorithm (FA), Butterfly Optimization Algorithm (BOA), Biogeography Based Krill Herd (BBKH),

Table 7.3 Mechanisms of the utilized optimization algorithms

Name of optimization algorithm	Mechanism
Particle Swarm Optimization (PSO)	Simulation of the social behavior of birds
Genetic Algorithm (GA)	Process of natural selection
	Genetic operators (crossover and mutation)
Differential Evolution (DE)	Differential-based mutation operations
	One-to-one competitive survival strategies
Firefly Algorithm (FA)	Idealized behavior of the flashing characteristics of fireflies
Butterfly Optimization Algorithm (BOA)	Food search and mating behavior of butterflies
Biogeography Based Krill Herd (BBKH)	Idealized swarm behavior of krill
Grey Wolf Optimization (GWO)	Leadership hierarchy of grey wolfs
	Hunting mechanism
Multi-Objective Grey Wolf Optimizer (MOGWO)	Leader selection mechanism
	Archive maintenance mechanism

and Grey Wolf Optimization (GWO). To compare with single-objective optimization algorithms, Multi-Objective Grey Wolf Optimizer (MOGWO) is utilized to represent the performance of multi-objective optimization. These algorithms are meta-heuristic and nature-inspired. The mechanisms of these algorithms are shown in Table 7.3.

7.2.3 Model Framework

Figure 7.6 shows the model framework of optimized weighting device recognition method. The modeling steps are clarified as follow:

(a) Extract period of the original current time series.
(b) Extract statistical and entropy features of period time series for each instance.
(c) Divide 1793 instances into training set and testing set.
(d) Normalize the feature set and use the normalized data to train four base classifiers, including DT, KNN, SVM, and NBM.

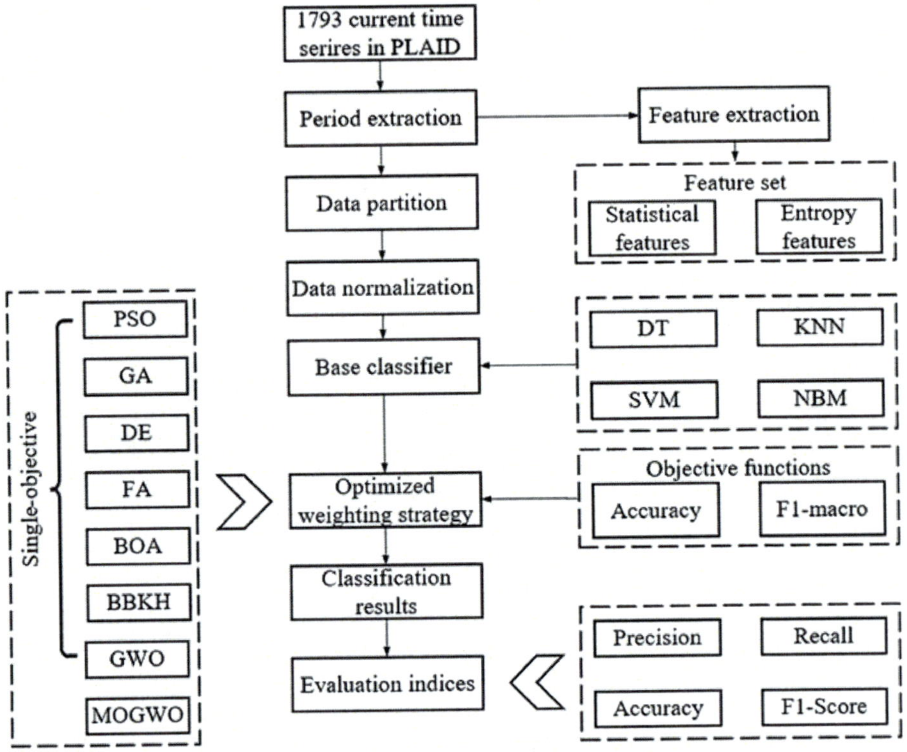

Fig. 7.6 Model framework of optimized weighting device recognition method

(e) Implement optimized weighting of base classifiers. The optimization algorithms include seven single-objective algorithms and one multi-objective algorithm. The objective functions are accuracy and F1-macro.
(f) Evaluate the classification results of base classifiers and optimized weighting ensemble models by calculating precision, recall, accuracy, and F1-Score. Compare the base classifiers and ensemble models.

7.2.4 *Analysis of Weighting Ensemble Device Recognition Model*

(1) *Analysis of base classifier*
The modeling processes of four base classifiers are totally different. They have different parameters and model structures. Therefore, it is necessary to further analyze the outcome of these base classifiers.
The well-formed DT has a tree structure. Since the DT in the study is too large to be shown. A tiny part of it is provided in Fig. 7.7. It can be observed that the instances are classified into different categories according to their features. Each split divides the feature into two parts and all splits would form a tree-like

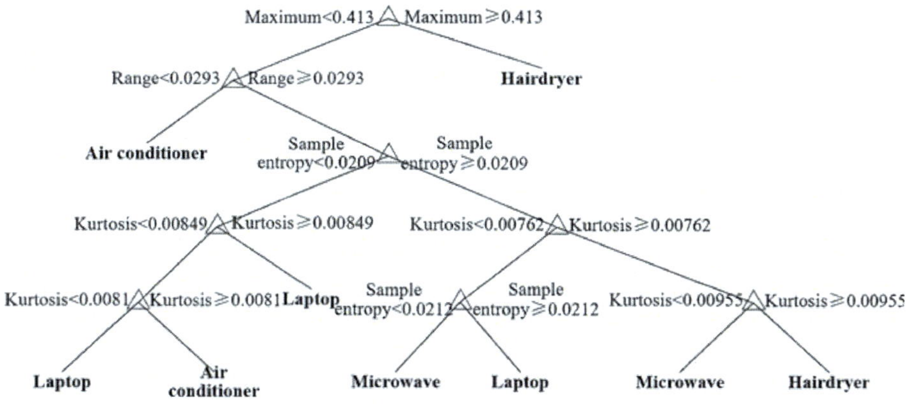

Fig. 7.7 Part of the well-formed DT

model. At the end of the tree, the categories would be given, so that the instances are successfully recognized.

The DT can estimate the importance of features by summing changes in the risk due to splits on every feature and dividing the sum by the number of branch nodes. Figure 7.8 shows the estimation results of feature importance. Kurtosis, sample entropy, and maximum are the three most important features in device recognition. However, the minimum, range, and median of instances are relatively not important in device recognition.

The hyperparameters of KNN, SVM, and NBM are automatically optimized by the Bayesian method. Therefore, the analysis of these three base classifiers mainly focuses on the optimization results of their hyperparameters. Taking

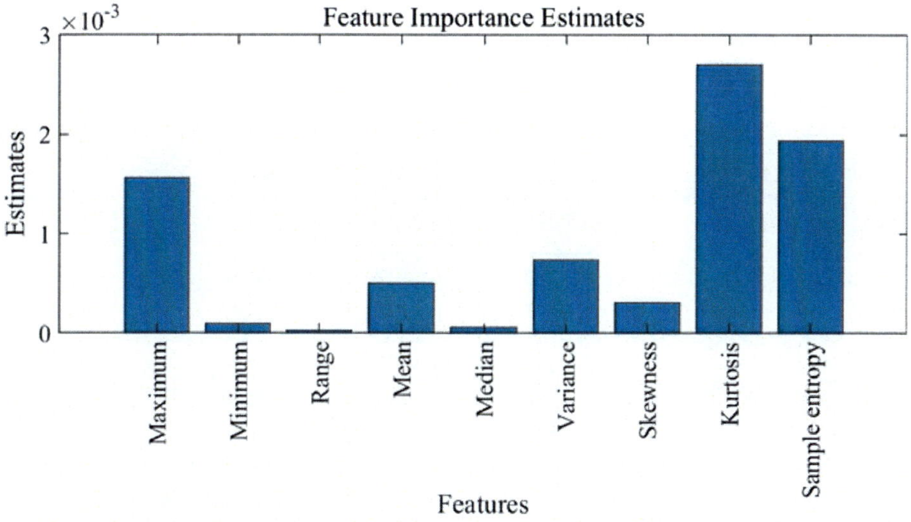

Fig. 7.8 Estimation results of feature importance obtained by DT

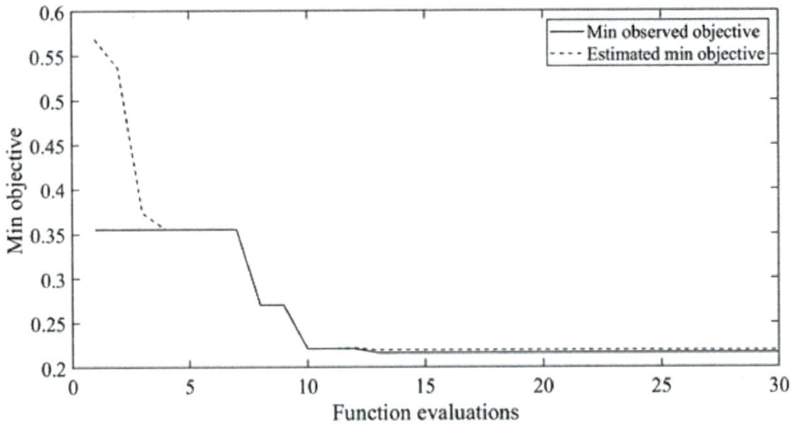

Fig. 7.9 Hyperparameter optimization process of the NBM

the NBM as an example, its optimization process is shown in Fig. 7.9. The y-axis represents min objective and the x-axis represents the number of function evaluations. It can be observed that the objective decreases as the number of function evaluations increases. Hence, the optimization of hyperparameters is significant. The final optimal hyperparameters for KNN, SVM NBM are listed in Table 7.4.

(2) *Convergence of optimization algorithms*
The chapter utilized eight optimization algorithms to achieve optimized weighting. Their convergence performances significantly affect the validity of comparison results. Therefore, it is necessary to analyze the convergence of optimization algorithms. The objective curves of MOGWO are shown in Fig. 7.10, which includes accuracy and F1-macro. According to Fig. 7.10, both objectives show increasing trends and finally become stable. The objective curves of seven single-optimization algorithms are shown in Fig. 7.11. During 10 iter-

Table 7.4 Optimized hyperparameters of KNN, SVM, and NBM

Model	Hyperparameter	Value
KNN	Number of nearest neighbors to find	3
	Distance metric	Euclidean
SVM	Kernel function	Gaussian
	Box constraint	50.14
	Kernel scale	0.0083637
NBM	Data distributions	Kernel smoothing density estimate
	Kernel smoothing window width	6.9672e-04

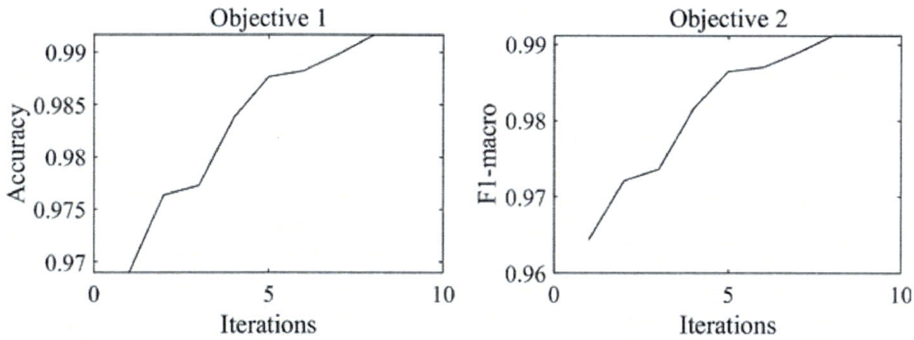

Fig. 7.10 Objective curves of MOGWO

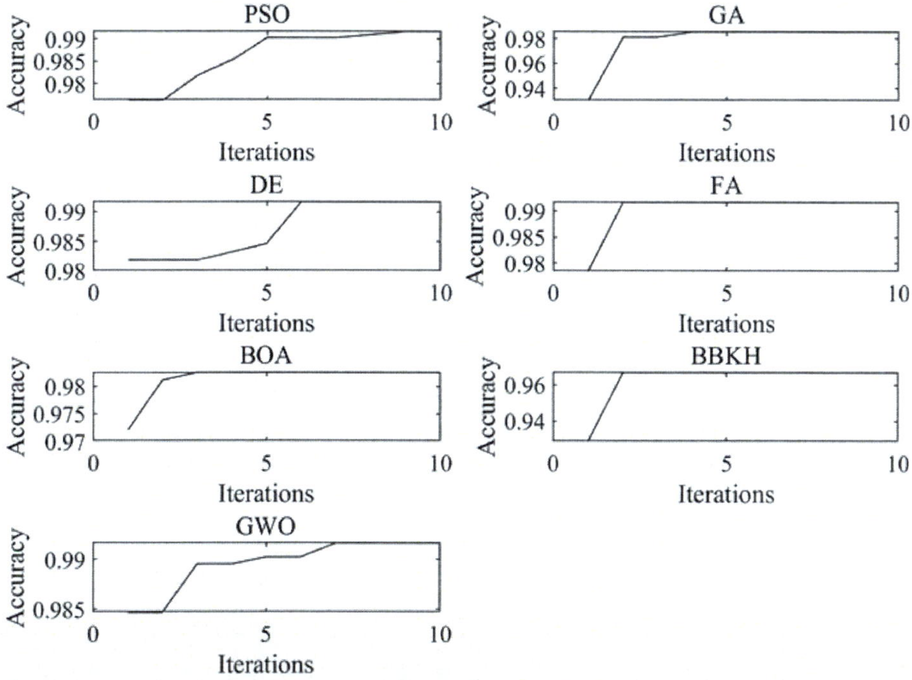

Fig. 7.11 Objective curves of seven single-optimization algorithms

ations, the accuracy of ensemble device recognition models based on single-objective optimization algorithms rapidly increases at the beginning and finally becomes stable. Therefore, all constructed optimization models are successfully converged.

The optimized variables are the combination weights of four base classifiers. The final obtained weights are listed in Table 7.5. Among them, each single-objective optimization algorithm obtains only one solution, while the MOGWO obtains four different solutions. In total, DT and KNN are given relatively high weights.

Table 7.5 Optimal combination weights of four base classifiers

Category	Optimization algorithm	DT	KNN	SVM	NBM
Single-objective	PSO	0.3412	0.4506	0.1601	0.0480
	GA	0.3120	0.4490	0.3150	−0.0760
	DE	0.4615	0.5706	0.0000	−0.0321
	FA	0.4550	0.4697	0.0462	0.0292
	BOA	0.4428	0.1531	0.2199	0.1842
	BBKH	0.6205	0.0144	0.5392	−0.1741
	GWO	0.4669	0.4579	0.0311	0.0441
Multi-objective	MOGWO	0.3899	0.4727	0.0821	0.0553
		0.3871	0.4815	0.0828	0.0486
		0.3880	0.4786	0.0826	0.0508
		0.3889	0.4756	0.0824	0.0530

NBM has the smallest weight. This is mainly caused by their different classification performance towards device recognition. Better classifiers are usually given higher weights to enhance their importance in ensemble model.

(3) *Classification results of base classifier and ensemble model*
The section analyzes the results of base classifiers and ensemble models. The confusion matrix charts of four base classifiers are shown in Figs. 7.12, 7.13, 7.14 and 7.15. Classes 1-11 represent air conditioner, compact fluorescent lamp, fridge, hairdryer, laptop, microwave, washing machine, bulb, vacuum, fan, and heater, respectively.
Table 7.6 shows the evaluation indices of four base classifiers. According to Figs. 7.12, 7.13, 7.14 and 7.15 and Table 7.6, it can be concluded that:

(a) The base classifiers have different classification performances towards different appliances. There is no classifier that is the best for all appliances. For instance, the precisions of DT, KNN, SVM, and NBM for samples of air conditioner are 0.6731, 0.7949, 0.6400, and 0.6250, respectively (KNN is the best). The precisions of DT, KNN, SVM, and NBM for heater are 0.9231, 0.9333, 1.0000, and 0.7647, respectively (SVM is the best). The precisions of DT, KNN, SVM, and NBM for vacuum are 1.0000, 1.0000, 0.9833, and 1.0000, respectively (SVM is the worst). This phenomenon is reasonable because these base classifiers have totally different characteristics, that is, they are heterogeneous. They are suitable for the recognition of different current time series. Besides, this poses great significance to ensemble learning. By combining base classifiers, the ensemble model can benefit from all classifiers and make sure it is the best for different classes.
(b) From the perspective of overall classification performance, DT and SVM are relatively better than KNN and NBM. The overall precision, recall, accuracy, and F1-Score of DT are 0.8604, 0.8403, 0.9782, and 0.8459,

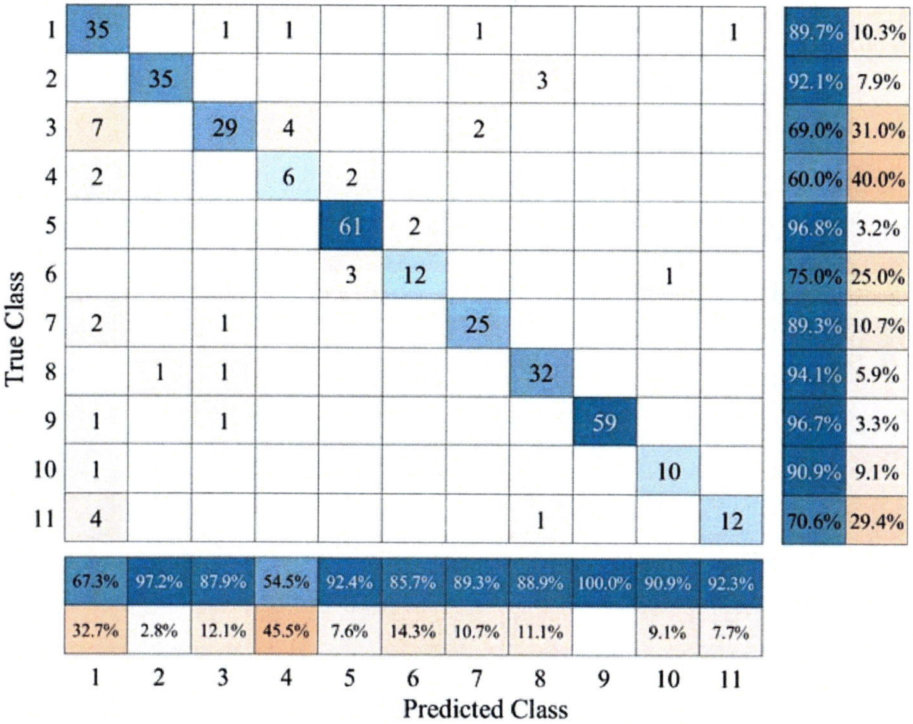

Fig. 7.12 Confusion matrix chart of DT in device recognition

respectively. The overall precision, recall, accuracy, and F1-Score of KNN are 0.8343, 0.8343, 0.9767, and 0.8313, respectively. The overall precision, recall, accuracy, and F1-Score of SVM are 0.8677, 0.8363, 0.9757, and 0.8475, respectively. The overall precision, recall, accuracy, and F1-Score of NBM are 0.7789, 0.7668, 0.9615, and 0.7680, respectively.

(c) Among four base classifiers, NBM has the worst classification performance, especially for appliances such as fridge, hairdryer, and washing machine. The precision, recall, accuracy, and F1-Score of NBM for hairdryer are 0.5000, 0.5000, 0.9721, and 0.5000, respectively. Due to sample imbalance, despite the high accuracy of NBM, its actual classification performance is not satisfactory. The other evaluation indices demonstrate that among 10 hairdryer instances, only half of them are correctly classified. This corresponds to the optimal combination weights as listed in Table 7.6. The NBM is given weights close to zero because of its poor classification performance.

To evaluate the classification results of ensemble models, the overall precision, overall recall, accuracy, and F1-macro are calculated. As shown in Fig. 7.16, it can be concluded that:

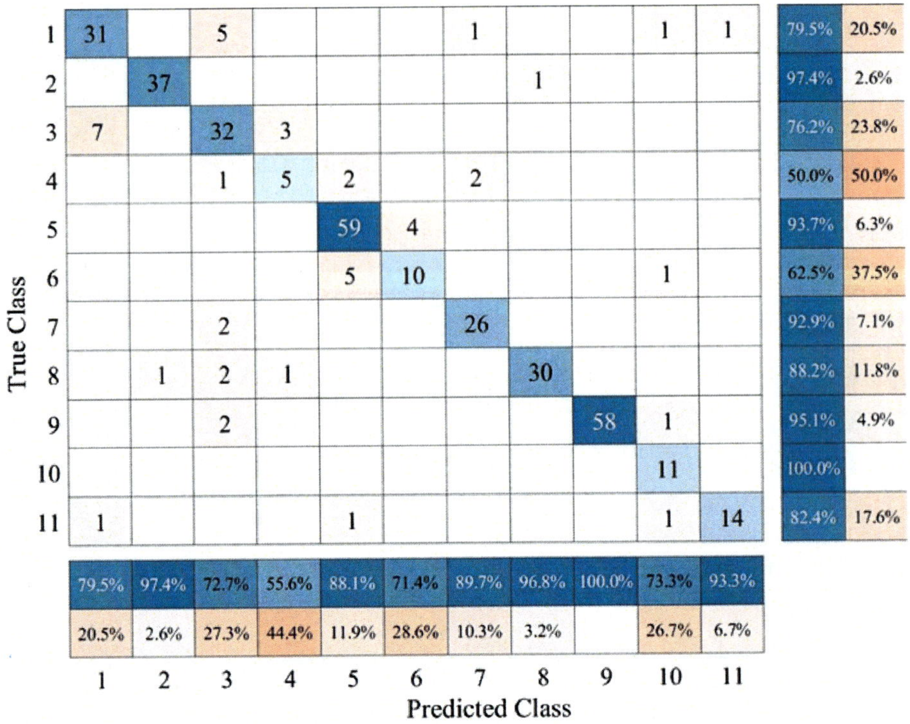

Fig. 7.13 Confusion matrix chart of KNN in device recognition

(a) MOGWO significantly outperforms seven single-objective optimization algorithms in terms of all indices. Its precision, recall, accuracy, and F1-macro are 0.9113, 0.9046, 0.9868, and 0.9061, respectively. While the precisions of PSO, GA, DE, FA, BOA, BBKH, and GWO are 0.9089, 0.9056, 0.9086, 0.9062, 0.8932, 0.8550, and 0.9029, respectively. They are not competitive when compared with MOGWO. It demonstrates that considering several objectives can effectively enhance classification performance. In the study, the superiority of MOGWO is guaranteed by two objective functions. During the optimization process, both accuracy and F1-macro of the ensemble model are improved. Therefore, the multi-objective ensemble model outperforms the single-objective ensemble models in device recognition.

(b) Apart from BBKH, the single-objective optimization algorithms have similar performance in ensemble device recognition. For instance, the accuracy indices of PSO, GA, DE, FA, BOA, and GWO are 0.9863, 0.9858, 0.9863, 0.9858, 0.9843, and 0.9853, respectively. It is mainly because the single-objective optimization of ensemble device recognition is a relatively simple task. The searching space of combination weights is limited in [0,1] so that the single-objective optimization can easily achieve the optimal value.

(c) BBKH obtains the worst classification performance. Its precision, recall, accuracy, and F1-macro are 0.8550, 0.8517, 0.9787, and 0.8496, respectively.

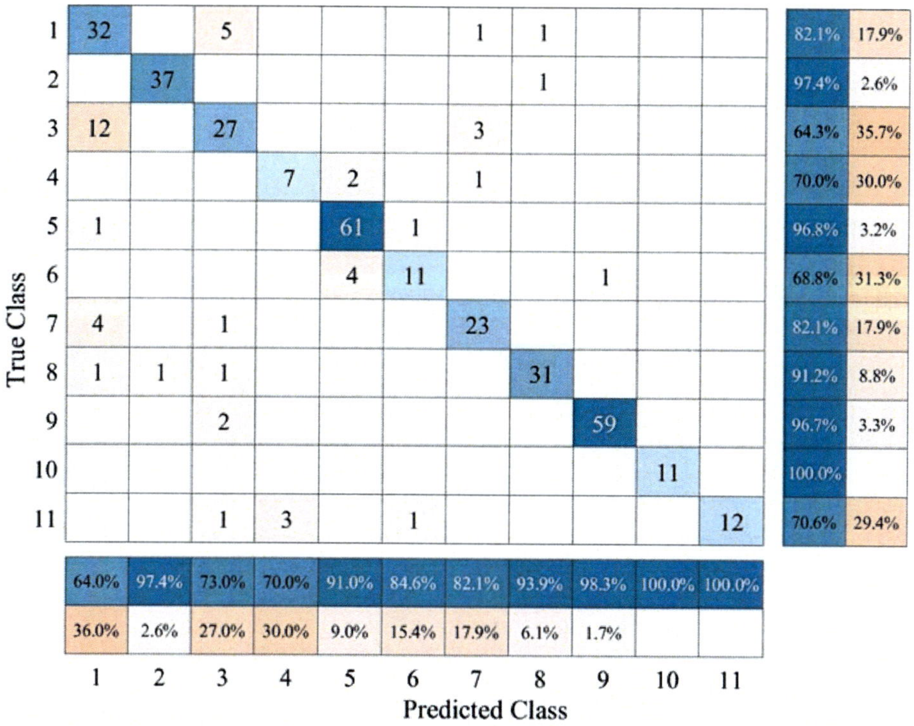

Fig. 7.14 Confusion matrix chart of SVM in device recognition

However, the result just demonstrates BBKH is worse than other optimization algorithms in the special case due to its own mechanism. It does not represent that BBKH cannot get satisfactory results in other applications.

7.3 Ensemble Device Recognition Method Based on Boosting Strategy

7.3.1 Theoretical Basis of Boosting Strategy

Boosting strategy is another frequently-used way to achieve ensemble learning. The algorithm is realized by changing the data distribution. It trains different weak classifiers and updates the weights of each training sample according to the classification error of the last time. Adaptive Boosting (AdaBoost) and Linear Programming Boosting (LPBoost) are two representative boosting algorithms. Their methodologies are presented as follows:

(1) *AdaBoost*
AdaBoost is a machine learning algorithm based on the idea of Boosting (Liu and Chen 2020). The input of AdaBoost is training set $T =$

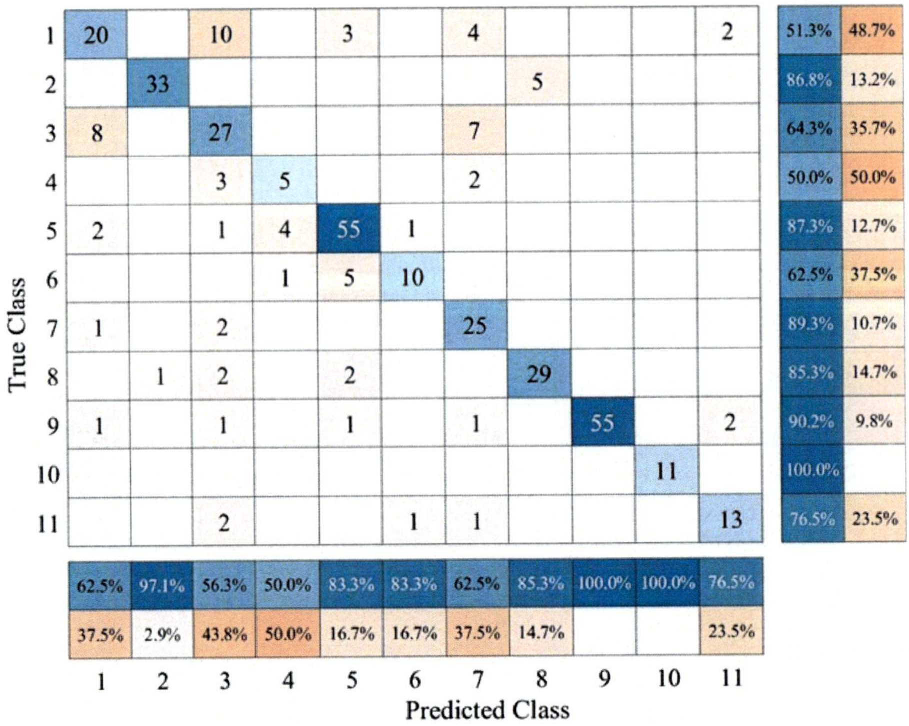

Fig. 7.15 Confusion matrix chart of NBM in device recognition

$(x_1, y_1), (x_2, y_2), \ldots, (x_N, y_N)$, where the range of y_i is $\{0, 1\}$. And the output is the final classifier $G(x)$. Its modeling theory is as follows:

(a) Firstly, initialize the weight distribution of weak classifiers as:

$$D_1 = (w_1 1, \ldots, w_1 i, \ldots, w_1 N), w_{1i} = \frac{1}{N} \tag{7.7}$$

(b) Then, update the weight distribution based on the training set. Use weight distribution to obtain base classifiers:

$$G_m(x) : X \to 0, +1 \tag{7.8}$$

(c) Calculate the classification error on the training set:

$$e_m = P(G_m(x) \neq y_i) = \sum\nolimits_{i=1}^{N} w_{mi} I(G_m(x_i) \neq y_i) \tag{7.9}$$

(d) Calculate the coefficient of $G_m(x)$:

$$\alpha_m = \frac{1}{2} \log \frac{1 - e_m}{e_m} \tag{7.10}$$

Table 7.6 Classification evaluation indices of four base classifiers

Classifier	Type of appliance	Precision	Recall	Accuracy	F1-Score
DT	Air conditioner	0.6731	0.8974	0.9415	0.7692
	Compact flourescent lamp	0.9722	0.9211	0.9889	0.9459
	Fridge	0.8788	0.6905	0.9526	0.7733
	Hairdryer	0.5455	0.6000	0.9749	0.5714
	Laptop	0.9242	0.9683	0.9805	0.9457
	Microwave	0.8571	0.7500	0.9833	0.8000
	Washing machine	0.8929	0.8929	0.9833	0.8929
	Bulb	0.8889	0.9412	0.9833	0.9143
	Vacuum	1.0000	0.9672	0.9944	0.9833
	Fan	0.9091	0.9091	0.9944	0.9091
	Heater	0.9231	0.7059	0.9833	0.8000
	Overall	0.8604	0.8403	0.9782	0.8459
KNN	Air conditioner	0.7949	0.7949	0.9554	0.7949
	Compact flourescent lamp	0.9737	0.9737	0.9944	0.9737
	Fridge	0.7273	0.7619	0.9387	0.7442
	Hairdryer	0.5556	0.5000	0.9749	0.5263
	Laptop	0.8806	0.9365	0.9666	0.9077
	Microwave	0.7143	0.6250	0.9721	0.6667
	Washing machine	0.8966	0.9286	0.9861	0.9123
	Bulb	0.9677	0.8824	0.9861	0.9231
	Vacuum	1.0000	0.9508	0.9916	0.9748
	Fan	0.7333	1.0000	0.9889	0.8462
	Heater	0.9333	0.8235	0.9889	0.8750
	Overall	0.8343	0.8343	0.9767	0.8313
SVM	Air conditioner	0.6400	0.8205	0.9304	0.7191
	Compact flourescent lamp	0.9737	0.9737	0.9944	0.9737
	Fridge	0.7297	0.6429	0.9304	0.6835
	Hairdryer	0.7000	0.7000	0.9833	0.7000
	Laptop	0.9104	0.9683	0.9777	0.9385
	Microwave	0.8462	0.6875	0.9805	0.7586
	Washing machine	0.8214	0.8214	0.9721	0.8214
	Bulb	0.9394	0.9118	0.9861	0.9254
	Vacuum	0.9833	0.9672	0.9916	0.9752
	Fan	1.0000	1.0000	1.0000	1.0000
	Heater	1.0000	0.7059	0.9861	0.8276
	Overall	0.8677	0.8363	0.9757	0.8475

(continued)

Table 7.6 (continued)

Classifier	Type of appliance	Precision	Recall	Accuracy	F1-Score
NBM	Air conditioner	0.6250	0.5128	0.9136	0.5634
	Compact flourescent lamp	0.9706	0.8684	0.9833	0.9167
	Fridge	0.5625	0.6429	0.8997	0.6000
	Hairdryer	0.5000	0.5000	0.9721	0.5000
	Laptop	0.8333	0.8730	0.9471	0.8527
	Microwave	0.8333	0.6250	0.9777	0.7143
	Washing machine	0.6250	0.8929	0.9499	0.7353
	Bulb	0.8529	0.8529	0.9721	0.8529
	Vacuum	1.0000	0.9016	0.9833	0.9483
	Fan	1.0000	1.0000	1.0000	1.0000
	Heater	0.7647	0.7647	0.9777	0.7647
	Overall	0.7789	0.7668	0.9615	0.7680

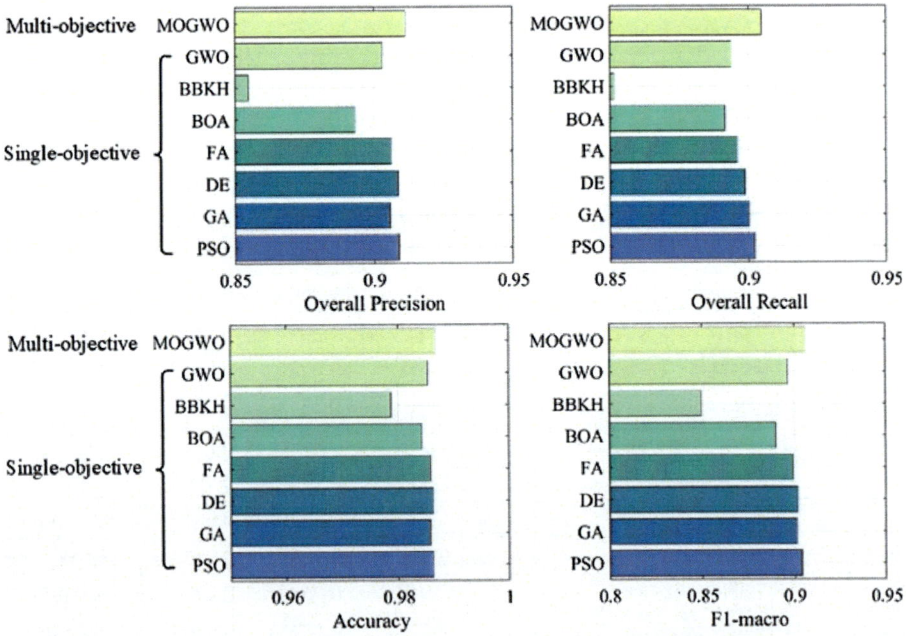

Fig. 7.16 Overall precision, recall, accuracy, and F1-macro of ensemble models

(e) Update the weight distribution of training set according to follow equations:

$$D_{m+1} = \left(w_{m+1,1}, \ldots, w_{m+1,i}, \ldots, w_{m+1,N}\right) \quad (7.11)$$

$$w_{m+1,i} = \frac{w_{mi}}{Z_m} \exp(-\alpha_m y_i G_m(x_i)) \quad (7.12)$$

where Z_m is normalization coefficient which makes D_{m+1} satisfy the probability distribution:

$$Z_m = \sum_{i=1}^{N} w_{mi} \exp(-\alpha_m y_i G_m(x_i)) \quad (7.13)$$

(f) The final classifier is defined as follows:

$$f(x) = \sum_{m=1}^{N} \alpha_m G_m(x) \quad (7.14)$$

(2) *LPBoost*

LPBoost is another manifestation of the Boosting algorithm. It is a supervised multi-classifier ensemble algorithm which maximizes the soft interval between different categories of training samples Define the number of iterations as $t = 1, 2, \ldots, T$ the modeling process of LPBoost algorithm is as follows:

(a) Normalize the weights of training samples $w_m^t(i)$ as follows:

$$w_m^t(i) = \frac{w_m^t(i)}{\sum_{i=1}^{v} w_m^t(i)} \quad (7.15)$$

(b) Calculate the classification error of the m-th weak classifier ε_m^t:

$$\varepsilon_m^t = \sum_{i=1}^{v} c_i, \quad \forall c_i = \begin{cases} w_m^t(i)\left|f_m^t(x_j) - y_j\right| f_m^t(x_j) \neq y_j \\ 0 \; f_m^t(x_j) = y_j \end{cases} \quad (7.16)$$

where x_j represents the collection of several samples x_i, y_j represents the values of weak classifiers. If $\varepsilon_m^t = 0$ or $\varepsilon_m^t \geq 0.5$, the training process is stopped.

(c) Calculate the weights a_m^t of weak classifiers $f_m^t(x_i)$ as follows:

$$a_m^t = \ln \frac{1}{\beta_m^t}, \quad \beta_m^t = \frac{\varepsilon_m^t}{1 - \varepsilon_m^t} \quad (7.17)$$

(d) Update the weights of training samples:

$$w_{m+1}^t(i) = \frac{w_m^t(i)}{z_m^t} \times \left(\beta_m^t\right)^{1-e_m^t(i)}$$
$$e_m^t(i) = \begin{cases} 0 \ \mathrm{f}_t(x_j) = y_j \\ 1 \ \mathrm{f}_t(x_j) \neq y_j \end{cases} \quad (7.18)$$

where z_m^t is the normalization coefficient. Its function is to guarantee $w_{m+1}^t(i)$ obey a distribution, that is, makes the weights of all training samples limited in [−1, 1].

(e) Return to step (a) and train the next weak classifier $f_{m+1}^t(x_i)$ until M weak classifiers are trained or the training termination condition is satisfied.

(f) Finally, combine M weak classifiers to get the result of LPBoost:

$$f_{LpBoost}(x) = \sum_{m=1}^{M} a_m f_m(x_i) \quad (7.19)$$

7.3.2 Model Framework

Figure 7.17 shows the model framework of boosting the ensemble device recognition method. The modeling steps are clarified as follow:

(a) Extract period of the original current time series.
(b) Extract statistical and entropy features of period time series for each instance.
(c) Divide 1793 instances into the training set and testing set.
(d) Normalize the feature set and use the normalized data to train two boosting ensemble models, namely AdaBoost-tree and LPBoost-tree.
(e) Evaluate the classification results of base classifiers and optimized weighting ensemble models by calculating precision, recall, accuracy, and F1-Score.

7.3.3 Analysis of Boosting Ensemble Device Recognition Model

(1) *Model parameters*
To avoid the error caused by manual parameter setting, the hyperparameters of AdaBoost-tree and LPBoost-tree are determined by the Bayesian optimization method. Their optimization processes are shown in Figs. 7.18 and 7.19. The y-axis represents min objective and the x-axis represents the number of function evaluations. During 25 evaluations, the main objective is greatly

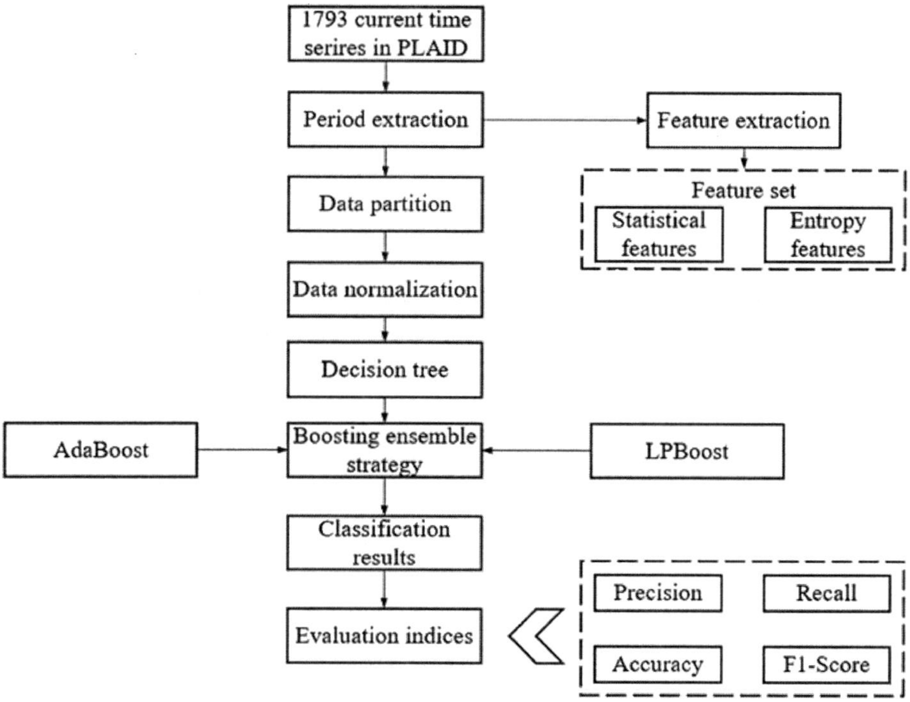

Fig. 7.17 Model framework of boosting ensemble device recognition method

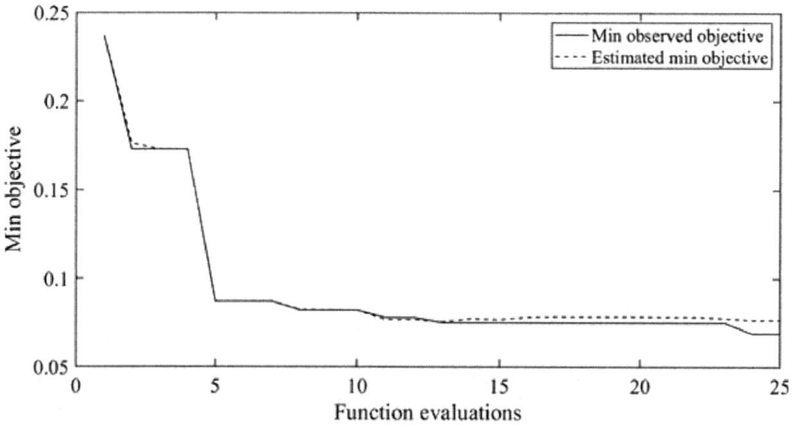

Fig. 7.18 Hyperparameter optimization process of the AdaBoost-tree

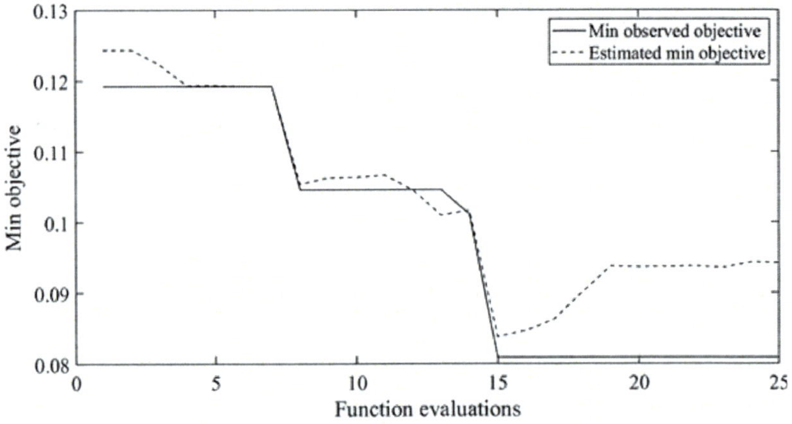

Fig. 7.19 Hyperparameter optimization process of the LPBoost-tree

reduced. Therefore, the Bayesian optimization can successfully determine optimal hyperparameters of boosting device recognition models.

The final hyperparameters obtained by the Bayesian optimization method are shown in Table 7.7. Three hyperparameters are determined, including the number of ensemble learning cycles, learning rate, and maximal number of decision splits. They are important hyperparameters that significantly affect the classification performance of ensemble model. Among them, the number of ensemble learning cycles determines the total number of decision trees that would be utilized to construct an ensemble model. Learning rate determines the fitting performance. High learning rate may lead to suboptimal and low learning rate may result in overfitting in some cases.

(2) *Feature importance*

According to Table 7.7, the trained AdaBoost-tree and LPBoost-tree model contains 498 and 46 decision trees in total. Each decision tree has an estimation result of feature importance. Figures 7.20 and 7.21 visualize the estimation

Table 7.7 Hyperparameters of AdaBoost-tree and LPBoost-tree models

Boosting strategy	Optimal hyperparameter	Value
AdaBoost	Number of ensemble learning cycles	498
	Learning rate	0.9522
	Maximal number of decision splits	52
LPBoost	Number of ensemble learning cycles	46
	Learning rate	1
	Maximal number of decision splits	57

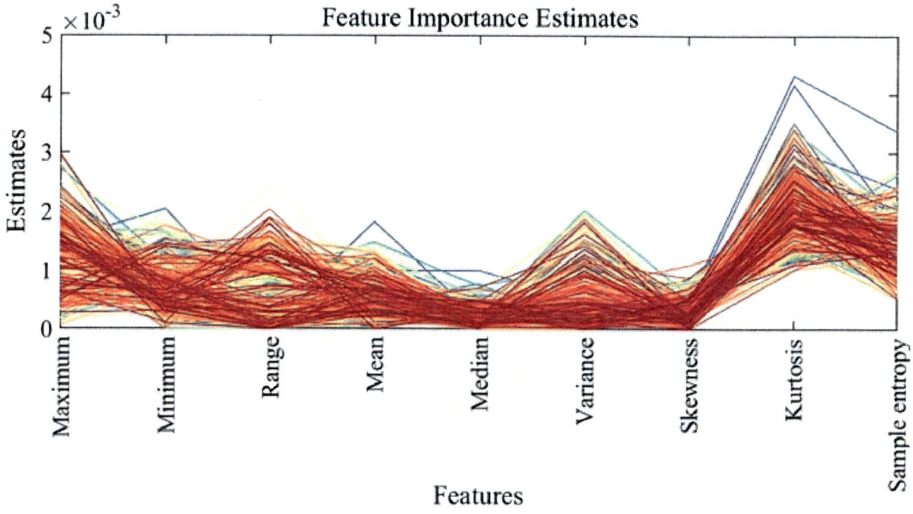

Fig. 7.20 Estimation results of feature importance obtained by AdaBoost-tree

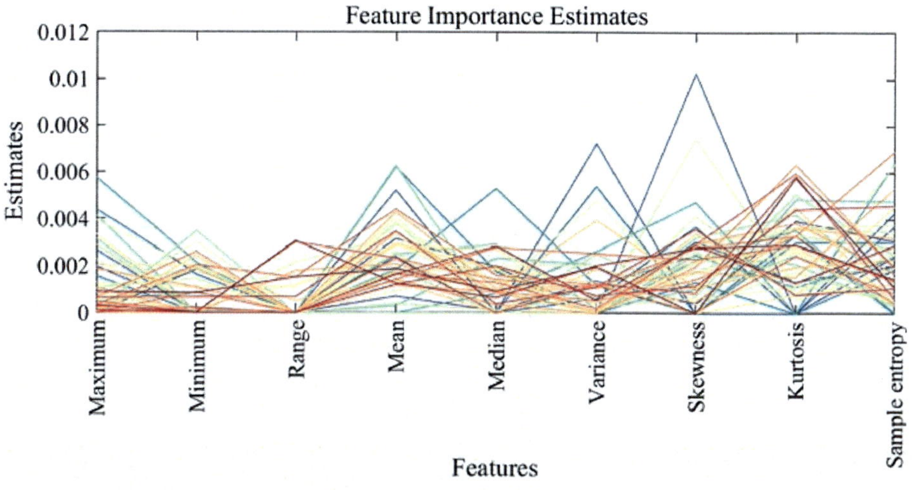

Fig. 7.21 Estimation results of feature importance obtained by LPBoost-tree

results of all constructed decision trees. Based on the results of AdaBoost-tree, it is obvious that kurtosis is the most important feature for device recognition. The overall estimates of kurtosis are much higher than other features. However, as for the results of LPBoost-tree, the importance rank of feature is not that obvious, except that the estimates of skewness are noticeably higher than other features at times.

To quantitatively analyze the overall estimation results of all decision trees, the average estimates are calculated. The calculation results of AdaBoost-tree and LPBoost-tree are listed in Tables 7.8 and 7.9. For the AdaBoost-tree model,

Table 7.8 Average feature importance estimates of AdaBoost-tree model

Feature	Importance estimate	Rank
Maximum	0.001347	3
Minimum	0.000715	4
Range	0.000556	6
Mean	0.000595	5
Median	0.000189	9
Variance	0.000492	7
Skewness	0.000310	8
Kurtosis	0.002065	1
Sample entropy	0.001350	2

Table 7.9 Average feature importance estimates of LPBoost-tree model

Feature	Importance estimate	Rank
Maximum	0.001152	6
Minimum	0.000799	8
Range	0.000362	9
Mean	0.002379	2
Median	0.001074	7
Variance	0.001213	5
Skewness	0.002078	4
Kurtosis	0.002655	1
Sample entropy	0.002345	3

maximum, kurtosis, and sample entropy are the three most important features. Their estimates are 0.001347, 0.002065, and 0.001350, respectively. For the LPBoost-tree model, mean, kurtosis, and sample entropy are the three most important features. Their estimates are 0.002379, 0.002655, and 0.002345, respectively. This is reasonable because kurtosis and sample entropy mainly evaluates the normality and complexity of appliance current data. Different appliances have totally different waveforms. Therefore, kurtosis and sample entropy are both useful features for device recognition.

(3) *Classification results of boosting ensemble model*
Table 7.10 shows the classification evaluation indices of AdaBoost-tree and LPBoost-tree for all classes. Based on the three most important features, the classification results of AdaBoost-tree and LPBoost-tree are visualized in Figs. 7.22 and 7.23, respectively. According to Table 7.10 and Figs. 7.22 and 7.23, it can be concluded that:

(a) The boosting ensemble device recognition models have different classification performances towards different appliances. For the AdaBoost-tree model, most of the appliances are correctly recognized, such as hairdryer,

Table 7.10 Classification evaluation indices of AdaBoost-tree and LPBoost-tree

Classifier	Type of appliance	Precision	Recall	Accuracy	F1-Score
AdaBoost-tree	Air conditioner	0.8750	0.8974	0.9749	0.8861
	Compact flourescent lamp	0.9744	1.0000	0.9972	0.9870
	Fridge	0.8444	0.9048	0.9694	0.8736
	Hairdryer	1.0000	0.8000	0.9944	0.8889
	Laptop	0.9692	1.0000	0.9944	0.9844
	Microwave	1.0000	0.9375	0.9972	0.9677
	Washing machine	0.9655	1.0000	0.9972	0.9825
	Bulb	1.0000	0.9118	0.9916	0.9538
	Vacuum	1.0000	0.9508	0.9916	0.9748
	Fan	1.0000	1.0000	1.0000	1.0000
	Heater	0.9444	1.0000	0.9972	0.9714
	Overall	0.9612	0.9457	0.9914	0.9518
LPBoost-tree	Air conditioner	0.9118	0.7949	0.9694	0.8493
	Compact flourescent lamp	0.9730	0.9474	0.9916	0.9600
	Fridge	0.8000	0.9524	0.9666	0.8696
	Hairdryer	0.7778	0.7000	0.9861	0.7368
	Laptop	0.9524	0.9524	0.9833	0.9524
	Microwave	0.8571	0.7500	0.9833	0.8000
	Washing machine	0.9333	1.0000	0.9944	0.9655
	Bulb	0.9412	0.9412	0.9889	0.9412
	Vacuum	1.0000	0.9672	0.9944	0.9833
	Fan	0.9167	1.0000	0.9972	0.9565
	Heater	0.9412	0.9412	0.9944	0.9412
	Overall	0.9095	0.9042	0.9863	0.9051

bulb, vacuum, and fan. However, the precision of fridge is relatively lower than other appliances. This may be because the current time series of fridge at work are more complex than those of other devices, but the statistical characteristics are not so different that they are difficult to identify. For the LPBoost-tree model, the classification results of fridge, hairdryer, and microwave are not satisfactory. But its recognition ability of vacuum is better than the AdaBoost-tree model. Therefore, it is difficult to determine which stacking ensemble model is the best when considering different appliances.

(b) From the perspective of overall classification performance, the AdaBoost-tree model significantly outperforms the LPBoost-tree model. The overall precision, recall, accuracy, and F1-Score of AdaBoost-tree are 0.9612, 0.9457, 0.9914, and 0.9518, respectively. The overall precision, recall, accuracy, and F1-Score of LPBoost-tree are 0.9095, 0.9042, 0.9863, and

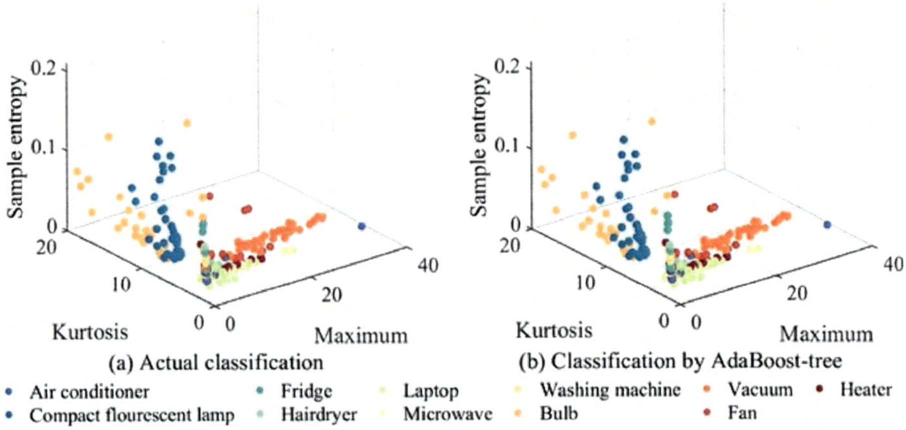

Fig. 7.22 Relationship between AdaBoost-tree classification results and the three most important features (maximum, kurtosis, sample entropy)

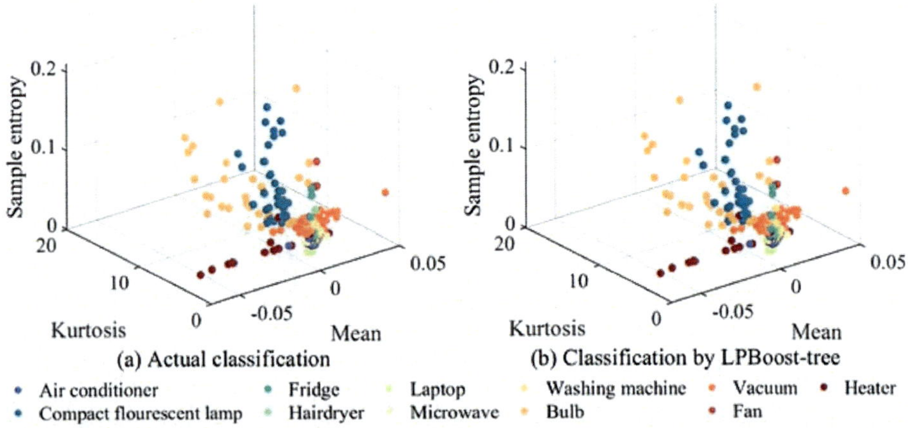

Fig. 7.23 Relationship between LPBoost-tree classification results and the three most important features (mean, kurtosis, sample entropy)

0.9051, respectively. The superiority of the AdaBoost-tree model depends not only on the algorithm itself but also on the total number of weak classifiers. Compared with the LPBoost-tree model which contains 46 weak classifiers, the AdaBoost-tree model contains 498 weak classifiers in total. Therefore, its accuracy and robustness are much higher than the LPBoost-tree model.

The confusion matrix charts of the AdaBoost-tree and LPBoost-tree device recognition models are shown in Figs. 7.24 and 7.25. Classes 1-11 represent air conditioner, compact flourescent lamp, fridge, hairdryer, laptop, microwave, washing machine, bulb, vacuum, fan, and heater, respectively. According to

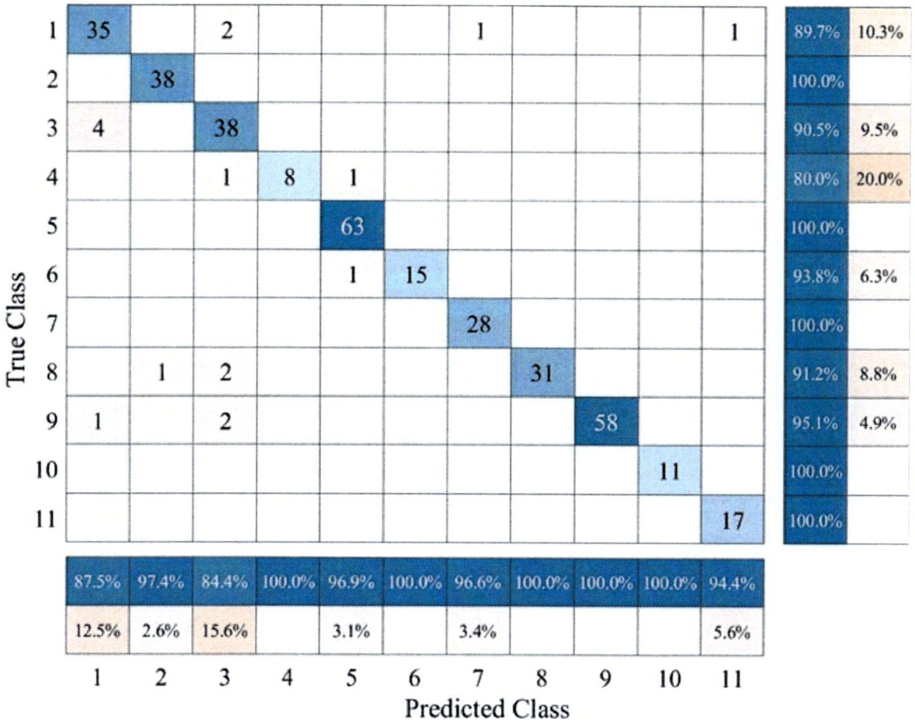

Fig. 7.24 Confusion matrix chart of AdaBoost-tree in device recognition

Figs. 7.24 and 7.25, it can be observed that the AdaBoost-tree model has very competitive and satisfactory performance. Most instances are correctly recognized. As for the LPBoost-tree model, 10 instances have been misclassified as fridge. And the recognition of hairdryer is also not satisfactory.

7.4 Experiment Analysis

In the section, the above device recognition models are thoroughly compared to evaluate their differences. The compared models include four base classifiers (DT, KNN, SVM, and NBM), the best weighting ensemble model (MOGWO), and two boosting ensemble models (AdaBoost and LPBoost).

7.4.1 Comparative Analysis of Classification Performance

To further evaluate the classification performance of these models, receiver operating characteristic (ROC) for each output class is calculated. The results are shown in Fig. 7.26. The more each ROC curve hugs the left and top edges of the plot, the

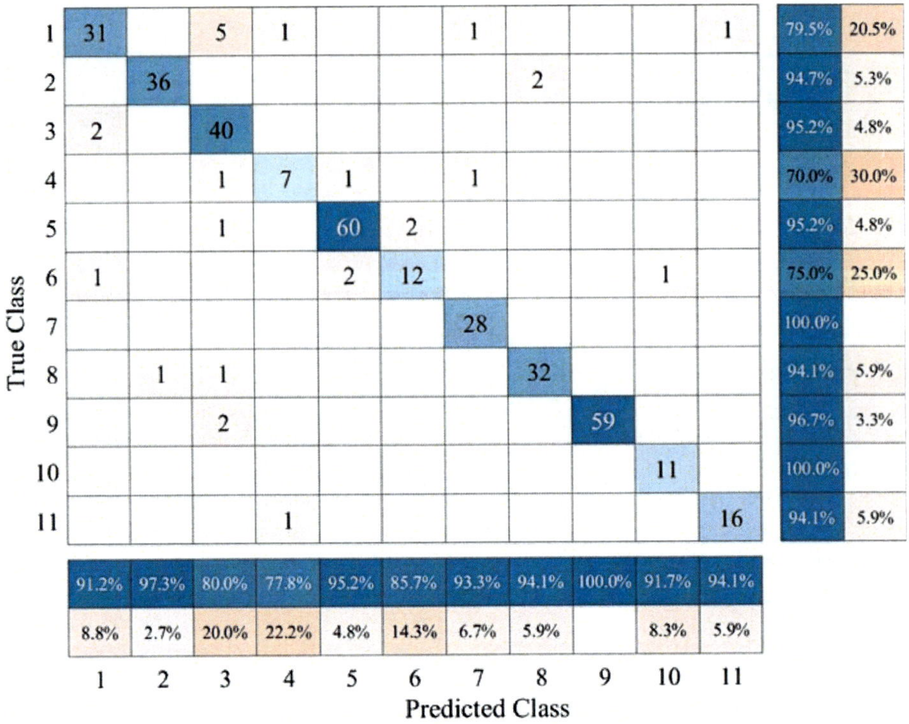

Fig. 7.25 Confusion matrix chart of LPBoost-tree in device recognition

better the classification. To quantitively analyze the ROC of classifiers, Area Under roc Curve (AUC) is provided in Table 7.11. In general, the AUC value ranges from 0.5 to 1. Larger AUC value represents better classification performance. If the AUC value is only 0.5, it demonstrates that the classifier is similar to random selection. Besides, the AUC improvement percentage index is utilized to characterize the difference between two models. The calculation formula is shown as follows:

$$P_{AUC}=\frac{AUC_2-AUC_1}{AUC_1}\times 100\% \tag{7.20}$$

The AUC improvement percentages of MOGWO for four base classifiers are shown in Table 7.12. The AUC improvement percentages of two boosting strategies for DT are shown in Table 7.13.

According to Fig. 7.26 and Tables 7.11, 7.12 and 7.13, it can be concluded that:

(a) Among all compared models, the AdaBoost-tree model is the most suitable for device recognition. The overall AUC value of AdaBoost-tree model is 0.9943, while the overall AUC values of DT, KNN, SVM, NBM, MOGWO, LPBoost-tree are 0.9509, 0.9578, 0.9795, 0.9692, 0.9868, and 0.9901, respectively. The main classification errors of the AdaBoost-tree model are concentrated on air conditioner and laptop. This is because of the inner characteristics of dataset.

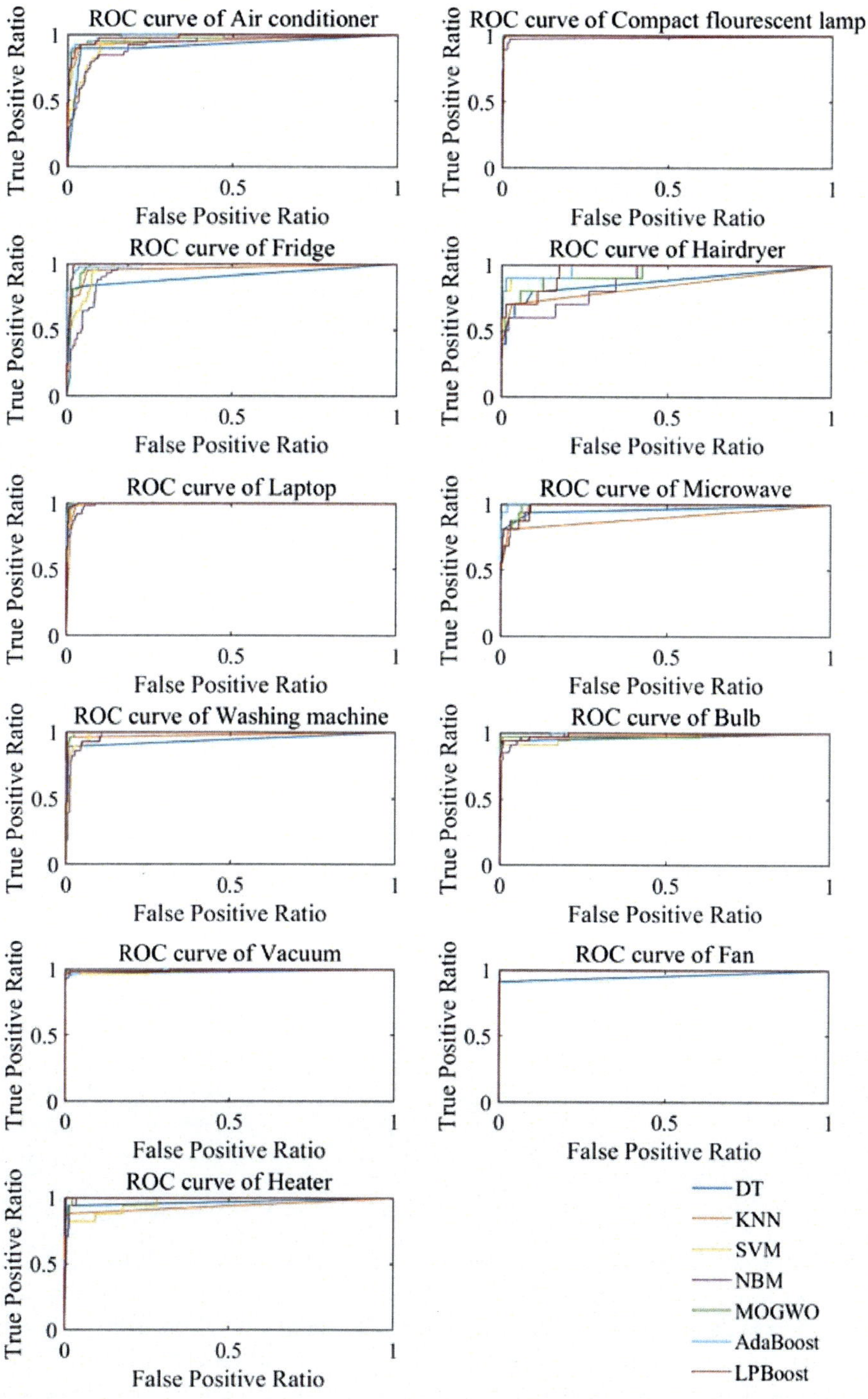

Fig. 7.26 ROC curves of compared base classifiers and ensemble models

Table 7.11 AUC values of compared base classifiers and ensemble models

Type of appliance	Base classifiers				Ensemble model		
	DT	KNN	SVM	NBM	MOGWO	AdaBoost	LPBoost
Air conditioner	0.9196	0.9526	0.9563	0.9268	0.9753	0.9896	0.9822
Compact flourescent lamp	0.9983	0.9997	0.9992	0.9836	1.0000	0.9994	0.9991
Fridge	0.9025	0.9610	0.9709	0.9504	0.9868	0.9914	0.9929
Hairdryer	0.8718	0.8404	0.9550	0.8785	0.9367	0.9762	0.9536
Laptop	0.9962	0.9910	0.9954	0.9902	0.9976	0.9994	0.9978
Microwave	0.9580	0.9001	0.9854	0.9840	0.9903	0.9985	0.9838
Washing machine	0.9402	0.9780	0.9851	0.9809	0.9959	0.9994	0.9984
Bulb	0.9699	0.9844	0.9704	0.9860	0.9823	0.9917	0.9905
Vacuum	0.9836	0.9917	0.9898	0.9886	0.9940	0.9940	0.9946
Fan	0.9531	0.9986	1.0000	1.0000	0.9997	1.0000	0.9997
Heater	0.9665	0.9384	0.9670	0.9924	0.9966	0.9981	0.9985
Overall	0.9509	0.9578	0.9795	0.9692	0.9868	0.9943	0.9901

Table 7.12 AUC improvement percentages of MOGWO for four base classifiers

Type of appliance	DT	KNN	SVM	NBM
Air conditioner	6.0556	2.3890	1.9943	5.2395
Compact flourescent lamp	0.1683	0.0328	0.0820	1.6627
Fridge	9.3330	2.6807	1.6322	3.8249
Hairdryer	7.4445	11.4558	−1.9202	6.6210
Laptop	0.1427	0.6629	0.2209	0.7501
Microwave	3.3761	10.0314	0.4993	0.6481
Washing machine	5.9215	1.8314	1.0953	1.5290
Bulb	1.2736	−0.2206	1.2217	−0.3763
Vacuum	1.0570	0.2302	0.4224	0.5425
Fan	4.8924	0.1177	−0.0261	−0.0261
Heater	3.1144	6.1950	3.0594	0.4159
Overall	3.7804	3.0307	0.7487	1.8173

The recognition of air conditioner and laptop is relatively difficult. To solve this problem, several methods can be tried. The first is to extend the dataset and use data sources other than current. The second is to consider other feature selection methods. In the study, the features are mainly statistical features. Some other feature extraction algorithms such as principal component analysis and autoencoder can be tried. The third is to adopt deep learning techniques to further enhance the complexity of classification model, making the classifier more powerful than shallow models.

Table 7.13 AUC improvement percentages of two boosting strategies for DT

Type of appliance	AdaBoost	LPBoost
Air conditioner	7.6065	6.8049
Compact flourescent lamp	0.1109	0.0780
Fridge	9.8490	10.0071
Hairdryer	11.9803	9.3837
Laptop	0.3203	0.1642
Microwave	4.2320	2.6914
Washing machine	6.2887	6.1854
Bulb	2.2440	2.1227
Vacuum	1.0570	1.1130
Fan	4.9198	4.8924
Heater	3.2746	3.3102
Overall	4.5695	4.1226

(b) Ensemble device recognition methods are effective in improving classification performance. All ensemble models outperform four base classifiers. MOGWO improves the overall AUC values of DT, KNN, SVM, NBM by 3.7804%, 3.0307%, 0.7487%, and 1.8173%, respectively. AdaBoost and LPBoost improve the overall AUC values of DT by 4.5695% and 4.1226%, respectively. The main reason is the combination of several weak classifiers makes ensemble models significantly more reliable. Hence, it is recommended to employ ensemble learning when actual application of device recognition.

(c) No model is absolutely superior when considering different appliances. Although the AdaBoost-tree model has the highest overall AUC value, it does not outperform other classifiers in the recognition of fridge and fan. Taking the recognition result of fan as an example, the AUC value of the AdaBoost-tree model for fan is 0.9940, while the AUC values of the MOGWO and LPBoost-tree model are 0.9940 and 0.9946, respectively. These results may be caused by the imbalanced dataset. Among the whole PLAID, the total number of fridge instances is 210. It is apparently higher than that of hairdryer instances (90). Therefore, methods can be utilized to solve imbalanced dataset in actual application and further improve the accuracy of device recognition. Common solutions include extending dataset, resampling dataset, and manually generating samples. Besides, when applied to other datasets and features, the classification results may be different. In application, it is wise to accordingly select and tune models to achieve satisfactory device recognition performance.

7.4.2 Conclusion

The chapter analyzes the ensemble learning methods in device recognition. Device recognition is essentially a multi-classification problem based on multi-dimensional features. In the chapter, the current data of 1793 instances are utilized. It includes 11 different devices in total and the models need to classify these devices. To generate useful features from current time series, period extraction and feature extraction methods are proposed. Nine different features are considered, including maximum, minimum, range, mean, median, variance, skewness, kurtosis, and sample entropy. Based on these extracted features, four base classifiers (DT, KNN, SVM, NBM), eight optimized weighting ensemble models (PSO, GA, DE, FA, BOA, BBKH, and GWO), and two boosting ensemble models (AdaBoost and LPBoost) are established. Based on the experimental results, the conclusion can be drawn as follows:

(a) The extracted features can properly represent the characteristics of original current time series. Among them, kurtosis and sample entropy show higher importance estimates.
(b) Different base classifiers show diverse classification performance towards different devices. DT and SVM are relatively better than KNN and NBM in terms of overall results.
(c) In optimized weighting ensemble models, the multi-objective algorithm significantly outperforms single-objective algorithms. Considering several objectives can consider multiple aspects of the model and effectively enhance classification performance.
(d) Among all ensemble models, the AdaBoost-tree model is the most recommended for device recognition. It has the most accurate classification performance for all 11 types of devices.

References

Aladesanmi, E. J., & Folly, K. A. J. I. P. (2015). Overview of non-intrusive load monitoring and identification techniques. *IFAC PapersOnLine, 48*(30), 415–420.

Breiman, L. (1996). Bagging predictors. *Machine learning, 24*(2), 123–140.

Buddhahai, B., Wongseree, W., & Rakkwamsuk, P. (2018). A non-intrusive load monitoring system using multi-label classification approach. *Sustainable Cities and Society, 39,* 621–630.

Chang, H. H., Lian, K. L., Yi-Ching, S., & Lee, W. J. J. (2014). Power-spectrum-based wavelet transform for nonintrusive demand monitoring and load identification. *IEEE Transactions on Industry Applications, 50*(3), 2081–2089.

Chang, H. H., Lin, L. S., Chen, N., & Lee, W J J I To I A. (2013). Particle-swarm-optimization-based nonintrusive demand monitoring and load identification in smart meters. *IEEE Transactions on Industry Applications, 49*(5), 2229–2236.

Du, L., Restrepo, J. A., Yang, Y., Harley, R. G., & Habetler, T G J I To S G. (2013). Nonintrusive, self-organizing, and probabilistic classification and identification of plugged-in electric loads. *IEEE Transactions on Smart Grid, 4*(3), 1371–1380.

Gohari, M., & Eydi, A. M. (2020). Modelling of shaft unbalance: Modelling a multi discs rotor using K-Nearest neighbor and decision tree algorithms. *Measurement, 151,* 107253. https://doi.org/10.1016/j.measurement.2019.107253.

Hastie, T., Rosset, S., Zhu, J., & Zou, H. (2009). Multi-class adaboost. *Statistics and its Interface, 2*(3), 349–360.

Himeur, Y., Alsalemi, A., Bensaali, F., & Amira, A. (2020). Robust event-based non-intrusive appliance recognition using multi-scale wavelet packet tree and ensemble bagging tree. *Applied Energy, 267,* 114877. https://doi.org/10.1016/j.apenergy.2020.114877.

Liu, H., & Chen, C. (2020). Spatial air quality index prediction model based on decomposition, adaptive boosting, and three-stage feature selection: A case study in China. *Journal of Cleaner Production, 265,* 121777. https://doi.org/10.1016/j.jclepro.2020.121777.

Michael, B., & Jiirgen, V. (2004). Genetic algorithm for pattern detection in nialm systems. *IEEE SMC, 4,* 3462.

Nalmpantis, C., & Vrakas, D. (2018). Machine learning approaches for non-intrusive load monitoring: From qualitative to quantitative comparation. *Artificial Intelligence Review, 2,* 1–27.

Sun, L., Zhou, K., & Yang, S. (2020). An ensemble clustering based framework for household load profiling and driven factors identification. *Sustainable Cities and Society, 53,* 101958. https://doi.org/10.1016/j.scs.2019.101958.

Thorsten, J. (2006). Training linear SVMs in linear time. Paper presented at the Proceedings of the ACM SIGKDD International Conference on Knowledge Discovery and Data Mining.

Yang, C. C., Soh, C. S., & Yap, V. V. (2017). A non-intrusive appliance load monitoring for efficient energy consumption based on Naive Bayes classifier. *Sustainable Computing: Informatics and Systems, 14,* 34–42. https://doi.org/10.1016/j.suscom.2017.03.001.

Chapter 8
Smart Non-intrusive Device Recognition Based on Deep Learning Methods

8.1 Introduction

The development of technology and science has been very rapid recently, the lifestyle of people has undergone tremendous changes. The ever-changing electronic products not only bring convenience to humans but also increase energy consumption, which posed big challenges to the power management system in the new era. The traditional electrical monitoring system only records the total power consumption of users and cannot provide equipment-level electrical power consumption information, which is not conducive to the energy consumption management and equipment failure detection of the users (Zoha et al. 2012). Non-intrusive device recognition (Wang and Zheng 2012) is a new type of intelligent device monitoring technology, which recognizes the types of the indoor device by collecting and analyzing the power data of the general interface. In this way, users can know the operating status of the indoor device, and when an electrical fault occurs, it is helpful to accurately locate the faulty equipment. On this basis, further statistics on the energy consumption of each device can provide users with personalized energy-saving solutions. The classic non-intrusive device identification framework usually includes three steps: data collection, feature extraction, and device classification (Buddhahai et al. 2018).

In recent years, deep learning (LeCun et al. 2015; Schmidhuber 2015) technology has made breakthroughs in many fields such as speech (Hinton et al. 2012), image (Dong et al. 2016; Litjens et al. 2017) and natural language processing (Aldabbas et al. 2021; Collobert et al. 2011). In 2012, *Hinton* and his students build a convolutional neural network on the current largest image database ImageNet and achieve a significant reduction in the top 5 error rate, which dropped from 26 to 15% (Hinton et al. 2012). The network automatically clusters the images in the training video and successfully realizes the recognition of cats without external interference. In the field of speech recognition, compared to the Gaussian mixture model (GMM) used in traditional acoustic models, deep learning models have achieved an error rate reduction of about 30%. Deep learning is considered to be the smart learning method closest

H. Liu et al., *Smart Device Recognition*,
https://doi.org/10.1007/978-981-33-4925-4_8

to the human brain so far, and its emergence has brought the possibility of breakthroughs in many traditional technologies. Deep learning networks usually consist of a series of hidden layers, the Convolutional Neural Networks (CNN) (Zhao et al. 2018), Auto-Encoder (AE) (Liu et al. 2017), Sparse Auto-Encoder (SAE) (Zhang et al. 2018), Restricted Boltzmann Machines (RBM) (Odense and Edwards 2016), etc. are common hidden layer components

Extremely strong feature extraction ability is one of the important reasons for the success of deep learning. The introduction of it is expected to improve non-intrusive device recognition. At present, there have been some achievements of deep learning applicated in device recognition. (Wu et al. 2020a) establish a database of various types of devices through load decomposition and then classify the decomposed waveforms through a CNN model to achieve equipment recognition. (Faustine and Pereira 2020a) proposed a device identification method based on Recursive Graph (RG) technology and CNN. The raw data of the devices can be processed to generate a Weighted Recursive Graph (WRG), which contains the characteristic information of the devices. The device classification can be realized by using CNN to recognize it. In addition, (Faustine and Pereira 2020b) also proposed a multi-label learning method based on CNN. The method first disaggregates the current features into active and reactive components by using the Fryze power theory and then realizes the conversion of the current signal to the image through the Euclidean distance similarity function. Finally, the CNN network is used to identify these images. As computer hardware and deep learning algorithms continue to develop, the accuracy of deep learning-based device recognition models will be further improved.

8.2 Deep Learning Device Recognition Method Based on Load Sequence Input

A series of electrical quantities including active power, reactive power, harmonics, etc., will be generated while the electrical device is running, which contains the characteristic information of the electrical device. By mining the equipment feature information in these sequences and classifying them, the non-intrusive device identification task can be realized. The power data during the operation of the device is used as the original feature to achieve non-intrusive device identification, and the results of different models will be given.

8.2.1 *Non-intrusive Device Identification Based on RNN Network*

1. *Algorithm introduction*
   Recurrent Neural Network (RNN) (Zhang et al. 2019; Ren et al. 2019) is a type of neural network, which is good at mining the characteristics of time series. RNN is different from ordinary feedforward neural network in that it has a recursive structure. In the RNN model, the output at the current moment not only depends on the input at the current moment but also is related to the input at all historical moments. In theory, RNN has the ability to model arbitrary length sequences, so it is very suitable for processing sequence data. Figure 8.1 shows the simplest unidirectional RNN structure with a hidden layer. RNN will process the input sequence in turn. At each moment, RNN will update the model parameters based on the input information at the current time and the previous time.
   Defining W_{ih} as the connection matrix from the input layer to the hidden layer, W_{hh} as the connection matrix between the adjacent moments of the hidden layer, and W_{ho} as the connection matrix from the hidden layer to the output layer. The state vector of the hidden layer of RNN at time k is denoted by h_k (Wu et al. 2020b):

$$h_k = f_h(w_{ih}x_k + w_{hh}h_{k-1} + b_h) \tag{8.1}$$

where b_h is the bias vector while f_h is the nonlinear activation function. In RNN, f_h is usually a sigmoid or tanh function. Therefore, the output at time k (o_k) can be calculated by the following formula:

$$o_k = f_o(w_{ho}h_k + b_o) \tag{8.2}$$

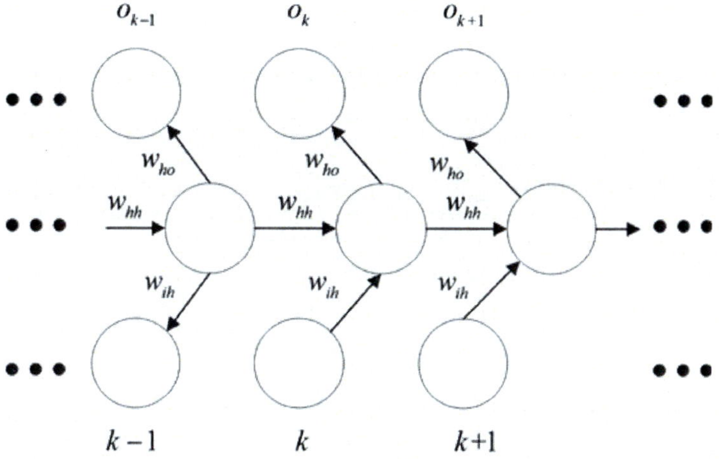

Fig. 8.1 The structure of the RNN with a hidden layer

2. *Model and experiment*

   In this section, the RNN is used to realize the device identification, and the parameters of the model are shown in Table 8.1. The PLAID (Gao et al. 2014) data set is used as the experimental data. A total of 1074 samples with 11 types of devices is contained in the PLAID dataset. Before data, the data in the dataset is preprocessed. After excluding the abnormal data in the data set, five electrical devices were selected as identification objects, namely Air Conditioner (AC), heater, laptop, vacuum, and Washing Machine (WM). The active power sequence of each device during steady-state operation is used as the model input, and the final output is the identification result of each device. For each sample, the sampling frequency is 1 Hz, and the total length of each sample is 2000. To avoid data imbalance in the data set, the resampling technology is used to expand the number of samples of the device with a small sample size, and Table 8.2 shows the final sample quantity of each device.

   In the experiment, 70% of the samples of each electrical device are divided into the training set, while the rest is the test set. The training results are shown in Fig. 8.2.

   According to Fig. 8.2, the training accuracy of the model is close to 100%, which shows that the model training is successful. After inputting the test set into the model, the recognition results can be obtained. Table 8.3 shows the recognition results of the RNN model, and Fig. 8.3 shows its confusion matrix. In the confusion matrix, the abscissa represents the prediction result, and the ordinate represents the label value.

   As seen in Table 8.3 and Fig. 8.3, the model has good recognition accuracy, with an average accuracy rate of 90.07%. The average recall, average precision, and average F1-Score are 0.9329, 0.9067, and 0.9043 respectively. The model has the highest recognition accuracy of vacuum, reaching 100, and has the lowest recognition accuracy for air conditioner, with an accuracy rate of only 91.33%. In the recognition process, there were many errors in the recognition of the heater, a total of 24 samples of the heater were incorrectly identified as air conditioners. This shows that the model's ability to distinguish the heater needs to be further improved. In general, the model can complete basic electrical device identification tasks.

Table 8.1 The parameters of the model

The number of layers	3
Number of neurons in each layer	30\50\30
Learning rate	0.001
Number of iterations	50

Table 8.2 The number of samples of the device

The type of device	AC	Heater	Laptop	Vacuum	WM
The number of samples	200	200	200	200	200

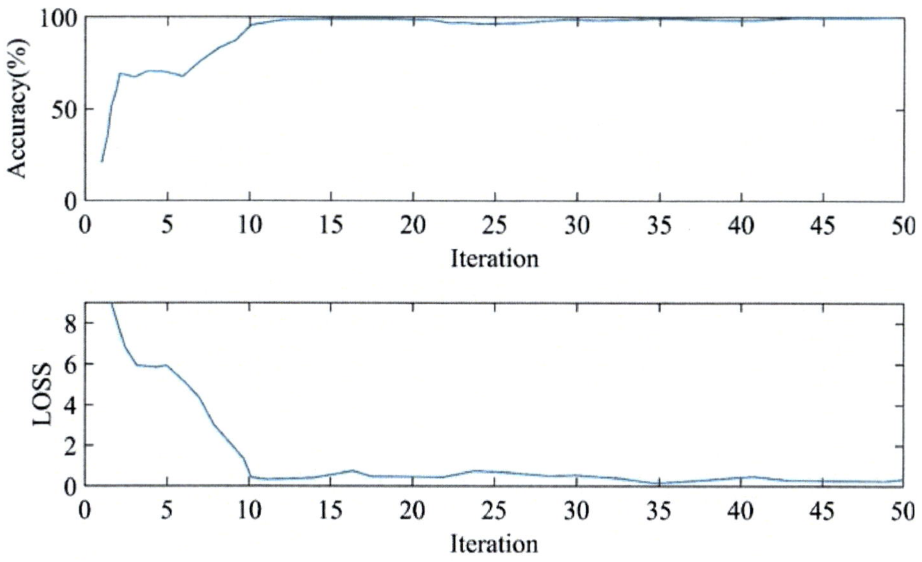

Fig. 8.2 The training results of the RNN model

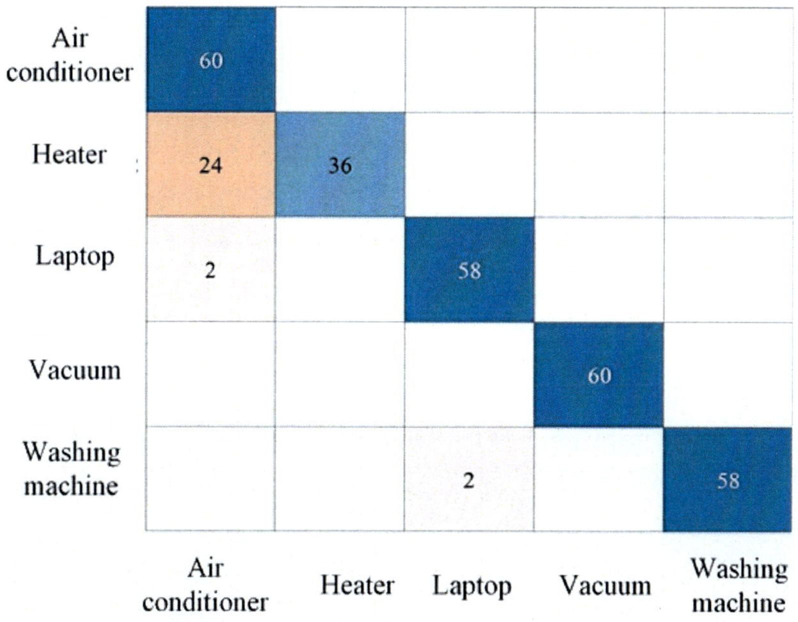

Fig. 8.3 The confusion matrix of the RNN model

Table 8.3 The recognition results of the RNN model

RNN		The type of device					
		Air conditioner	Heater	Laptop	Vacuum	Washing machine	Average
Evaluation index	Accuracy	0.9133	0.9200	0.9867	1.0000	0.9933	0.9067
	Recall	0.6977	1.0000	0.9667	1.0000	1.0000	0.9329
	Precision	1.0000	0.6000	0.9667	1.0000	0.9667	0.9067
	F1-Score	0.8219	0.7500	0.9667	1.0000	0.9831	0.9043

8.2.2 Non-intrusive Device Identification Based on LSTM Network

1. *Algorithm introduction*

Long Short-Term Memory network (LSTM) (Zhu et al. 2018) is one of the commonly used deep learning networks. LSTM is a variant of RNN, which inherits the advantages of RNN, and the structure of the hidden layer of the LSTM was improved (Sherstinsky 2020). By adding 4 multiplication units and multiple time series state storage units, the LSTM effectively prolongs the memory cycle of the network (Smagulova and James 2019). LSTM has an excellent performance in data processing, especially time-series processing. Figure 8.4 shows the structure of the LSTM network.

Four multiplication units form the hidden layer of the LSTM model, and the mathematical expressions of them are given in formulas 8.3 to 8.6, respectively (Mbatha and Bencherif 2020).

$$f_t = \sigma\psi_{f,t} = \sigma\left(w_f\left[h_{t-1}, x_t\right]^T + b_f\right) \tag{8.3}$$

The mathematical expression of the forget gate is given by formula 8.3. Where f_t is the output, σ is the activation coefficient, $\psi_{f,t}$ is the input of the neuron, w_f is the weight matrix, h_{t-1} is the output of the hidden layer at a time point, x_t is the input at the current time point, and b_f is the bias matrix.

$$i_t = \sigma\psi_{i,t} = \sigma\left(b_i + w_i\left[x_t, h_{t-1}\right]^T\right) \tag{8.4}$$

$$\tilde{c}_t = \tanh\psi_{\tilde{c},t} = \sigma\left(b_c + w_c\left[x_t, h_{t-1}\right]^T\right) \tag{8.5}$$

The mathematical expression of the input gate is given by formula 8.4 and 8.5. Where i_t is the scale factor saved to the state storage unit; $\tilde{c}_t$ is the modified state storage unit; w_c and w_i are the corresponding weight matrices, and b_c, b_i are the corresponding bias matrices.

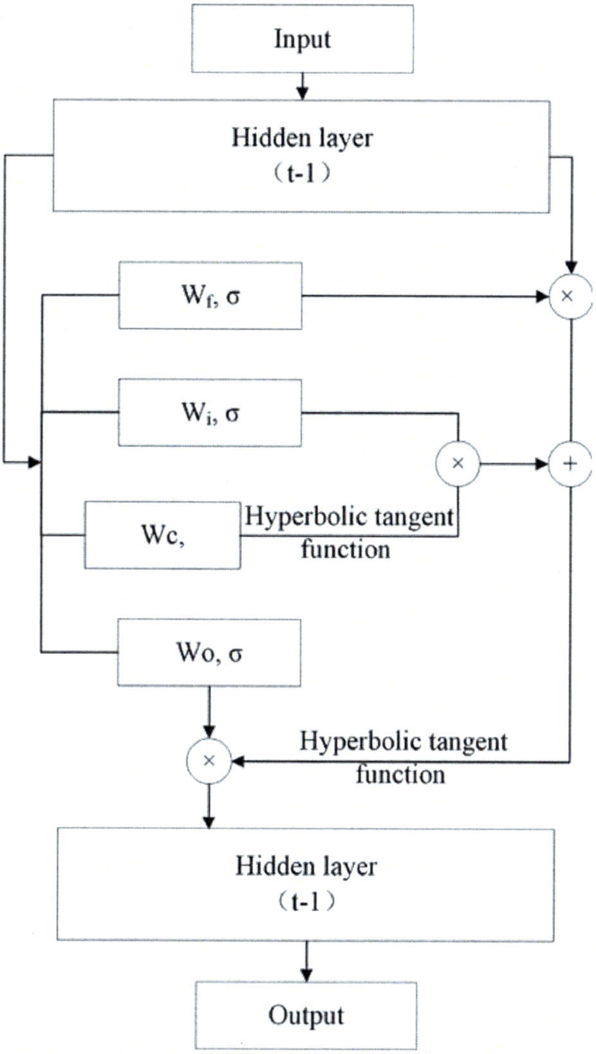

Fig. 8.4 The typical structure of the LSTM network

$$c_t = f_t c_{t-1} + i_t \tilde{c}_t \tag{8.6}$$

The mathematical expression of the update gate is given by formula 8.6. Where c_t represents the state storage unit, and $c_{t\text{-}1}$ represents the state storage unit at previous.

$$o_t = \sigma \psi_{o,t} = \sigma \left(b_o + w_o [x_t, h_{t-1}]^T \right) \tag{8.7}$$

$$h_t = \tanh(c_t) \bullet o_t \tag{8.8}$$

The mathematical expression of the update gate is given by formula 8.7 and 8.8 (Wen et al. 2019). Where o_t represents the output value after the output gate is modified; h_t represents the final hidden layer output; w_o represents the corresponding weight matrix, and b_o represents the bias matrix.

2. *Model and experiment*
   In this section, the LSTM network is constructed to identify the different devices, and the parameters of it are shown in Table 8.4. The input of the model is the active power data of the device during steady-state operation, and the type and number of the samples are consistent with the RNN model in Sect. 8.2.1.
   In the experiment, 70% of the samples of each electrical device are divided into the training set while the rest is the test set. The training results of the model are shown in Fig. 8.5.
   As shown in Fig. 8.5, the model's accuracy continues to rise during the iteration process, and eventually approaches 100%. There is no overfitting phenomenon in model training, which shows that the model is reliable. Table 8.5 shows the recognition results of the LSTM model, and Fig. 8.6 shows its confusion matrix. In the confusion matrix, the abscissa represents the prediction result, and the ordinate represents the label value.

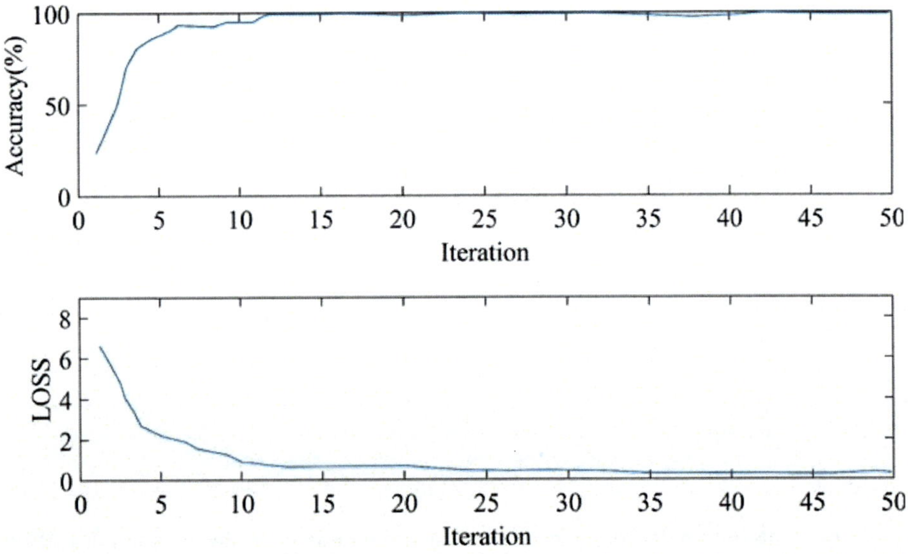

Fig. 8.5 The training results of the LSTM model

Table 8.4 The parameters of the model

The number of layers	3
Number of neurons in each layer	30\50\30
Learning rate	0.001
Number of iterations	50

Table 8.5 The recognition results of the LSTM model

LSTM		The type of device					
		AC	Heater	Laptop	Vacuum	Washing machine	Average
Evaluation index	Accuracy	0.9700	1.0000	0.9800	0.9867	0.9967	0.9667
	Recall	0.9180	1.0000	1.0000	0.9375	0.9836	0.9678
	Precision	0.9333	1.0000	0.9000	1.0000	1.0000	0.9667
	F1-Score	0.9256	1.0000	0.9474	0.9677	0.9917	0.9665

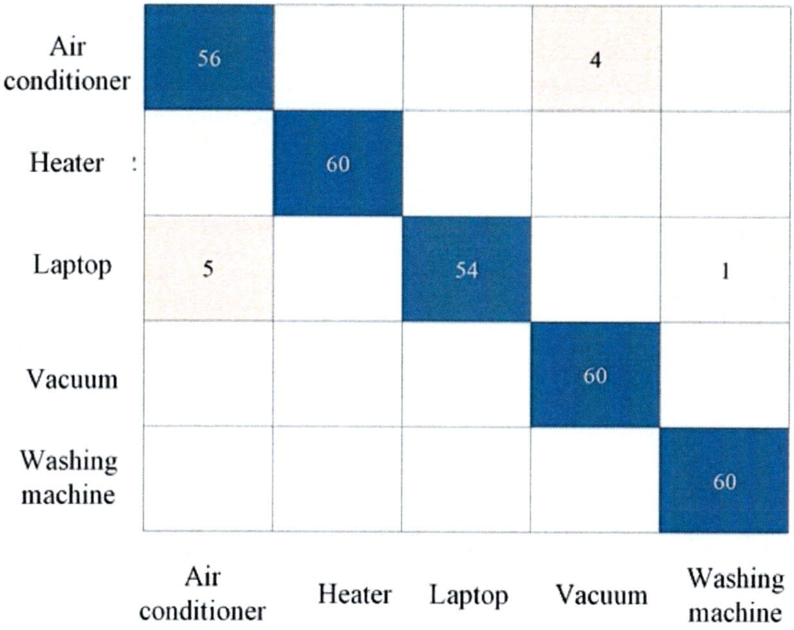

Fig. 8.6 The confusion matrix of the LSTM model

From Table 8.5, compared to the RNN model, the accuracy of the LSTM model has been significantly improved. The reason for this is that the structure of LSTM has better adaptability to time series. The average accuracy, average recall, average precision, and average F1-Score of the model are 0.9667, 0.9678, 0.9667, and 0.9665 respectively, which shows the model has high accuracy. The model has the highest recognition accuracy for heaters, which is 100%, and the lowest recognition accuracy for air conditioners, which is 97.00%. For all five electrical devices, the model can give high-precision recognition results, which shows that the model can efficiently complete the electrical device identification task.

8.2.3 *Non-intrusive Load Identification Based on GRU Network*

(1) *Algorithm introduction*
Compared with RNN, LSTM has long-term memory function and controllable memory ability, but the structure of the LSTM network which has multiple gate structures is too complicated to reduce model efficiency. In 2014, the Gated Recurrent Unit network (GRU) was developed which is a variant of the LSTM network. Figure 8.7 shows the structure of a typical GRU network hidden layer, compared with LSTM, GRU (Noh et al. 2020) simplifies the structure of LSTM. The GRU has fewer parameters and reduces its calculation (Pan et al. 2018). The GRU model has two gate control units, including the update gate, and the reset gate. The GRU network does not retain internal memory, which makes it more efficient than traditional LSTM models. The mathematical model of GRU is shown in the following formula (Pan et al. 2020):

$$z_t = \sigma\left(h_{t-1}W^{(z)} + x_t U^{(z)} + b_z\right) \tag{8.9}$$

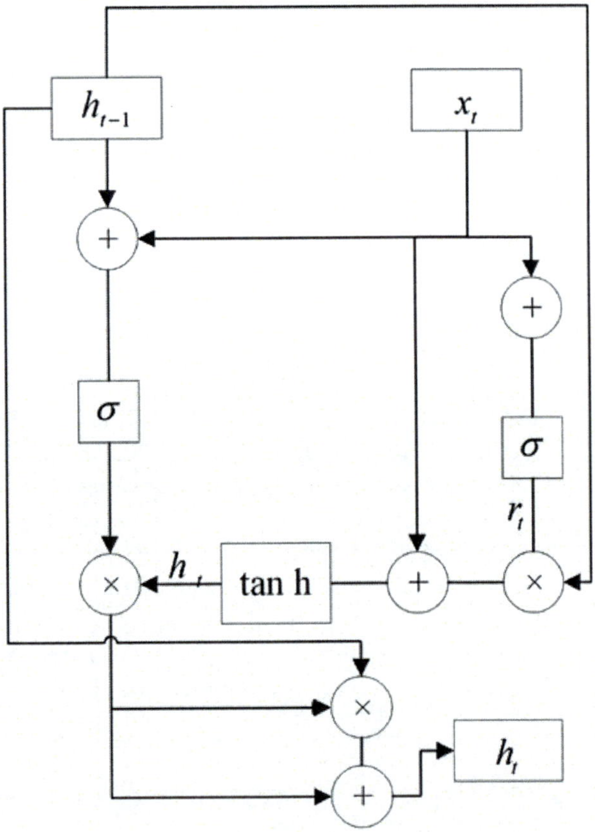

Fig. 8.7 The structure of the GRU neural network

Table 8.6 The parameters of the model

The number of layers	3
Number of neurons in each layer	30\50\30
Learning rate	0.001
Number of iterations	50

$$r_t = \sigma\left(h_{t-1}W^{(r)} + x_tU^{(r)} + b_r\right) \tag{8.10}$$

$$h_t^{'} = \tanh\left(r_t * W^{(h)}h_{t-1} + W^{(x)}x_t + b_h\right) \tag{8.11}$$

$$h_t = h_{t-1} - z_th_{t-1} + z_t * h_t^{'} \tag{8.12}$$

where z_t is the output of the network at time t, r_t is the input of the network at time t, z_t and r_t are the output of the update gate and reset gate respectively. $W^{(z)}$ is the weight matrice of the update gate output at t-1, $U^{(z)}$ is the weight matrice of the update gate input at t, and the results of the above two gates will be mapped to [0,1] through a sigmoid function. $h_t^{'}$ and h_t respectively represent the candidate hidden state and hidden state, which determines the proportion of information that is forgotten and saved in the hidden state h_{t-1} at the previous moment. $W^{(r)}$ is the weight matrice of the reset gate output at t-1 and $U^{(r)}$ is the weight matrice of the reset gate input at t. The candidate hiding result will be obtained through a tanh function. b_z,b_r and b_h represent the offset of update gate, reset gate, and candidate hidden state, respectively.

(2) *Model and experiment*

In this section, the GRU is used to realize the device identification. The GRU is simpler in structure than the LSTM model, which helps to improve the model efficiency. The parameters of the GRU model are given in Table 8.6, and the data used by the GRU model is consistent with the RNN model in Sect. 8.2.1 (the detailed information of the samples is shown in Table 8.2). The division of training and test sets is consistent with that in Sect. 8.2.1, the ratio is 7:3. The training results are shown in Fig. 8.8.

As shown in Fig. 8.8, the training accuracy is close to 100% at the 30th iteration. In the training process, there is no over-fitting phenomenon. The training of the model is successful. Using the test set to test the model performance, Table 8.7 shows the recognition results, and Fig. 8.9 shows its confusion matrix. In the confusion matrix, the abscissa represents the prediction result, and the ordinate represents the label value.

From Table 8.7, the average accuracy, average recall, average precision and average F1-Score are 0.9633, 0.9766, 0.9633, and 0.9630 respectively. The accuracy of the model is comparable to the LSTM model, but the GRU model has advantages in operating efficiency. The model has the highest recognition accuracy of the washing machine, reaching 100, and has the lowest recognition

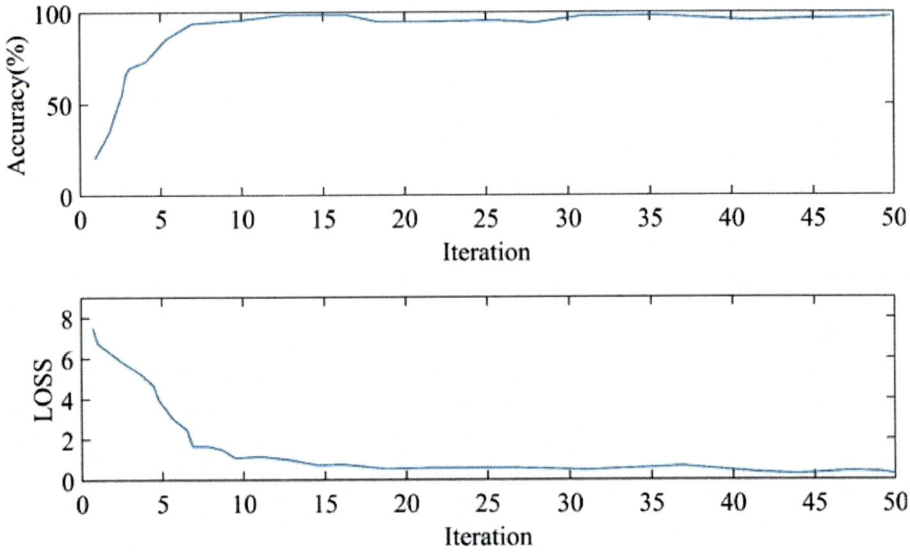

Fig. 8.8 The training results of the GRU model

Table 8.7 The recognition results of the GRU model

GRU		The type of device					
		AC	Heater	Laptop	Vacuum	Washing machine	Average
Evaluation index	Accuracy	0.9700	0.9867	0.9767	0.9933	1.0000	0.9633
	Recall	0.9180	0.9375	1.0000	0.9677	1.0000	0.9766
	Precision	0.9333	1.0000	0.8833	1.0000	1.0000	0.9633
	F1-Score	0.9256	0.9677	0.9381	0.9836	1.0000	0.9630

accuracy for air conditioner, with an accuracy rate of 97.00%. In general, the model can effectively distinguish all kinds of electrical devices, and the false recognition rate is extremely low.

8.3 Deep Learning Device Recognition Method Based on Graph Processing

8.3.1 Data Conversion

In Sect. 8.2, the original input of the device recognition model is the load series. The images can also be used as input to the model to realize device recognition. In this section, the original load sequence of the equipment is converted into a V-I image to realize device identification. The original data used in the model is consistent

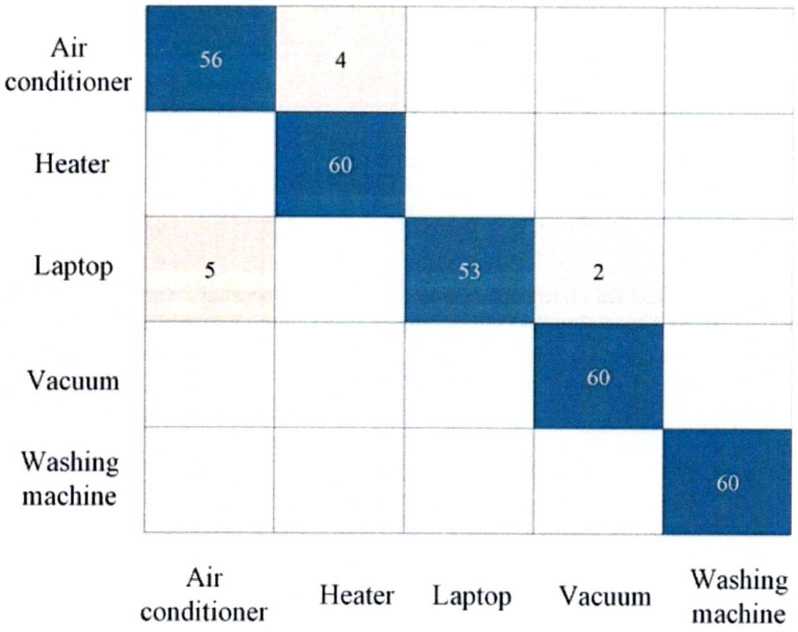

Fig. 8.9 The confusion matrix of the GRU model

with that in Sect. 8.2, and five electrical devices including air conditioner, heater, laptop, vacuum, and washing machine are selected as identification objects. For each device, extracting 2000 voltage signal points and current signal points during stable operation of the device, and the sampling frequency is 1 Hz. After the data were normalized to $[-1, 1]$, the current will be taken as the abscissa while the voltage will be taken as the ordinate. Figure 8.10 shows the V-I images of the five devices.

8.3.2 Non-intrusive Device Identification Based on CNN

1. *Algorithm introduction*
   Deep learning is a collection of algorithms, which belong to machine learning. It is intended to simulate the human brain to obtain the ability to analyze and learn (Cosgriff and Celi 2020). The deep learning network refers to the learning mechanism of the human brain to interpret the data, trying to establish a model for transmitting and processing information like brain neurons. In general, deep learning divides complex problems into steps, and each step is represented by a network layer. Under the same conditions, the deeper the network, the more parameters it needs.
   CNN (Li et al. 2021) has strong learning ability, it can realize the translation-invariant classification of input information, so it is also called "shift-invariant artificial neural network (SIANN)". It has been utilized in many fields (Tian

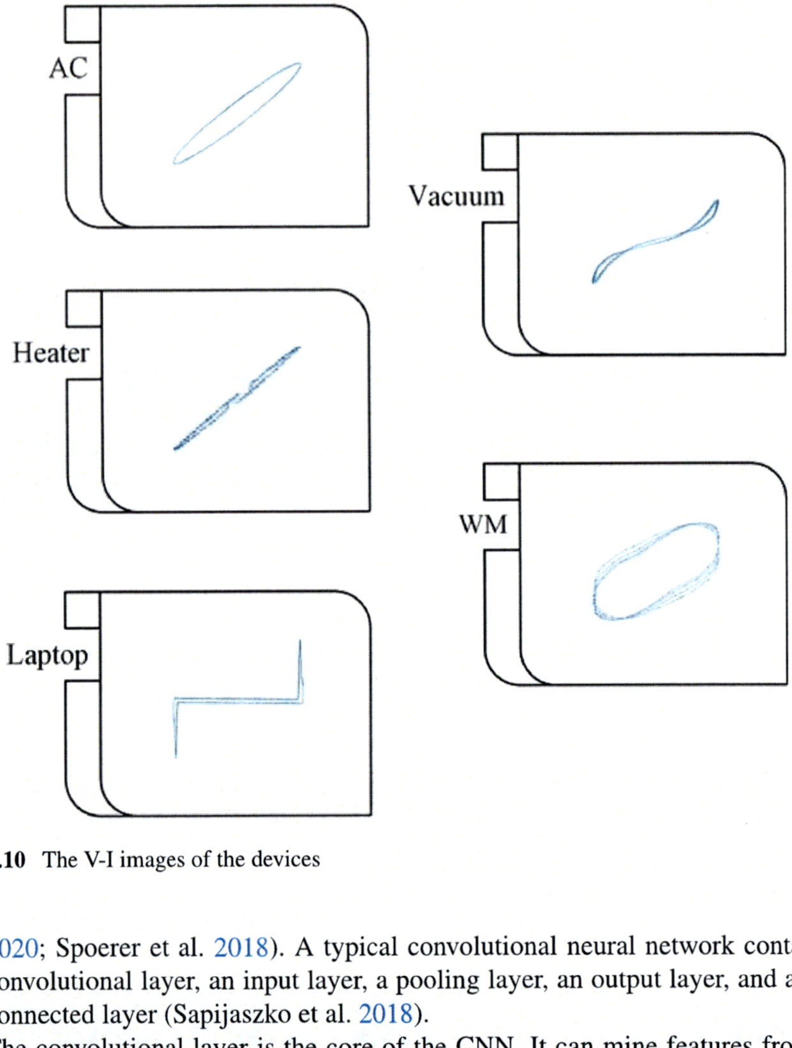

Fig. 8.10 The V-I images of the devices

2020; Spoerer et al. 2018). A typical convolutional neural network contains a convolutional layer, an input layer, a pooling layer, an output layer, and a fully connected layer (Sapijaszko et al. 2018).

The convolutional layer is the core of the CNN. It can mine features from the input data. When the model is working, the input features will be scanned regularly by the convolution kernel and the operation of the matrix elements will be performed in the receiving domain. The mathematical formula of the two-dimensional convolution kernel is as follows (Hou et al. 2018):

$$\begin{aligned} Z^{l+1}(i,j) &= [Z^l \otimes w^{l+1}](i,j) + b \\ &= \sum_{k=1}^{K_l} \sum_{x=1}^{f} \sum_{y=1}^{f} [Z_k^l(s_0 i + x, s_0 j + y) w_k^{l+1}(x,y)] + b(i,j) \in \{0,1,\ldots L_{l+1}\}, L_{l+1} \\ &= \frac{L_l + 2p - f}{s_0} + 1 \end{aligned} \tag{8.13}$$

Table 8.8 The structure of the CNN

Layer1	Input
Layer2	Convolutional layer 3 × 3
Layer3	Convolutional layer 3 × 3
Layer4	Max pooling
Layer5	Fully connected layer
Layer6	Output

where b represents the deviation vector, Z^l is the input of the $(l + 1)$-th layer, Z^{l+1} is the output of the $(l + 1)$-th layer. Assuming that the length and width of output are equal, L_{l+1} represents the size of $Z_{l+1}.Z(i, j)$ corresponds to the pixel in the i-th row and j-th column of the feature map, and K is the number of channels. f is the convolution kernel size, s_0 is the strided convolutions, and p is padding value.
The calculation method of one-dimensional convolution and three-dimensional convolution is similar to that of two-dimensional convolution. When the convolution kernel size f = 1, the strided convolutions and the value of padding are both 0, the cross-correlation calculation in the convolution layer is equivalent to matrix multiplication. From this, the mathematical expression of the fully connected layer can be derived.

2. *Model and experiment*
The V-I image processed in Sect. 8.3.1 is used as the model input, and the traditional CNN model is used to realize electrical device identification. The structure of the CNN used in this chapter are shown in Table 8.8, and the type and number of samples used in the model are shown in Table 8.2 in Sect. 8.2.1.
The network is mainly composed of two consecutive convolutional layers. As shown in Table 8.8, the first convolution layer contains 32 convolution kernels with 3 × 3 size, while the second convolution layer contains 64 convolution kernels with 3 × 3 size, and the step size of them is 1. The pooling layer uses the maximum pooling method. The number of neurons in the fully connected layer contains is 5, corresponding to five devices. The input layer size is 200 × 200 × 3, which is compatible with the image size used. Seventy percent of the images are used as the training set. The learning rate is set to 0.001 during training, and the number of iterations is 50. The model training process is given by Fig. 8.11. It can be seen from Fig. 8.11 that there is no over-fitting phenomenon during the training process. Table 8.9 shows the recognition results, and Fig. 8.12 shows its confusion matrix of the CNN model. In the confusion matrix, the abscissa represents the prediction result, and the ordinate represents the label value.
It can be obtained from Table 8.9 that the average accuracy, average recall, average precision, and average F1-Score are 0.9267, 0.9463, 0.9627, and 0.9284 respectively. From the confusion matrix in Fig. 8.12, during the model recognition process, the vacuum and the air conditioner are confused. There are 15 samples of the vacuum that were incorrectly identified as the air conditioner.

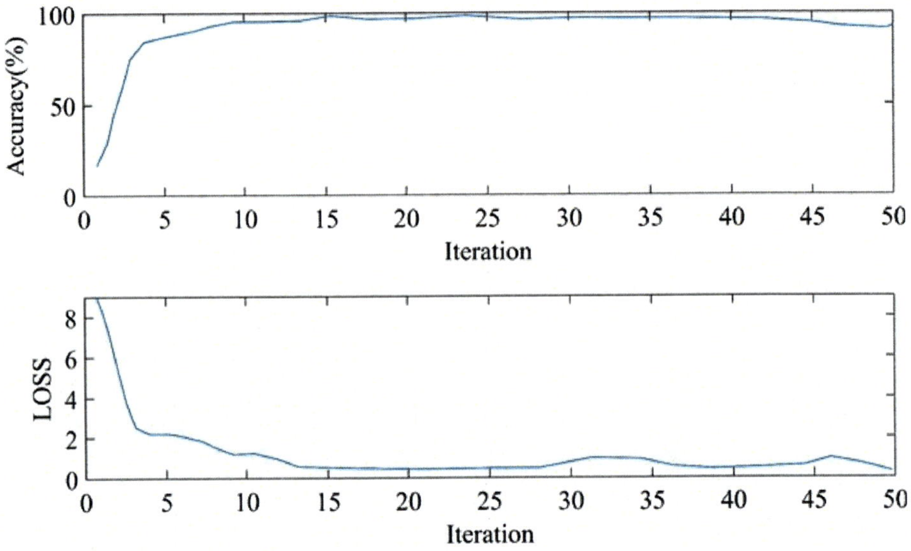

Fig. 8.11 The training results of the CNN model

Table 8.9 The recognition results of the CNN model

CNN		The type of device					
		AC	Heater	Laptop	Vacuum	Washing machine	Average
Evaluation index	Accuracy	0.9267	1.0000	0.9833	0.9500	0.9933	0.9267
	Recall	0.7317	1.0000	1.0000	1.0000	1.0000	0.9463
	Precision	1.0000	1.0000	0.9167	0.7500	0.9667	0.9267
	F1-Score	0.8451	1.0000	0.9565	0.8571	0.9831	0.9284

8.3.3 Non-intrusive Device Identification Based on AlexNet

1. *Algorithm introduction*
   AlexNet is a typical deep convolutional network and has achieved excellent results in the field of image classification (Zhao et al. 2019; Han et al. 2017). Compared with the traditional convolutional network, the structure of AlexNet is more complicated. It contains 650,000 neurons in total. Figure 8.13 shows its structure.
   AlexNet has a huge amount of parameters. In order to better train the network, some adjustments to the network were made by (Krizhevsky et al. 2012). The first part is the improvement of the activation function. Traditional convolutional networks usually choose logarithmic functions, tanh functions, arctan functions, etc. as their activation functions, but in deep models, these functions often encounter the problem of gradient disappearance. To overcome this problem, AlexNet uses a rectified linear unit as its activation function, called the ReLU

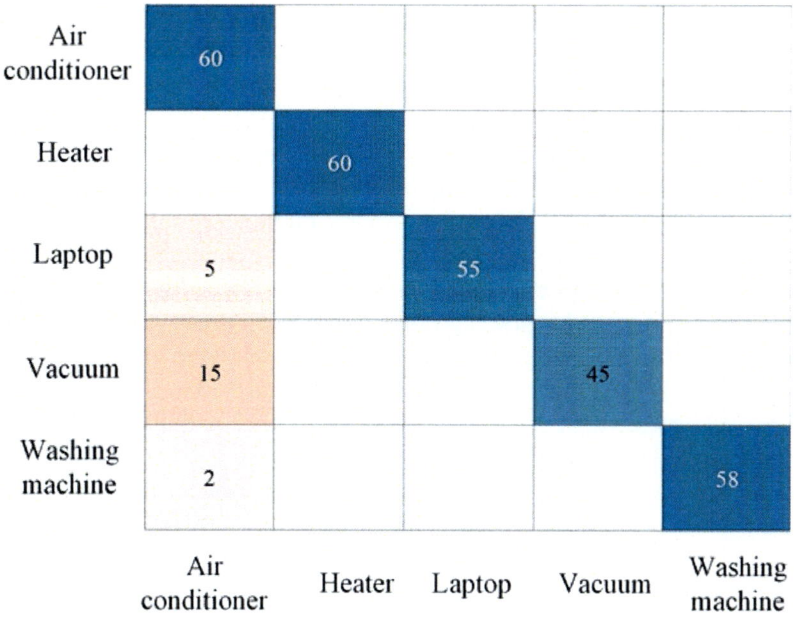

Fig. 8.12 The confusion matrix of the CNN model

Fig. 8.13 The structure of AlexNet

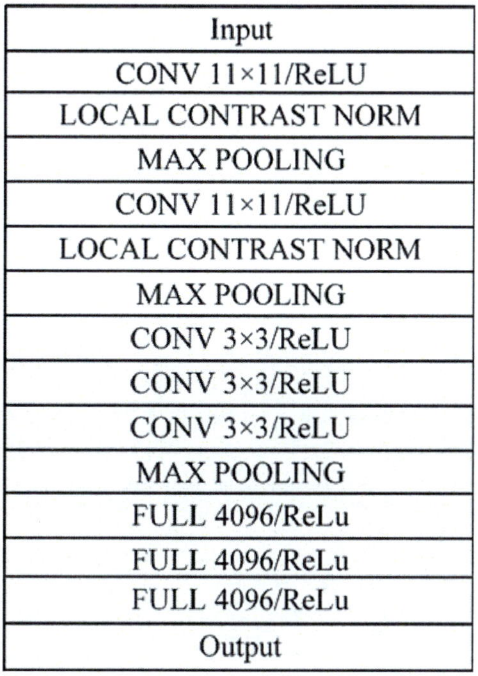

layer. The definition of the ReLU layer is as follows:

$$ReLU(x) = \max(x, 0) \tag{8.14}$$

In addition, the method of dropout is used to avoid overfitting during model training. In this method, only a part of the neurons is trained in each iteration, which forces the neurons to cooperate. Convolution technology provides a way for models to extract features from images, and the parameter sharing feature of the convolutional network reduces the complexity. In addition, the network also uses a pooling layer for pooling operations, which greatly reduces the feature dimension. This also reduces the difficulty of training the model. AlexNet won the championship in the ImageNet competition, and the accuracy of AlexNet was about 10% higher than the best method on the benchmark data set at that time.

2. Model and experiment
   To improve the recognition accuracy, a device recognition model based on AlexNet was established. Compared with the CNN model in Sect. 8.3.2, AlexNet has a deeper network structure. The experimental data used in this model and the partitioning method of the training set and test set are the same as those in *Sect.* 8.3.2. The size of the device V-I images was adjusted to 227 × 227 × 3. The training process is given by Fig. 8.14.
   The structure of the AlexNet network is more complicated, and only using experimental data sets may lead to incomplete model training. Therefore, the transfer learning method (Lu et al. 2015; Taylor and Stone 2009) is introduced, and the network data set ImageNet is first tried to pre-train it, and then the experimental data is used for the second training to complete the model training process. The

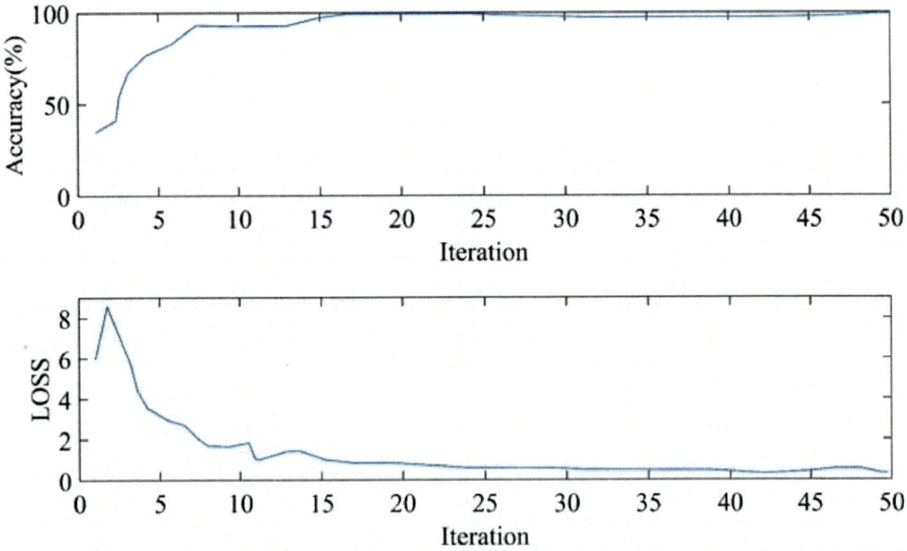

Fig. 8.14 The training results of the AlexNet model

learning rate and the number of iterations during training are 0.001 and 50 respectively. Table 8.10 shows the recognition results of the RNN model, and Fig. 8.15 shows its confusion matrix. In the confusion matrix, the abscissa represents the prediction result, and the ordinate represents the label value.

From Table 8.10 and Fig. 8.15, the AlexNet model has a higher recognition accuracy than the CNN model in Sect. 8.3.2, which shows that increasing the depth of the convolutional network can improve the accuracy of the model. However, the accuracy of the model has not improved much, the average accuracy, average recall, average precision, and average F1-Score are 0.9433, 0.9458, 0.9433, and 0.9435 respectively. Using a deeper convolutional network may improve the model's accuracy.

Table 8.10 The recognition results of the AlexNet model

AlexNet		The type of device					
		AC	Heater	Laptop	Vacuum	Washing machine	Average
Evaluation index	Accuracy	0.9500	0.9533	0.9900	1.0000	0.9933	0.9433
	Recall	0.8358	0.9259	0.9672	1.0000	1.0000	0.9458
	Precision	0.9333	0.8333	0.9833	1.0000	0.9667	0.9433
	F1-Score	0.8819	0.8772	0.9752	1.0000	0.9831	0.9435

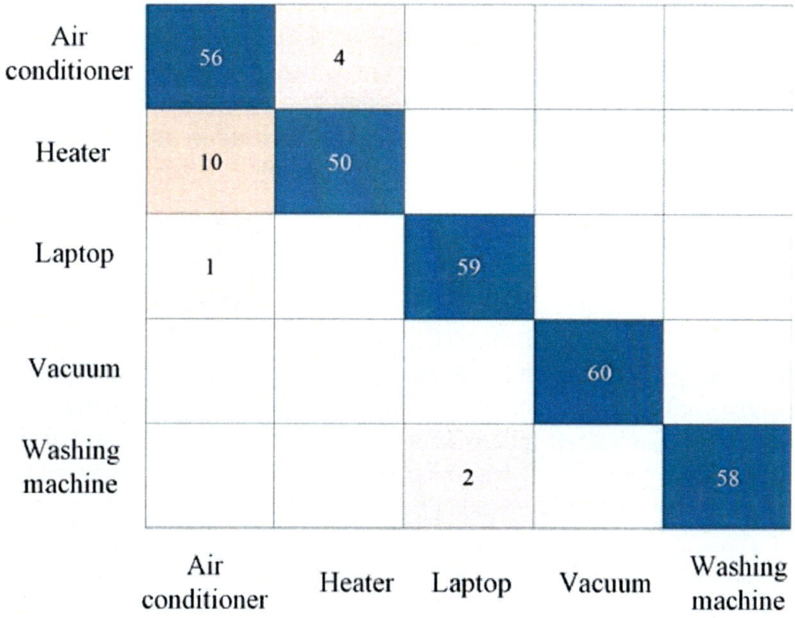

Fig. 8.15 The confusion matrix of the AlexNet model

8.3.4 Non-intrusive Device Identification Based on GoogLeNet

1. *Algorithm introduction*

   Traditional convolutional networks have achieved good results in many application fields, but for some more complex situations, there are still problems that affect the performance of the network, such as incomplete feature extraction (Du et al. 2020). Therefore, when faced with more complex practical problems, researchers improve the model performance by expanding the scale of the model, that is, expanding the depth and width of the network. However, increasing the network scale will cause the problem of the increase of network parameters, making the calculation more difficult. If the data set used in the experiment is limited, it will more easily lead to overfitting. In addition, the deeper the network is, the easier it is to cause the gradient disappearance problem. And it is difficult to optimize the model (Lee et al. 2018). To solve the above problems, the GoogLeNet was proposed.

   GoogLeNet (Khan et al. 2019) was proposed in the ImageNet Challenge in 2014, along with the VGG network (Mahdianpari et al. 2018; Milella et al. 2019). The innovation of GoogLeNet lies in the structure of the network, which increases the depth of the network while improving the computational efficiency. Simulate the structure of human brain neuron stacking, by clustering sparse matrices into relatively dense sub-matrices, the computational performance can be improved (Tang et al. 2017). GoogLeNet uses this idea to innovatively build a sparse and high-performance basic neuron structure, which is called the Inception network structure. The Inception network structure has gone through multiple versions such as Inception v1, Inception v2, Inception v3, and Inception v4. Each of these versions has its advantages but also has corresponding disadvantages (Serbanescu et al. 2020). The GoogLeNet used in this chapter is Inception v1, and the native structure of Inception v1 is given by Fig. 8.16.

   By placing multiple convolution kernels and pooling layers in the convolutional neural network in parallel on the same layer, the size after convolution and pooling can be the same, which increases the width of the network. Different convolution kernels can mine every detail feature in the input, and the pooling operation is mainly used to reduce parameters. Therefore, the Inception structure does not need to manually determine whether to add a convolutional layer or a pooling layer, and the network can determine whether it needs what parameters. However, this native Inception still has a big flaw: the number of feature maps outputted by the above layers is large. If the number of network layers continues to increase, the model will become very complicated and difficult to train and optimization. In addition, the 5×5 convolution kernel may cause the feature map thickness to be too large due to excessive calculation.

   To solve the above problems, the researchers optimized the network and proposed a new Inception network structure. In the new structure, before the 3×3 convolution kernel and the 5×5 convolution kernel, the 1×1 convolution kernel

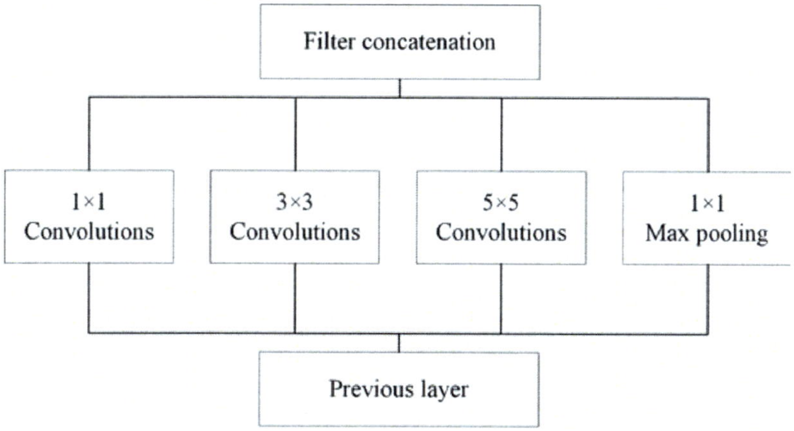

Fig. 8.16 The native structure of Inception v1

is added and it is also added after the max pooling, which not only reduces the dimension but also greatly reduces the numbers of the parameter. This structure is named Inception v1, and the structure of the Inception v1 is given in Fig. 8.17. The entire GoogLeNet consists of 22 layers, containing 9 layers of Inception v1 layers, and the structure of it is shown in Table 8.11. To avoid the problem of gradient disappearance caused by the overly complex network structure, the GoogLeNet added two auxiliary classifiers in the middle layer, so that the gradient

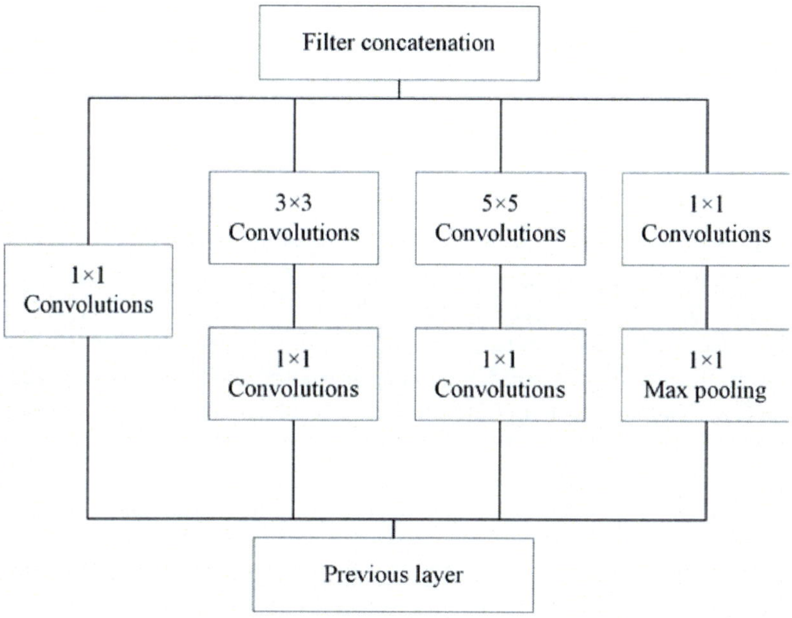

Fig. 8.17 The structure of the Inception v1

Table 8.11 The structure of the GoogLeNet

Type	Size/ stride	Depth	#1 × 1	#3 × 3 reduce	#3 × 3	#5 × 5 reduce	#5 × 5
Convolution	7 × 7/2	1					
Max pool	3 × 3/2	0					
Convolution	3 × 3/1	2		64	192		
Max pool	3 × 3/2	0					
Inception (3a)		2	64	96	128	16	32
Inception (3b)		2	128	128	192	32	96
Max pool	3 × 3/2	0					
Inception (4a)		2	192	96	208	16	48
Inception (4b)		2	160	112	224	24	64
Inception (4c)		2	128	128	256	24	64
Inception (4d)		2	112	144	288	32	64
Inception (4e)		2	256	160	320	32	128
Max pool	3 × 3/2	0					
Inception (5a)		2	256	160	320	32	128
Inception (5b)		2	384	192	384	48	128
Avg pool	7 × 7/1	0					
Dropout (40%)		0					
Linear		1					
Softmax		0					

signal of the network can be backpropagated, which is of great help to the training of the model.

2. *Model and experiment*

   In this section, GoogLeNet is used to identify the device. The GoogLeNet has a complex structure with 22 layers. The data used in this section is consistent with that in Sect. 8.3.2, The input image size was adjusted to 224 × 224 × 3 to fit the input of the network. The training results are illustrated in Fig. 8.18.
   To make up for the shortcomings of insufficient data, transfer learning technology is used, and the implementation method is the same as that in Sect. 8.3.3. Table 8.12 shows the recognition results. Figure 8.19 shows its confusion matrix. In the confusion matrix, the abscissa represents the prediction result, and the ordinate represents the label value.
   Owes to the deeper structure, the device recognition model based on GoogLeNet has higher accuracy. Concluded from Table 8.12, the average accuracy, average recall, average precision, and average F1-Score reach 0.9767, 0.9774, 0.9767, and 0.9766 respectively. From the perspective of the confusion matrix, the model has a good resolution for all kinds of devices.

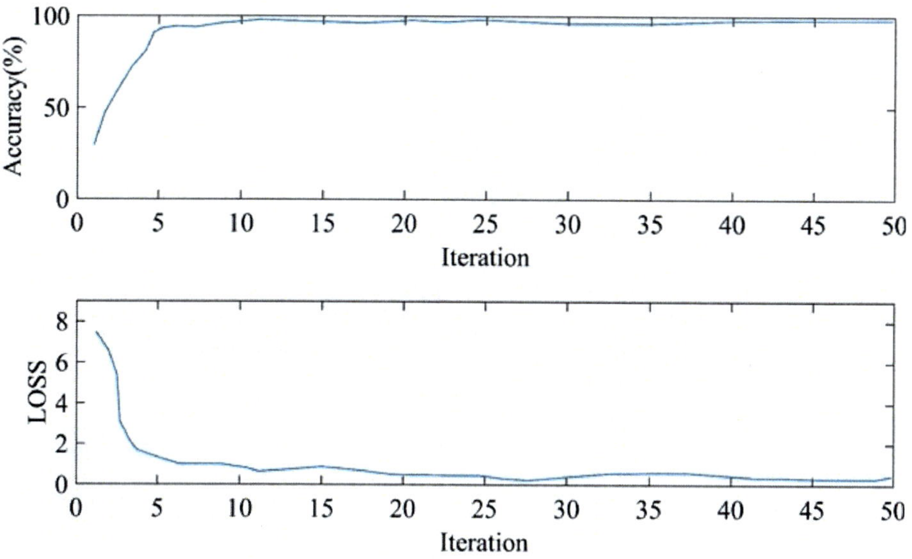

Fig. 8.18 The training results of the GoogLeNet model

Table 8.12 The recognition results of the GoogLeNet model

GoogLeNet		The type of device					
		AC	Heater	Laptop	Vacuum	Washing machine	Average
Evaluation index	Accuracy	0.9833	1.0000	0.9900	0.9867	0.9933	0.9767
	Recall	0.9825	1.0000	0.9672	0.9375	1.0000	0.9774
	Precision	0.9333	1.0000	0.9833	1.0000	0.9667	0.9767
	F1-Score	0.9572	1.0000	0.9752	0.9677	0.9831	0.9766

8.4 Experiment Analysis

8.4.1 Experimental Analysis of the Load Sequence-Based Device Recognition

In Sect. 8.2, RNN, LSTM, and GRU are used to build device recognition models respectively. Table 8.13 shows the recognition results of the three models, Fig. 5.20 shows its histogram for a more intuitive comparison.

The following conclusions can be drawn from Table 8.13 and Fig. 8.20:

(a) In general, all models can achieve device identification. The RNN model has the lowest recognition accuracy, 90.67%. The performance of the LSTM model and the GRU model are comparable, and the average accuracy of them is 96.67% and 96.33% respectively. The LSTM model has an advantage inaccuracy, but the GRU model is more efficient due to its simple structure. In actual tasks, the

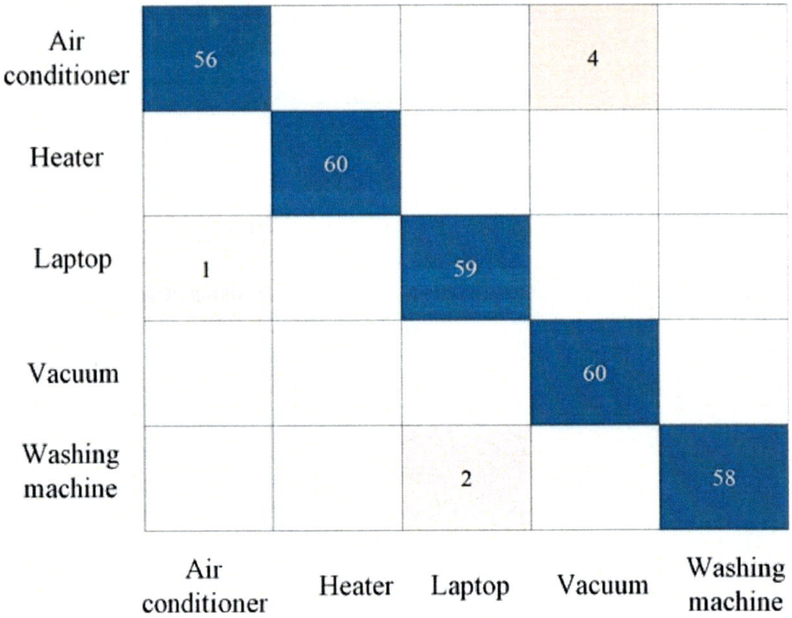

Fig. 8.19 The confusion matrix of the GoogLeNet model

Table 8.13 The results of the three models in Sect. 8.2

MODEL		Evaluation index			
		Accuracy	Recall	Precision	F1
RNN (Model1)	AC	0.9133	0.6977	1.0000	0.8219
	Heater	0.9200	1.0000	0.6000	0.7500
	Laptop	0.9867	0.9667	0.9667	0.9667
	Vacuum	1.0000	1.0000	1.0000	1.0000
	WM	0.9933	1.0000	0.9667	0.9831
	Average	0.9067	0.9329	0.9067	0.9043
LSTM (Model2)	AC	0.9700	0.9180	0.9333	0.9256
	Heater	1.0000	1.0000	1.0000	1.0000
	Laptop	0.9800	1.0000	0.9000	0.9474
	Vacuum	0.9867	0.9375	1.0000	0.9677
	WM	0.9967	0.9836	1.0000	0.9917
	Average	0.9667	0.9678	0.9667	0.9665
GRU (Model3)	AC	0.9700	0.9180	0.9333	0.9256
	Heater	0.9867	0.9375	1.0000	0.9677
	Laptop	0.9767	1.0000	0.8833	0.9381
	Vacuum	0.9933	0.9677	1.0000	0.9836
	WM	1.0000	1.0000	1.0000	1.0000
	Average	0.9633	0.9766	0.9633	0.9630

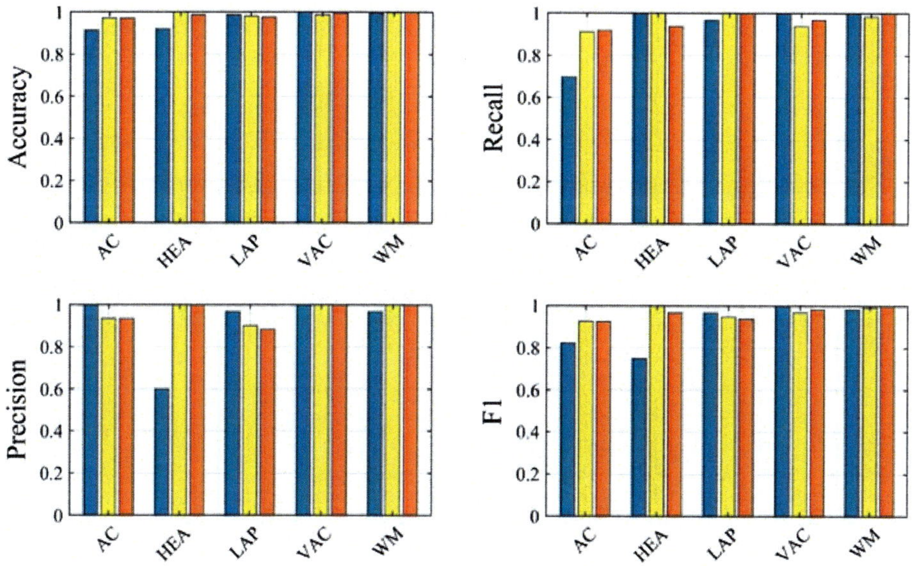

Fig. 8.20 The results of the three models in Sect. 8.2 (The blue, yellow, and red correspond to the RNN model, LSTM model, and GRU model respectively, and AC, HEA, LAP, VAC, WM correspond to the air conditioner, heater, laptop, vacuum, and washing machine respectively)

final model can be determined according to the equipment configuration and accuracy requirements.

(b) The devices with the lowest accuracy of the three model recognition rates are air conditioners. This shows that compared to other equipment, the characteristics of air conditioners are more difficult to identify. To improve the recognition accuracy of air conditioners, the use of higher performance algorithms or the addition of auxiliary features can be considered.

(c) From the results of the three models, it can be found that the best recognition devices of different models are different. The RNN model has the highest recognition accuracy for vacuum, while the best recognition device for the LSTM model and the GRU model are heater and washing machine, respectively. This shows that different models have their unique advantages, and the use of integrated algorithms to integrate them is expected to make the model more accurate.

8.4.2 Experimental Analysis of the Graph Processing Based Device Recognition

Three different deep neural networks are used to build device recognition models in Sect. 8.3, including CNN, AlexNet, and GoogLeNet. The recognition results of them are given in Table 8.14 and Fig. 8.21.

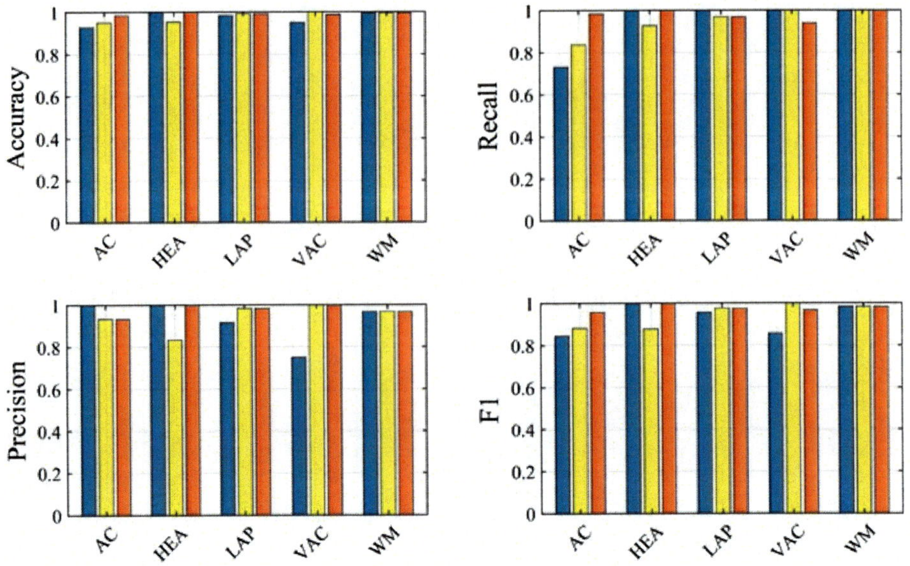

Fig. 8.21 The results of the three models in Sect. 8.3 (The blue, yellow, and red correspond to the CNN model, AlexNet model, and GoogLeNet model respectively, and AC, HEA, LAP, VAC, WM correspond to the air conditioner, heater, laptop, vacuum, and washing machine respectively)

Table 8.14 The results of the three models in Sect. 8.3

Model		Evaluation index			
		Accuracy	Recall	Precision	F1
CNN (Model1)	AC	0.9267	0.7317	1.0000	0.8451
	Heater	1.0000	1.0000	1.0000	1.0000
	Laptop	0.9833	1.0000	0.9167	0.9565
	Vacuum	0.9500	1.0000	0.7500	0.8571
	WM	0.9933	1.0000	0.9667	0.9831
	Average	0.9267	0.9463	0.9267	0.9284
AlexNet (Model2)	AC	0.9500	0.8358	0.9333	0.8819
	Heater	0.9533	0.9259	0.8333	0.8772
	Laptop	0.9900	0.9672	0.9833	0.9752
	Vacuum	1.0000	1.0000	1.0000	1.0000
	WM	0.9933	1.0000	0.9667	0.9831
	Average	0.9433	0.9458	0.9433	0.9435
GoogLeNet (Model3)	AC	0.9833	0.9825	0.9333	0.9572
	Heater	1.0000	1.0000	1.0000	1.0000
	Laptop	0.9900	0.9672	0.9833	0.9752
	Vacuum	0.9867	0.9375	1.0000	0.9677
	WM	0.9933	1.0000	0.9667	0.9831
	Average	0.9767	0.9774	0.9767	0.9766

According to the results, the following conclusions can be drawn:

(a) The average accuracy of the three models of CNN, AlexNet, and GoogLeNet are 92.67%, 94.33%, and 97.67% respectively. The GoogLeNet model has the highest recognition accuracy, and the lowest is the CNN model. In general, all models can successfully identify the five devices in the data set.
(b) According to the experimental results, the best recognition objects of the three models are different. The device with the highest recognition accuracy of the CNN model and the GoogLeNet model is the heater, while the AlexNet model is the vacuum. This shows that networks of different structures have different recognition capabilities for devices.
(c) From the experimental results, it can be found that as the network structure deepens, its accuracy rate gradually improves. Compared with the CNN model which has a simple structure, the GoogLeNet model which has a complicated structure has a significant improvement in recognition accuracy. This shows that deepening the depth of the model is a feasible way to make the model more accurate. However, the complexity of the model structure will inevitably lead to a reduction in model efficiency and an increase in the difficulty of training, which needs to be selected according to actual needs.
(d) Compared with the model in Sect. 8.2, the device recognition model based on deep networks and pictures has certain advantages in improving accuracy. However, deep networks have higher requirements for hardware, which will increase the cost of the system.
(e) In summary, deep learning brings a new path to device recognition. With the development of hardware and the advancement of deep learning algorithms, it is expected to push the accuracy of device recognition models to a higher level.

References

Aldabbas, H., Bajahzar, A., Alruily, M., Qureshi, A. A., Latif, R. M. A., & Farhan, M. (2021). Google Play Content Scraping and Knowledge Engineering using Natural Language Processing Techniques with the Analysis of User Reviews. *J Intell Syst, 30*(1), 192–208. https://doi.org/10.1515/jisys-2019-0197.

Buddhahai, B., Wongseree, W., & Rakkwamsuk, P. (2018). A Non-Intrusive Load Monitoring System Using Multi-Label Classification Approach. *Sustainable Cities and Society, 39,* 621–630.

Collobert, R., Weston, J., Bottou, L., Karlen, M., Kavukcuoglu, K., & Kuksa, P. (2011). Natural Language Processing (Almost) from Scratch. *J Mach Learn Res, 12,* 2493–2537.

Cosgriff, C. V., & Celi, L. A. (2020). Deep learning for risk assessment: all about automatic feature extraction. *British Journal of Anaesthesia, 124*(2), 131–133.

Dong, C., Loy, C. C., He, K. M., & Tang, X. O. (2016). Image Super-Resolution Using Deep Convolutional Networks. *IEEE Transactions on Pattern Analysis and Machine Intelligence, 38*(2), 295–307. https://doi.org/10.1109/tpami.2015.2439281.

Du, B. L., He, Y. G., He, Y. Z., Duan, J. J., & Zhang, Y. R. (2020). Intelligent Classification of Silicon Photovoltaic Cell Defects Based on Eddy Current Thermography and Convolution Neural

Network. *IEEE Transactions on Industrial Informatics, 16*(10), 6242–6251. https://doi.org/10.1109/tii.2019.2952261.
Faustine, A., & Pereira, L. (2020a). Improved Appliance Classification in Non-Intrusive Load Monitoring Using Weighted Recurrence Graph and Convolutional Neural Networks. *Energies, 13*(13), 15. https://doi.org/10.3390/en13133374.
Faustine, A., & Pereira, L. (2020b). Multi-Label Learning for Appliance Recognition in NILM Using Fryze-Current Decomposition and Convolutional Neural Network. *Energies, 13*(16), 4154. https://doi.org/10.3390/en13164154.
Gao J, Giri S, Kara EC, Bergés M (2014) Plaid: a public dataset of high-resoultion electrical appliance measurements for load identification research: demo abstract. In: proceedings of the 1st ACM Conference on Embedded Systems for Energy-Efficient Buildings, 2014. pp 198–199
Han, X. B., Zhong, Y. F., Cao, L. Q., & Zhang, L. P. (2017). Pre-Trained AlexNet Architecture with Pyramid Pooling and Supervision for High Spatial Resolution Remote Sensing Image Scene Classification. *Remote Sens, 9*(8), 22. https://doi.org/10.3390/rs9080848.
Hinton, G., Deng, L., Yu, D., Dahl, G. E., Mohamed, A. R., Jaitly, N., et al. (2012). Deep Neural Networks for Acoustic Modeling in Speech Recognition. *IEEE Signal Processing Magazine, 29*(6), 82–97. https://doi.org/10.1109/msp.2012.2205597.
Hou, C., He, Y., Jiang, X., & Pan, J. (2018). Deep Convolutional Neural Network Based on Two-Stream Convolutional Unit. *Laser & Optoelectronics Progress, 55*(2), 021005. https://doi.org/10.3788/lop55.021005.
Khan, R. U., Zhang, X., & Kumar, R. (2019). Analysis of ResNet and GoogleNet models for malware detection. *Journal of Computer Virology and Hacking Techniques, 15*(1), 29–37. https://doi.org/10.1007/s11416-018-0324-z.
Krizhevsky, A., Sutskever, I., & Hinton, G. (2012). ImageNet Classification with Deep Convolutional Neural Networks. *Advances in Neural Information Processing Systems, 25*(2), 1097–1105.
LeCun, Y., Bengio, Y., & Hinton, G. (2015). Deep learning. *Nature, 521*(7553), 436–444. https://doi.org/10.1038/nature14539.
Lee, S. G., Sung, Y., Kim, Y. G., & Cha, E. Y. (2018). Variations of AlexNet and GoogLeNet to Improve Korean Character Recognition Performance. *J Inf Process Syst, 14*(1), 205–217. https://doi.org/10.3745/jips.04.0061.
Li G, Zhang M, Li J, Lv F, Tong G (2021) Efficient densely connected convolutional neural networks. Pattern Recognition 109
Litjens, G., Kooi, T., Bejnordi, B. E., Setio, A. A. A., Ciompi, F., Ghafoorian, M., et al. (2017). A survey on deep learning in medical image analysis. *Medical Image Analysis, 42,* 60–88. https://doi.org/10.1016/j.media.2017.07.005.
Liu, W. F., Ma, T. Z., Xie, Q. S., Tao, D. P., & Cheng, J. (2017). LMAE: A large margin Auto-Encoders for classification. *Signal Process, 141,* 137–143. https://doi.org/10.1016/j.sigpro.2017.05.030.
Lu, J., Behbood, V., Hao, P., Zuo, H., Xue, S., & Zhang, G. Q. (2015). Transfer learning using computational intelligence: A survey. *Knowledge-Based Syst, 80,* 14–23. https://doi.org/10.1016/j.knosys.2015.01.010.
Mahdianpari, M., Salehi, B., Rezaee, M., Mohammadimanesh, F., & Zhang, Y. (2018). Very Deep Convolutional Neural Networks for Complex Land Cover Mapping Using Multispectral Remote Sensing Imagery. *Remote Sens, 10*(7), 21. https://doi.org/10.3390/rs10071119.
Mbatha, N., & Bencherif, H. (2020). Time Series Analysis and Forecasting Using a Novel Hybrid LSTM Data-Driven Model Based on Empirical Wavelet Transform Applied to Total Column of Ozone at Buenos Aires, Argentina (1966-2017). *Atmosphere, 11*(5), 20. https://doi.org/10.3390/atmos11050457.
Milella, A., Marani, R., Petitti, A., & Reina, G. (2019). In-field high throughput grapevine phenotyping with a consumer-grade depth camera. *Computers and Electronics in Agriculture, 156,* 293–306. https://doi.org/10.1016/j.compag.2018.11.026.

Noh, J., Park, H. J., Kim, J. S., & Hwang, S. J. (2020). Gated Recurrent Unit with Genetic Algorithm for Product Demand Forecasting in Supply Chain Management. *Mathematics, 8*(4), 14. https://doi.org/10.3390/math8040565.

Odense, S., & Edwards, R. (2016). Universal Approximation Results for the Temporal Restricted Boltzmann Machine and the Recurrent Temporal Restricted Boltzmann Machine. *J Mach Learn Res, 17,* 1–21.

Pan, E. T., Mei, X. G., Wang, Q. D., Ma, Y., & Ma, J. Y. (2020). Spectral-spatial classification for hyperspectral image based on a single GRU. *Neurocomputing, 387,* 150–160. https://doi.org/10.1016/j.neucom.2020.01.029.

Pan, F. L., Li, J., Tan, B. D., Zeng, C. L., Jiang, X. F., Liu, L., et al. (2018). Stacked-GRU Based Power System Transient Stability Assessment Method. *Algorithms, 11*(8), 10. https://doi.org/10.3390/a11080121.

Sapijaszko G, Mikhael WB, Ieee (2018) An Overview of Recent Convolutional Neural Network Algorithms for Image Recognition. In: 2018 Ieee 61st International Midwest Symposium on Circuits and Systems. Midwest Symposium on Circuits and Systems Conference Proceedings. Ieee, New York, pp 743–746

Schmidhuber, J. (2015). Deep learning in neural networks: An overview. *Neural Networks, 61,* 85–117. https://doi.org/10.1016/j.neunet.2014.09.003.

Serbanescu, M.-S., Manea, N. C., Streba, L., Belciug, S., Plesea, I. E., Pirici, I., et al. (2020). Automated Gleason grading of prostate cancer using transfer learning from general-purpose deep-learning networks. *Romanian Journal of Morphology and Embryology, 61*(1), 149–155. https://doi.org/10.47162/rjme.61.1.17.

Sherstinsky, A. (2020). Fundamentals of Recurrent Neural Network (RNN) and Long Short-Term Memory (LSTM) network. *Physica D: Nonlinear Phenomena, 404,* 28. https://doi.org/10.1016/j.physd.2019.132306.

Smagulova, K., & James, A. P. (2019). A survey on LSTM memristive neural network architectures and applications. *Eur Phys J-Spec Top, 228*(10), 2313–2324. https://doi.org/10.1140/epjst/e2019-900046-x.

Spoerer, C. J., McClure, P., & Kriegeskorte, N. (2018). *Recurrent Convolutional Neural Networks: A Better Model of Biological Object Recognition Front Psychol, 9*(2), 1551. https://doi.org/10.3389/fpsyg.2018.01695.

Tang, P. J., Wang, H. L., & Kwong, S. (2017). G-MS2F: GoogLeNet based multi-stage feature fusion of deep CNN for scene recognition. *Neurocomputing, 225,* 188–197. https://doi.org/10.1016/j.neucom.2016.11.023.

Taylor, M. E., & Stone, P. (2009). Transfer Learning for Reinforcement Learning Domains: A Survey. *J Mach Learn Res, 10,* 1633–1685.

Tian, Y. H. (2020). Artificial Intelligence Image Recognition Method Based on Convolutional Neural Network Algorithm. *IEEE Access, 8,* 125731–125744. https://doi.org/10.1109/access.2020.3006097.

Wang, Z. Y., & Zheng, G. L. (2012). Residential Appliances Identification and Monitoring by a Nonintrusive Method. *Ieee Transactions on Smart Grid, 3*(1), 80–92. https://doi.org/10.1109/tsg.2011.2163950.

Wen, J. H., Liu, Y. S., Shi, Y., Huang, H. R., Deng, B., & Xiao, X. P. (2019). A classification model for lncRNA and mRNA based on k-mers and a convolutional neural network. *BMC Bioinformatics, 20*(1), 14. https://doi.org/10.1186/s12859-019-3039-3.

Wu, X., Jiao, D., & Du, Y. (2020a). Automatic Implementation of a Self-Adaption Non-Intrusive Load Monitoring Method Based on the Convolutional Neural Network. *Processes, 8*(6), 20. https://doi.org/10.3390/pr8060704.

Wu, Z., Rincon, D., & Christofides, P. D. (2020b). Process structure-based recurrent neural network modeling for model predictive control of nonlinear processes. *Journal of Process Control, 89,* 74–84. https://doi.org/10.1016/j.jprocont.2020.03.013.

Zhang, L., Lu, Y. P., Wang, B. J., Li, F. Z., & Zhang, Z. (2018). Sparse Auto-encoder with Smoothed Regularization. *Neural Processing Letters, 47*(3), 829–839. https://doi.org/10.1007/s11063-017-9668-5.

Zhao, J. F., Mao, X., & Chen, L. J. (2018). Learning deep features to recognise speech emotion using merged deep CNN. *IET Signal Process, 12*(6), 713–721. https://doi.org/10.1049/iet-spr.2017.0320.

Zhao, X. Y., Dong, C. Y., Zhou, P., Zhu, M. J., Ren, J. W., & Chen, X. Y. (2019). Detecting Surface Defects of Wind Tubine Blades Using an Alexnet Deep Learning Algorithm. *IEICE Trans Fundam Electron Commun Comput Sci, E102A*(12), 1817–1824. https://doi.org/10.1587/transfun.E102.A.1817.

Zhu, X. X., Li, L. X., Liu, J., Li, Z. Y., Peng, H. P., & Niu, X. X. (2018). Image captioning with triple-attention and stack parallel LSTM. *Neurocomputing, 319,* 55–65. https://doi.org/10.1016/j.neucom.2018.08.069.

Zoha, A., Gluhak, A., Imran, M. A., & Rajasegarar, S. (2012). Non-Intrusive Load Monitoring Approaches for Disaggregated Energy Sensing: A Survey. *Sensors, 12*(12), 16838–16866. https://doi.org/10.3390/s121216838.

Chapter 9
Potential Applications of Smart Device Recognition in Industry

9.1 Introduction

Electricity has penetrated every aspect of our production and life from scratch. Electric energy has been the most widely used and important energy in the modern economy. Meanwhile, the traditional power industry has changed to a highly intensive, highly knowledgeable, and highly technical direction. The power grid has become one of the largest and most complex machines in the world. Thanks to the continuous promotion of science and technology to the power grid, the Ubiquitous Electric Internet of Things (UEIOT) has been seen everywhere in the world.

Users yearn for the reliability, security, economy, and stability of the power system. However, in the whole process of production, transmission, distribution, and use of electric energy, there is still a lack of information about its operation. The traditional monitoring system often needs a lot of hardware equipment. The on-line installation and measurement of sensors and other equipment not only requires lots of investment but also requires a lot of money and time in the installation and maintenance stage. In addition, because the traditional power monitoring system adopts the "intrusive" technology design, the power users need to be temporarily cut off during installation and maintenance, which is easy to cause user dissatisfaction or other economic losses, and it is often uneasy to achieve the needs of system optimization, energy conservation, fault detection, and analysis. All of these urgently need to improve the monitoring level of the power system.

In addition, in today's energy shortage, the entire international community is calling for the conservation of energy and reduction of emission, and building an energy-saving society (Omer 2008). Especially for countries with large population bases and scarce per capita resources, it is of great practical significance to reduce power loss and improve power transmission efficiency to build an "energy-saving society", promote efficient use of various energies and alleviate the global energy crisis (Bose 2010).

At the same time, in various industrial and military fields, the use of various complex electromechanical systems (such as ship systems) supported by the electric

H. Liu et al., *Smart Device Recognition*,
https://doi.org/10.1007/978-981-33-4925-4_9

power system has promoted the leap of human industrial civilization, and they have become an important pillar in various fields. However, for the stability of complex electromechanical systems, the failure of each subsystem is crucial to the stability of electromechanical systems. How to effectively carry out a real-time diagnosis of electromechanical equipment has become a research hotspot. Compared with the identification of household electrical equipment, fault identification, and diagnosis of electromechanical equipment becomes more difficult and important. The application of smart device recognition technology undoubtedly provides great technical support for the identification and fault diagnosis of complex electromechanical systems.

Recently, with the rapid progress of global industrialization civilization, the unreasonable utilization of earth resources has caused great global environmental problems (Briggs 2003). Environmental pollution has become the focus of attention all over the world. In the relevant studies of environmental pollution, urban environmental pollution is becoming more and more serious, which has a huge impact on people's physical and mental health. Although most countries have the determination to solve the problem of environmental pollution, the immaturity of environmental monitoring technology poses no small challenge to environmental governance. Therefore, it is necessary to monitor the pollutants in the environment space in real-time. At the same time, the identification of pollution sources through "non-intrusive" means is also the key to control environmental pollution.

Energy and the environment are important issues of close concern in today's world, and the progress of science and technology should be conducive to energy management and environmental protection. In the background of UEIOT, the potential application of smart device recognition technology in the industry also plays an important role, and its application fields are shown in Fig. 9.1.

9.2 Electric and Energy Applications

9.2.1 Overview of Electricity and Energy

(1) *The composition of the power system*

The power system is actually a power production and consumption system. The system includes power plants, transmission lines, power supply and distribution stations, and power supply. The function of electric power system is to generate new energy by using primary energy in nature, and then convert this new energy into electric energy through power generation device. Finally, the generated electricity is transmitted, transformed and distributed to all users. In this process, it is necessary to configure the corresponding information and control system in each link of the power system in order to realize the regulation, control, and protection of electric energy. Finally, through these means, users are guaranteed to obtain a high-quality electric energy experience. Figure 9.2 shows the composition of the power system.

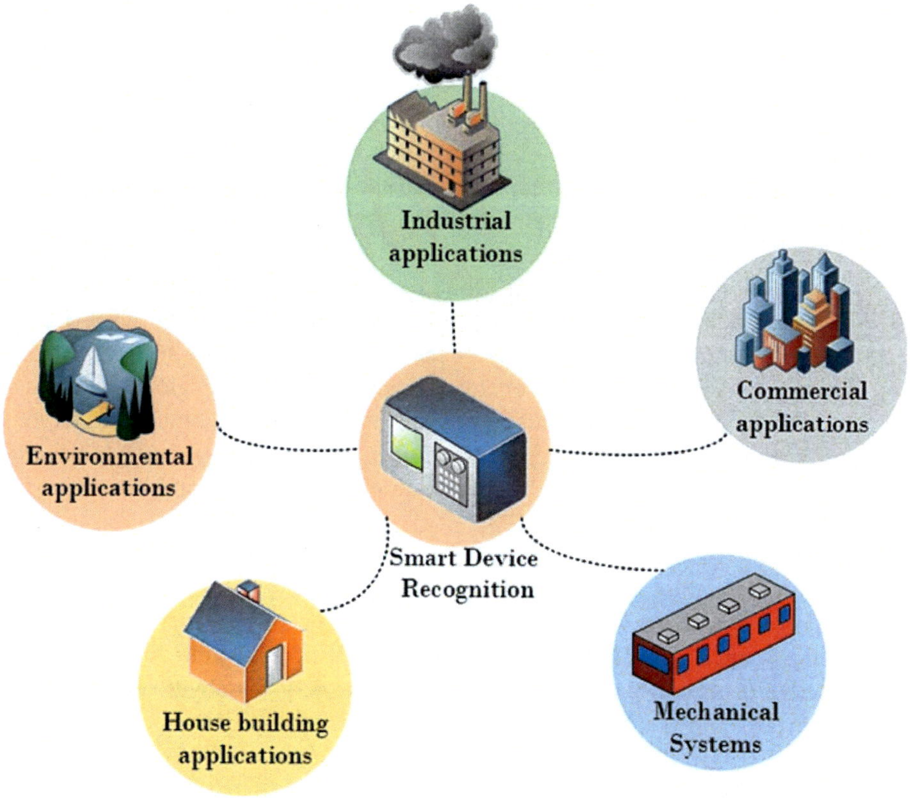

Fig. 9.1 The application field of smart device recognition

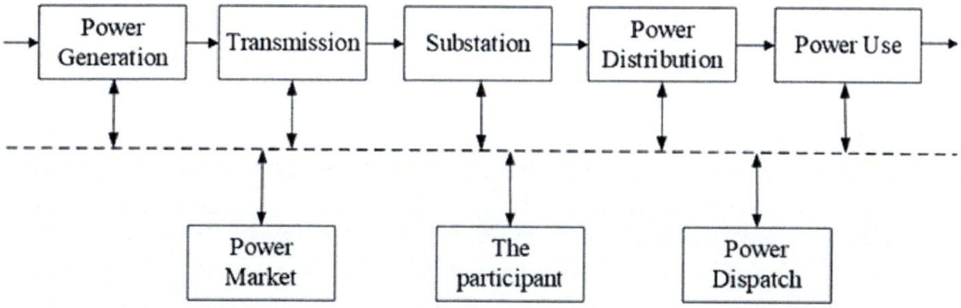

Fig. 9.2 The composition of the power system

As shown in Fig. 9.2, there are five main parts of the power system, namely, generation, transmission, transformation, distribution, and electricity consumption. Affected by the power market and participants, relevant dispatching is carried out according to demand. The electric power system is the main source of power to support social and economic development. Its operation mode is affected by external demand and changes dynamically due to the criterion differences in different countries, regions, and stages of economic development.

(2) *Global energy crisis*

Energy and environmental issues are major challenges facing the world today (Jacobson and Delucchi 2011; Delucchi and Jacobson 2011; Fang and Zeng 2007). After long-term development and the use of fossil energy, it tends to be exhausted. Carbon emission intensifies and causes climate change (Panwar et al. 2011), which becomes a hard constraint to the high-quality and sustainable development of the social economy. Moreover, the frequent occurrence of geopolitical events driven by energy causes constant regional wars (Colgan 2014), resulting in the instability of the international situation. In the face of the increasingly serious fossil energy dilemma and the many global problems it causes, the international community needs to cooperate and innovate, promote the development of large-scale clean and low-carbon energy industry, accelerate the energy transformation, and meet the development demand in a clean and green way.

At present, traditional fossil energy is still the main force of global energy supply, and with technological innovation and breakthroughs, clean energy will take the leading position (Resch et al. 2008; Pleßmann et al. 2014). From the perspective of energy resource endowment, the global supply of renewable energy resources far exceeds the development needs of human society, with hydro energy, terrestrial wind energy, and solar energy resources exceeding 10 billion kW, 1 trillion kW, and 14 100 trillion kW respectively. From the point of energy development, global consumption of oil, natural gas, coal, although from 3.8956 billion tons of oil equivalent in 1965 to 2017, 11.5094 billion tons of oil equivalent, but the weight has dropped from 94% to 85%. The proportion of renewable energy in the global energy consumption structure is increasing, which cannot be separated from the rapid development of the electric power industry.

In 2017, forty percent of the world's electricity comes from primary sources, global power generation increased by about 2.8%, 49% of the growth came from renewable energy, and the proportion of renewable energy in the power generation structure increased by 8.4% from 7.4%. From the perspective of transmission carrier, uneven distribution of resources determines the need for trans-regional transmission of energy. Ships, pipelines, and railways, as the main carriers of fossil energy transmission, take a long time, have many links, cause serious pollution, and have high transportation costs. Comparatively speaking, clean energy is transmitted in the form of electric energy, which is fast, environmentally friendly, efficient, and economical.

In recent years, a lot of studies and engineering practices have been carried out in the fields of the smart power grid, clean energy, and interconnection of large power grids, both internal and external. The European supergrid initiative, the US Department of Energy's Grid 2030 initiative, and China's strong smart Grid construction have provided important infrastructure to advance the global energy Internet.

In the future, with the progress of power supply, power grid, energy storage, and information and communication technology, it is obvious that electric energy still plays an indispensable role in the present and future energy composition.

(3) *Smart grid*

Nowadays, smart grid is becoming a new product of power system in the era of artificial intelligence. Through the integration of physical and digital technologies, smart grid has become a new power system that is more economical, safe, environmentally friendly and reliable. It is of vital importance to the mitigation of energy problems and environmental pollution problems.

As a kind of comprehensive power engineering that integrates various technologies, the smart power grid is still a new type of power system in essence. Therefore, when exploring its operation mechanism, the research should start from the main technical components of the smart power grid. Figure 9.3 shows the main technical components of a smart grid.

At present, smart grid research includes Advanced metering infrastructure (AMI), Advanced Distribution Operation (ADO), Advanced Transmission Operation (ATO), and Advanced Asset Management (AAM). Among them, to authorize user is the primary function of the AMI. By connecting systems and loads, AMI helps users support grid operations. The self-healing function is implemented by ADO. ATO is vital for reducing the risk of downtime. In addition, the ATO emphasizes blocking management. The improvement of power operation and the improvement of asset use efficiency are based on the cooperation between AMM and AMI, ADO and ATO.

The specific interpretation of each abbreviation is Automatic meter reading (AMR), Alternating current/direct current (AC/DC), Fast simulation and modeling (FSM), Distributed energy resource (DER), International Standardization Organization (ISO), Energy management system (EMS), Measurement data management system (MDMS).

Smart grid technology is aimed at many things. On the one hand, it can provide high-quality electricity and improve the utilization rate of renewable energy. On the other hand, it can reduce the exploitation of fossil energy and reduce environmental pollution. Finally, smart grid realizes green electricity consumption and reduces the production cost of the original power system.

A smart grid emphasizes two-way interaction feedback between the power supply grid and users. On the one hand, for users, they are able to know the real-time power supply information of the grid, real-time electricity price, power outage information, and the corresponding power use situation, so as to monitor and adjust their power consumption status in real time and better complete power consumption management. On the other hand, for the energy provider, it is helpful to realize the safe and efficient transmission of electricity by establishing a sound and comprehensive information interaction mechanism to understand the electricity consumption information of users in real time and optimize resource allocation. Interaction and coordination can not only meet the needs of both sides, but also enhance the economy and stability of the power grid, efficient

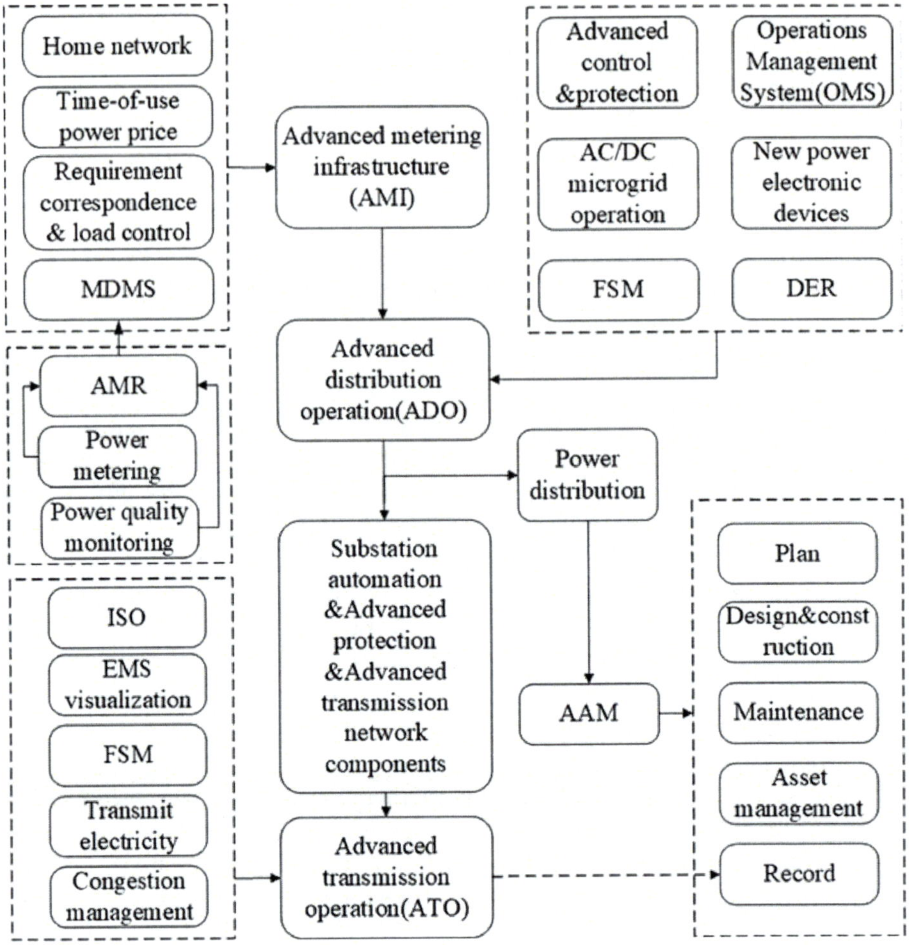

Fig. 9.3 The technical composition of smart grid

transmission of energy signals, reduce energy waste, and bring convenience to users and save residential electricity costs.

9.2.2 *Home Energy Management*

Home Energy Management System (HEMS) is an intelligent network control system. HEMS is based on a smart grid, which integrates all the power and electrical equipment in a home. It is the application of smart grid in the household electricity side, which can realize the control and management of the household electricity. It is the application of a smart power grid in the household power consumption side (Zhou et al. 2016). The household energy management system can promote the active participation of residents in electricity management, improve power supply safety and energy efficiency, and play a key role in promoting energy conservation, emission reduction, and environmental pollution reduction. The intelligent equipment identification based on non-intrusive technology can help the household energy management (Biansoongnern and Plungklang 2016; Ruano et al. 2019), which can be divided into the identification of household electrical equipment and fault detection of electrical equipment.

(1) *Electrical equipment identification*

For residential users, the residential power users can inquire about the power consumption and running status of each household electrical appliance in real time by installing an intelligent device identification device. Users with a strong sense of energy saving can optimize their electricity costs according to the information, so as to achieve the purpose of saving electricity and electricity costs. By optimizing the running parameters of electric appliances and adjusting the running state of electric appliances, the economic benefits of household electricity consumption can be greatly improved. At the same time, combined with the improvement strategy pushed by the electric power company and personalized energy-saving regulation measures such as dynamic electricity price, the energy-saving ratio is expected to reach nearly 20% under the condition of meeting the electricity demand and ensuring the comfort of users.

In fact, the identification of electrical equipment is essentially the collection and treatment of electrical equipment load. The common loads of home users can be divided into three types according to their characteristics: the first type is resistive loads. These are mainly electric kettles, electric heaters, electric rice cookers, etc. The characteristic of this kind of load is that harmonic content is less, and its transient process is generally relatively gentle. The second type of load is the capacitive load. This kind of load basically has a laptop, LCD TV to wait. The characteristics of this kind of load are that it generally has a start and stop power supply, high harmonic content, and obvious impulse current when switching and cutting. The third type of load is the perceptual load. This kind of load mainly has an air conditioner, hairdryer, washing machine, electric fan, fluorescent lamp, and so on. The characteristic of this kind of load is that these loads are equipped with coils or motors, so there is usually an impulse current when switching off, and this kind of load may have a high harmonic content. These three types of loads are usually supplied through a distribution circuit so that data collection can be done through a detection device installed at the incoming line of the residential distribution circuit.

The non-intrusive load identification algorithm is the most effective in current equipment identification applications. The non-intrusive load identification algorithm is based on the user's electricity consumption information to identify and decompose the user's detailed household electricity consumption. In the process of identification, the total collected data is processed and analyzed, and finally the information of various electrical appliances is obtained. The basic principle of this method is as follows: Firstly, the characteristic parameters of various common household appliances should be measured, and the characteristic parameter model of household appliances commonly used in the family should be established. Then, the calculated characteristic parameters are compared with the characteristic parameters model of household appliances, which can be used to identify the household load. The characteristic parameters of household load include two kinds: steady-state and transient parameters. Steady-state characteristic parameters include load steady-state current, harmonic power, and load voltage. Transient characteristic parameters include transient changes of voltage and current of load switch.

Non-intrusive load identification algorithm based on steady-state load value uses steady-state characteristic parameters to complete load identification calculation. The non-intrusive load identification monitoring method based on load transient characteristic parameters is used to identify load by measuring transient characteristic parameters. Based on the load identification method based on steady-state characteristics and transient characteristics, the application framework of non-intrusive monitoring and identification technology for typical resident electrical appliances is shown in Fig. 9.4.

As shown in Fig. 9.4, by reading the voltage and current signals coming from the total incoming line, and after processing with non-intrusive technology, the information such as the type, duration, and quantity of the electrical appliance can be output finally, that is, the intelligent identification of the electrical equipment can be completed.

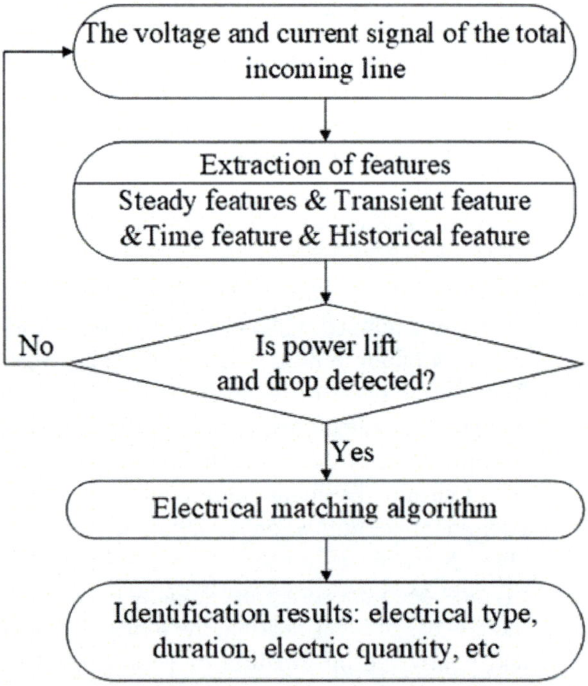

Fig. 9.4 The general framework of electrical equipment identification

For power systems, smart device recognition helps power companies to understand different load categories, and information such as the working status, power consumption rules, and current power consumption of the power equipment. Intelligent device identification is helpful to determine dynamic electricity price more scientifically, so as to encourage friendly interaction between users and power grid companies to realize load peak shifting and peak load shifting and improve load coefficient. Finally, it improves the economy and reliability of power system operation (such as saving energy and reducing loss, extending the service life of power equipment, reducing investment and maintenance cost of power equipment, etc.).

(2) *Fault detection of electrical equipment*

For buildings, no matter industrial buildings or civil buildings, all kinds of electrical equipment (such as refrigerators, air conditioners, hot water bottles, etc.) constitute an important part of indoor energy consumption. In real life, electrical appliances will fail due to their service life or other objective reasons in the process of use and often lead to useless consumption of electrical energy, or even the destruction of family circuits. Therefore, if the working state of the equipment in the building can be real-time monitoring, and can detect its abnormal conditions in a timely manner, for the energy-saving and safety of electricity are crucial.

NILM algorithm can realize state monitoring and energy consumption identification of electrical equipment through the power measurement of electrical equipment. However, the performance of the device may degrade over time. Equipment performance degradation is often manifested by abnormal energy consumption and abnormal running in the NILM monitoring environment. With the continuous improvement of technology, smart electricity meters have begun to be put into abnormal detection of electrical equipment at the household level. (Rashid et al. 2019) detected abnormal behavior of equipment through the NILM monitoring. In the experiment, the normal operation of electrical appliances was trained, then the open, recognized and available NILM technology was used to obtain the power track of electrical appliances, and finally, the abnormal detection of air conditioners and refrigerators in residential buildings was attempted.

In the application of anomaly detection, to improve the ability of the intelligent device identification system with anomaly detection function, it is necessary to extract the characteristic lines of the device during the anomaly. Because the failure is not frequent, it can only be simulated by experimental means. In a short time, the fault is injected into the electrical apparatus, and the characteristics of the fault circuit, such as current and voltage, are extracted to capture the abnormal behavior of the electrical apparatus. To popularize application, the characteristic extraction experiment of abnormal behavior is also carried out for electrical appliances with low work energy efficiency and close to their service life. In conclusion, by creating such a database with abnormal behavior

of NILM, intelligent device identification is expected to have more and better applications in anomaly detection of electrical equipment in buildings.
Based on the above two sections, the intelligent device identification technology based on the NILM system is expected to monitor the abnormal behavior and energy consumption of electrical equipment, so as to realize effective energy management.

9.2.3 Fault Discrimination of User's Electric Line

In the power system, the so-called "short circuit" refers to the connection between phases or between phase and ground (or neutral line) other than the normal operation of the power system. Very large currents flow when abnormal connections (short circuits) occur between phases or between phases and ground. The current value is much higher than the rated current and depends on the electrical distance between the short circuit point and the power supply. A short circuit is a low resistance short connection between conductive parts of different potentials, which is equivalent to the power supply being connected directly by the wire into a closed circuit without passing through the load.

As is known to all, in daily life, the accident of the short circuit of an electric circuit often happens. Usually, this is a kind of serious circuit fault that should be avoided as far as possible, it will cause the circuit to burn down due to too much current and cause fire. Therefore, it is very important to distinguish the fault of the user's electric line, and the non-intrusive intelligent equipment identification technology will play an important role.

(1) *Brief introduction of non-intrusive power user fault discrimination technology*

The basis of non-intrusive power fault discrimination technology for users is to collect the current and voltage signals collected by the equipment, process the current and voltage signals through the corresponding algorithm, extract the characteristic parameters in the signals, and judge the faults and fault types in the lines through the identification algorithm.
At present, the main signal extraction algorithms include loaded online decomposition processing algorithm, non-parametric Cumulative Sum (CUSUM) variable point monitoring algorithm, sliding window bilateral CUSUM variable point detection algorithm, and weighted CUSUM bilateral variable point monitoring algorithm.
The low-voltage distribution network generally adopts the radial type and tree-trunk type topological structure. In the radial type topological structure, the power point is connected with the subordinate bus or equipment by a special line, and there is no other load on the line, which is more suitable for places with large equipment capacity and high-grade load. In the trunk type topology, each outlet is led out by the high-voltage power supply bus, and the line is connected with other loads. The advantage of trunk type topology is to save the use of

switching equipment and the use of lines, but if the mainline fails, the influence range is larger. In practical application, the hybrid topological structure of radial type and trunk type is generally adopted.

There are many protection and control devices for low-voltage distribution networks, including circuit breakers, fuses, surge protectors, isolation switches, etc. Circuit breakers are most commonly used among home users. After the short circuit occurs, if the circuit breaker does not operate, the short circuit accident will generally produce a periodic current component and a transient aperiodic current component. In severe cases, the peak value of the short circuit current will be close to the sum of the two, which will affect the power grid.

The user-side power grid is generally equipped with a current-limiting circuit breaker. The role of the circuit breaker is to limit and close the short-circuit current when the short-circuit occurs, so as to reduce the impact caused by the short-circuit. After a short circuit occurs, the current limiting circuit breaker first starts the proper action, which takes about 3 ms. The arc resistance increases rapidly to limit or cut off the short circuit current, which is generally less than 10 ms. After the current limiting mode circuit breaker, the amplitude of short circuit current is generally smaller than that of the periodic component. Finally, according to the set short-circuit occurrence criterion, the short-circuit can be detected.

(2) *Non-intrusive short-circuit accident positioning method*

Managers are concerned not only about the number of a short circuit in the building but also about the location of the accident. The location of the short circuit accident under the non-intrusion architecture (the location of the short circuit accident) is judged, that is, the location algorithm of the short circuit accident mainly adopts the following algorithms.

A. Initial positioning method for distribution hierarchy

The size of the short circuit current is related to the level of the power network in which the short circuit occurs. The higher the distribution level in which the short circuit occurs, the greater the amplitude of the short circuit current, because the higher level of the distribution line impedance is smaller. Based on the above principles, the hierarchical positioning method is adopted in the determination process as follows:

- Short-circuit criterion for high-level distribution level.
  Calculate the short circuit current value of all high-rise distribution positions, such as general distribution cabinet, floor distribution cabinet, central air conditioning distribution cabinet, power distribution cabinet, etc., and take the minimum value as the short-circuit criterion of high-rise distribution level. When the sudden current of the short circuit is greater than this value, it is considered that the short circuit of high distribution level occurs.
- Short-circuit criterion at the end of the room.
  At the end of the room, there are the most electrical appliances, and it is also the most prone to short-circuit. The short-circuit current of all the rooms is

calculated, and the maximum value is taken as the criterion of short-circuit at the end of the room. When the short circuit mutation current is less than this value, it is considered that a short circuit may occur at the end of the room.

B. Initial location method of the loop based on power drop.

By modeling the operating power of the important circuit, the short circuit can be located according to the power drop after the short circuit occurs. The loop positioning method based on the power drop is as follows:

- Important power loop positioning method.
  Establish a list of important power circuits to be monitored, including basic information of the circuit and operating power. After the occurrence of a short circuit, calculate the power drop, and traverse the list. If a circuit matching the power drop is found, a short circuit accident may occur in the circuit. The disadvantage is that there may be several power circuits with similar operating power, and it is difficult to distinguish them by this method alone.
- If the power sag is less than 5 kW, there may be the following two situations: First, the short circuit occurs at the end of the room, and the power of the removed electrical appliance is very small, and most of them are single-phase power sag. In this case, the short circuit current mutation is not large, which can be combined to form the judgment of the short circuit at the end of the room. Second, the short circuit position occurs in the higher distribution level, but the short circuit with small loads, so the performance is small power drop, but at this time the short circuit current mutation amount is large, can be combined with both to form a high distribution level short circuit judgment.

C. Precise positioning methods for multiplexing itemized meters
Most smart buildings realize itemized measurement. Smart electricity meters are adopted to collect electricity information of important circuits and summarize it to the main station of the central control system of the building. The multi-item metering meters can be reused to form an accurate positioning structure of the short circuit accident as shown in Fig. 9.5.

- Install the non-intrusive short circuit accident monitoring terminal at the main power supply to monitor the current at the main incoming line in real time. Once the short circuit is determined, send the short circuit information to the main station immediately, mainly including the abrupt current and DC component.
- After receiving the short-circuit accident alarm information from the terminal, the master station will start the function module of accurate fault positioning to communicate with the intelligent electricity meters installed on each distribution line (the information of electricity meters on the branch indicated by crude positioning of the terminal shall be read first). If the communication establishment between the master station and the intelligent

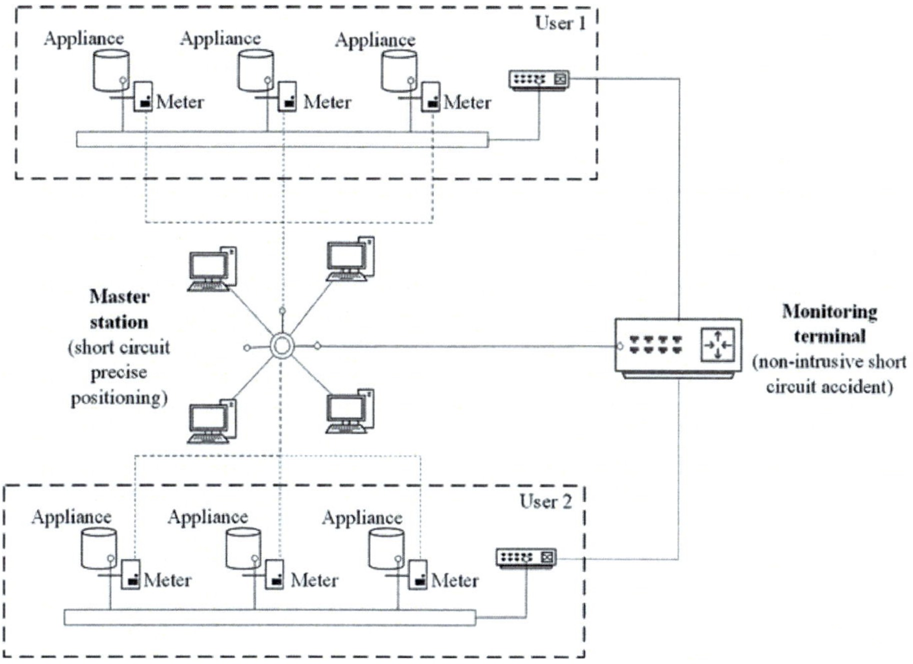

Fig. 9.5 Item metering electricity meter—shaped short—circuit accident precise positioning architecture

electricity meter fails, the master station will display the crude fault location result given by the terminal directly and list it as the suspect fault location.

- After the master station determines the short circuit location, it shall make statistics of the short circuit accident information and push it to the management personnel for on-site confirmation.

(3) *The scheme design of non-intrusive circuit fault identification device*
On the basis of completing the criterion of non-intrusive short circuit occurrence, setting method, and setting process, the scheme of non-intrusive internal circuit fault discrimination device for users is designed. The overall scheme design is shown in Fig. 9.6.

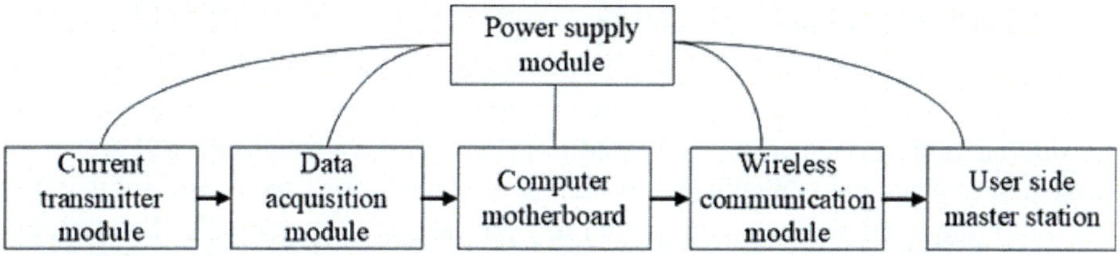

Fig. 9.6 Schematic diagram of the fault discrimination scheme for non-intrusive user internal circuit

Figure 9.6 is a schematic diagram of the non-intrusive internal circuit fault discrimination scheme for users, which includes the power supply module, current transmitter module, Data Acquisition module (DAQ), industrial PC motherboard and wireless communication module.

The function of the power supply module is to supply power to other modules of the monitoring device. The mainboard of the industrial PC is directly connected to the power frequency 220 V alternating current. The wireless communication module and current transmitter module are more suitable to choose 12 V DC voltage for power supply, so the current power supply module finally chooses 220 V to $\pm$ 12 V switching power supply.

The current transmitter module is responsible for converting large current signals into small voltage signals for easy detection. A wire with a high current pass through a hole in the current transmitter, and its electrical capacity is measured at its output.

The function of DAQ is to acquire the output voltage signal of the current transmitter.

The mainboard of IPC receives the voltage signal of the data acquisition module, and processes the corresponding data by using the corresponding fault detection algorithm. The IPC mainboard is the core module of the non-intrusive user internal circuit fault discrimination device. Through the IPC mainboard, the functions of user internal circuit fault signal monitoring, fault type determination, and fault waveform recording can be completed.

The mainboard of IPC communicates with the wireless communication module, which can transmit the fault to the user side master station. The master station summarizes and analyzes the energy efficiency and safety data of electricity consumption, analyzes the fault type, the location and causes of the fault, and finally gives the analysis report of the protection of the current fault action.

9.3 Complex Electromechanical System Applications

With advances in science and technology, the functional requirements of electromechanical equipment are increasingly complex, and the design with high precision and high efficiency is the foundation of producing high-performance complex electromechanical equipment. New design theory must be developed in mechanical design to promote the innovation of modern complex electromechanical equipment. The modern complex electromechanical system is composed of multiple subsystems such as machinery, power, information, and control, etc., and generates overall functions through the interaction among subsystems (also known as coupling). But at the same time, some strangeness also lurks. When the system is in some service environment, the coupling between subsystems that are not designed targets may be stimulated and destroy the operation of the system.

To ensure the safe, reliable, and healthy operation of machinery and equipment, it is of great significance to develop state monitoring and fault diagnosis technology for large and complex electromechanical systems. In fault diagnosis, whether the signal can be correctly analyzed and the effective fault characteristics can be obtained is the premise to analyze, identify and prevent possible faults, so the signal processing technology directly affects the effect of fault diagnosis. The fault model provides the basis for the diagnosis. The fault diagnosis based on the model has strong robustness and wide adaptability. The accurate model can be used to obtain highly reliable fault diagnosis conclusions.

9.3.1 Overview of Complex Electromechanical System Problem Analysis and Solution Research

The complex electromechanical system is the inevitable outcome of the progress of science and technology, and its design and development have important strategic significance for improving industrial competitiveness and strengthening national economic strength. The main feature of a complex electromechanical system is that it consists of a large number of mechanical systems and power systems, and the coupling relationship between them is very strong. Once the failure of a certain subsystem occurs, it may cause many problems such as the running function defect and function weakening of the whole system.

Scholars from all over the world have studied complex electromechanical systems from different research perspectives. Figure 9.7 shows some of the research results. The research field mainly involves complex electromechanical system design, application cases, and problem-solving.

Research on System Design: *Esteves* conducted an ecological design of complex electromechanical systems (Esteves 2014). *Wang* studied the dynamic similarity

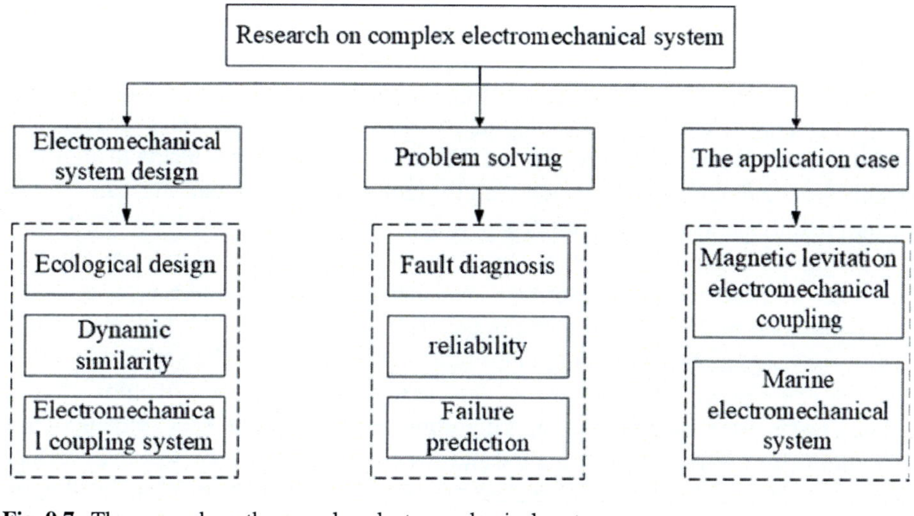

Fig. 9.7 The research on the complex electromechanical system

of complex electromechanical systems based on the bonding graph method (Wang 2010). *Ohki* studied the dynamics of a series of complex electromechanical coupling systems (Ohki 2011).

Research on the problem solving of systems: In terms of fault diagnosis, *Catelani* studies the factors affecting the failure of complex electromechanical systems and evaluates them (Catelani et al. 2010). In the field of reliability research, *Littlefield* et al. discussed the reliability synthesis problem of complex systems with failure-free data (Littlefield et al. 2012). Schumann et al. presented a method for predicting the reliability of systems by using faults and time-varying failure rates (Schumann et al. 2011). Based on multi-class data sources, *Anderson* et al. studied the modeling method of complex system reliability (Anderson-Cook et al. 2012).

Relevant scholars have also done many studies on complex electromechanical systems. The research on decoupling and coupling of complex electromechanical systems has always been a research hot spot. This research mainly introduces the research achievements in solving the coupling issues of complex electromechanical systems, improving system reliability, and reducing complexity.

9.3.2 Motor Fault Detection and Motor Energy Management

Motors provide power for fans, pumps, compressors, machine tools, and other equipment. They are indispensable power driving equipment in industrial production and play an important role in the industrial automation system.

The motor is composed of five independent and interrelated working systems: circuit system, magnetic circuit system, insulation system, mechanical system, and ventilation and heat dissipation system. Under the influence of various operating conditions, such as power supply, load, and environment, the performance of some structures and components in the above system will gradually deteriorate, which will cause various mechanical and electrical faults.

Motor failure will not only damage the motor itself, but also affect the normal operation of the whole system, and even endanger personal safety, causing huge economic losses. Therefore, it is of great theoretical significance and practical application value to carry out research on motor state monitoring and fault diagnosis technology and design and develop a motor fault diagnosis system that can meet the actual requirements of industrial production.

The motor is the main energy consumption equipment in industrial production, however, because of motor technology level backward, low level of energy efficiency, operation load, and rated load does not match, with fault operation, lack of advanced energy management method and other reasons, the current industrial production in the motor operation efficiency is low, the energy waste is serious, the motor system has a great energy-saving potential. Therefore, the research and development of low-cost motor on-line monitoring method and system technology is the key to implement motor energy management and realize the goal of small and medium-sized motor energy conservation.

In addition, the existing motor monitoring technology and means are backward, most of the production and use of motor manufacturers still use the voltmeter, ammeter, tachometer instrument for testing and maintenance, such as low working efficiency not only so, the intensity is big, and in the process of industrial energy-saving transformation, cannot provide sufficient effective motor running data, makes the improvement is lack of effective energy saving. Moreover, the layout of the motor and its control components has been determined by the motor manufacturer, which makes the installation of common monitoring devices heavy and easy to destroy the reasonable space arrangement of the equipment.

(1) A Non-intrusive Asynchronous Motor Rotor Initial Fault Detection Method
When rotor fault occurs to the asynchronous motor, it not only produces deviation in the mechanical structure, but also transmits fault information to signals such as air-gap flux, stator current, electromagnetic torque, and speed through electromagnetic coupling. Thus, a variety of fault detection methods are produced, such as electromagnetic torque detection method, stator residual voltage detection method, rotational speed detection method, vibration detection method, air gap flux detection method, stator voltage or current detection method, and so on. Among them, the electromagnetic torque or speed detection method requires the installation of an expensive torque or speed sensor, which not only increases the complexity of the system but also restricts its use scope due to the construction difficulty and site conditions. Stator residual voltage detection is an off-line detection method, which cannot be realized online. The air gap flux detection method needs to install a detection coil inside the motor, which not only takes a long time to stop but also is difficult to construct. Vibration signal detection is another effective method, but this method not only needs to add additional sensors but also is easy to be disturbed by other mechanical faults.
Besides magnetic flux, the stator current can best reflect the rotor fault, and the detection method based on this signal can be made into a non-intrusive method, which has the strengths of easy implementation and low cost and represents the future development direction of motor fault diagnosis.
To overcome the disadvantages of the above existing technologies, a non-intrusive rotor fault detection method for asynchronous motors is provided. Taking LabVIEW as the software development platform, the modular programming idea is adopted to realize the rotor fault online detection for asynchronous motors. The methods are as follows:
After the voltage and current signals of the motor are collected by the signal acquisition and conditioning equipment, the converted standard voltage/current signal is transmitted to the data acquisition card, and then the signal is transmitted to the computer through wireless communication. The software operating system on the computer mainly includes the functions of real-time signal acquisition, real-time algorithm analysis, and real-time display of results. At the same time, the software platform of the detection system is released to the cloud, allowing users to conduct remote detection and monitoring. The voltage

and current signals in the practical application can be collected by using the current clamp and hall sensor, so there is no need to intrude into the motor shell, which has strong convenience and practicability.

As shown in Fig. 9.8, a non-intrusive rotor initial fault detection method for asynchronous motors is characterized by the following steps:

Step 1: Configure the acquisition card driver before rotor fault detection, then create a signal acquisition task, and configure initial sampling parameters;

Step 2: After the data acquisition module completes the acquisition and processing of signals, the signal is displayed in real time through the waveform control in LabVIEW;

Step 3: Select the corresponding detection algorithm, including fast Fourier transform, Hilbert transform, and Park vector algorithm, calculate the voltage/current signal, and extract the fault feature;

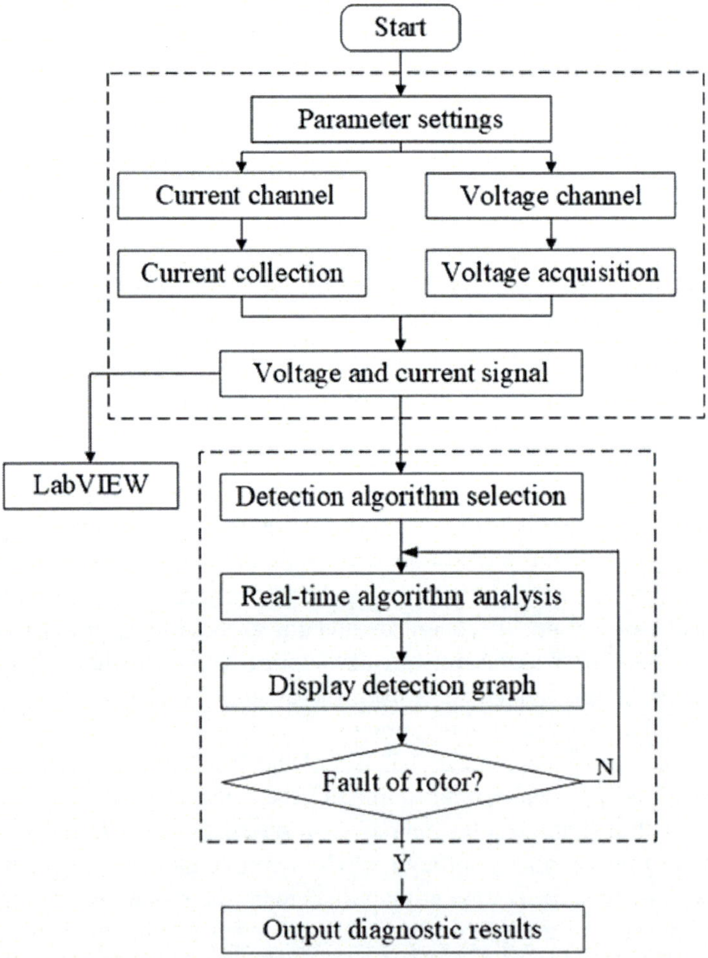

Fig. 9.8 Structure diagram of rotor fault detection method for non-intrusive asynchronous motor

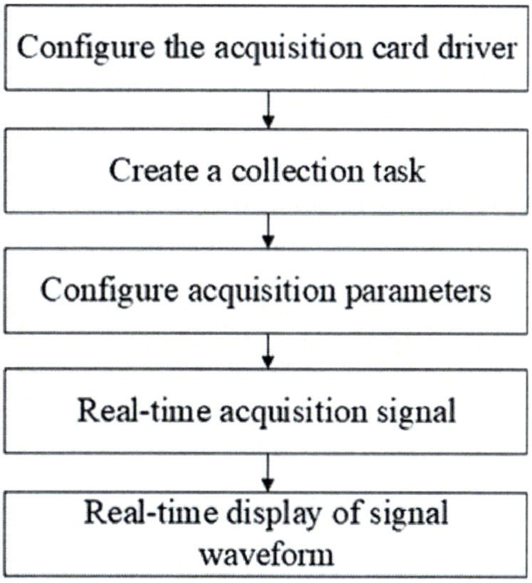

Fig. 9.9 Flow chart of DAQ module

Step 4: Judge whether the rotor is at fault according to the display result of waveform control and the detection principle of the algorithm adopted, output the diagnosis result, and timely feedback the detection result to the user.

In the flow chart of the data acquisition module shown in Fig. 9.9, before the signal acquisition, the driver of NI acquisition card should be configured, the DAQ Create Channel function should be used to create signal acquisition task, and the signal sampling frequency, sampling mode and sampling number of each Channel should be initialized. The acquisition process is mainly completed by DAQ Timing and DAQ Read and other functions. Among them, the DAQ Timing function sets the sampling frequency, the number of samples per channel, the sampling mode and the wiring terminal, and other parameters, while the DAQ Read function completes the function of reading data. Finally, a While loop structure is added outside the module to ensure the function of continuously collecting signals and displaying waveforms in real time.

The flow chart of rotor fault detection is shown in Fig. 9.10. The detection algorithm is selected through TAB control in LabVIEW. When one method is not ideal, another or several detection methods are used for collaborative detection, including fast Fourier transform, Hilbert transform, and Park vector algorithm. These algorithms have a small amount of computation to meet the requirements of on-line real-time detection.

Fast Fourier transform is realized by FFT spectrum sub-VI in LabVIEW. The realization of Hilbert transform needs to transform the data type of signal from the array data type whose content is waveforms to that whose content is a double-precision array. Then the square of the analytic signal module is obtained by using the calculation of "Hilbert transform" sub-VI and the sum of the square module. The DC component is then filtered by taking a negative number after two consecutive Hilbert transformations. Finally, FFT analysis is carried out to

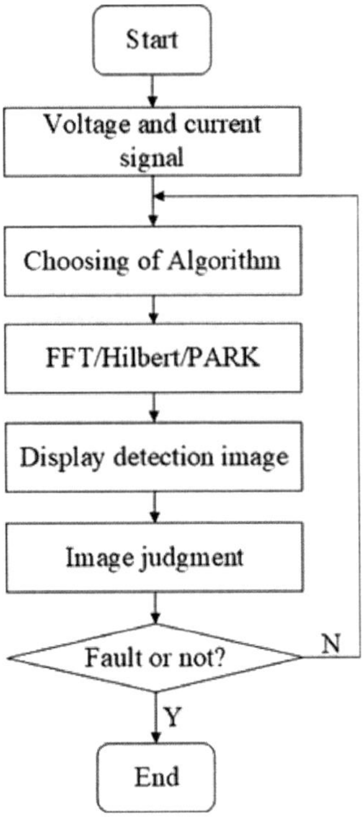

Fig. 9.10 Flow chart of rotor fault detection

obtain the spectrum of analytic signal square mode, and waveform control is used for real-time display.

The implementation of the Park vector algorithm requires the first use of two sub-VI, "index array" and "Waveform acquisition", to obtain the time interval of the waveform and return the waveform data value. Then, the three-phase stator current signal is transformed into the two-phase coordinate system according to the mathematical formula of PARK transformation, and the current signal I_d and I_q are obtained. Finally, the signal is bound into cluster elements by "binding" sub-VI and transmitted to the X-Y graph to display its trajectory shape in real time.

Web page publishing is carried out through the Web publishing tools in LabVIEW. Users only need to enter the given Web address to realize the function of remote online on the computer or mobile phone.

For this non-intrusive asynchronous motor rotor initial fault detection method, only follow the following steps when it is used:

Step 1: The inspector remotely logs into the login interface of the software operating system on the industrial tablet computer, inputs the user and password, and enters the operation interface.

Step 2: Create a signal acquisition task and set relevant acquisition parameters, such as sampling mode, sampling frequency, sampling points, etc.

Step 3: Press the start button of the software operating system to start signal detection.

Step 4: Select the corresponding detection algorithm through the TAB control of the interface.

Step 5: According to the graph displayed in the algorithm analysis module and combined with the corresponding detection principle of the algorithm, distinguish whether there is a rotor fault in the asynchronous motor.

(2) *Non-intrusive motor efficiency on-line detection device*

As mentioned above, the existing motor monitoring technology and means are relatively backward, and the motor system has great energy-saving potential. Therefore, the research and development of low-cost motor on-line monitoring method and system technology is the key to implement motor energy management and realize the goal of small and medium-sized motor energy conservation.

Therefore, aiming at the defects of motor state monitoring in existing technologies, a non-intrusive motor efficiency on-line monitoring system is proposed based on the technical idea of non-intrusive intelligent device identification. The system uploads the monitoring data to the remote monitoring computer and supports the control operation of the remote monitoring computer.

The non-intrusive motor efficiency online detection device is a monitoring device and method based on embedded computing and wireless communication technology. It supports distributed applications and has strong universality. The specific design and implementation framework are shown in Fig. 9.11.

As displayed in Fig. 9.11, the device is composed of a sensor unit, signal acquisition and processing unit, and a communication unit. The circuit board where the three units are located is connected together through cables to complete all functions of signal acquisition, acquisition, processing, and communication. The sensor unit is responsible for converting the line voltage and current of the motor to a weak voltage signal that can be easily measured. The signal acquisition and processing unit collects the weak voltage signal from the sensor unit, digitizes it, calculates the collected data, and sends the required monitoring results to the communication unit. The communication unit sends the output data of the signal acquisition and processing unit to the remote monitoring computer through a wireless network, and forwards the feedback control information of the monitoring computer to the signal acquisition and processing unit.

The details of the work are described in more detail below.

The sensor unit consists of AC-DC module, Hall voltage sensor and Hall current sensor. The AC-DC module supplies power to each sensor and provides the reference ground level of the sensor unit's output signal. The Hall voltage sensor is connected to the two-phase power cord of the induction motor to detect the line voltage of the motor. The hall current sensor's through-hole is threaded through the single-phase power line of the induction motor to detect the line current of the motor.

The signal acquiring and processing unit includes a conditioning circuit module, an analog-digital conversion circuit module, and a Digital Signal Processing

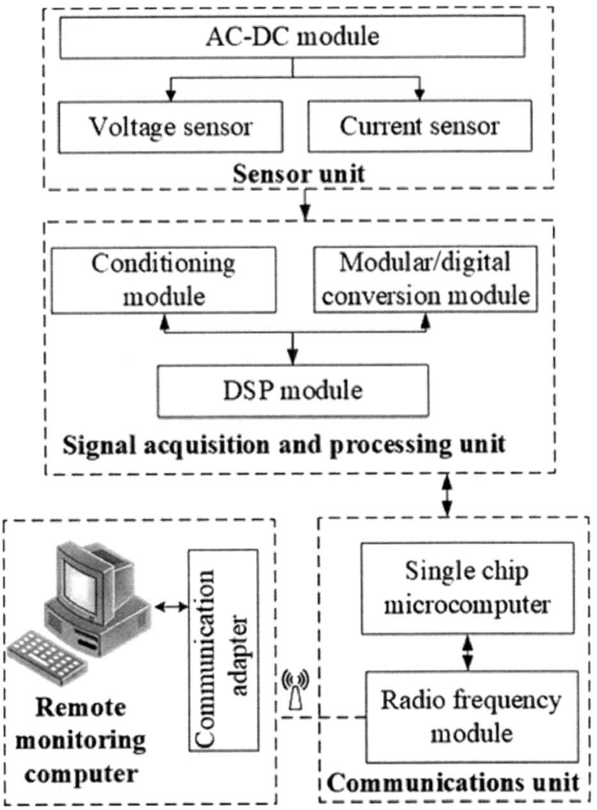

Fig. 9.11 Schematic diagram of non-intrusive motor efficiency monitoring device

(DSP) module, in which the conditioning circuit module is used to adjust the signal input by the sensor unit to the required voltage level. The analog to digital conversion circuit module continuously collects thousands of voltage signals from the conditioning circuit module at one frequency to form a digital sequence with equal time intervals. The DSP module analyzes and processes the above digital sequence and sends the results to the communication unit.

The conditioning circuit module includes a proportional circuit and an isolation circuit composed of an operational amplifier, in which the proportional circuit shares the level of the voltage signal output from the sensor and the signal acquisition and processing unit to scale the amplitude and lift the level of the input voltage signal. The isolation circuit is a voltage follower, which serves as a first-level buffer component before the input of the analog/digital conversion circuit module.

The core of the communication unit is the radio frequency module and the single-chip microcomputer, which runs the wireless protocol stack, controls the work of the radio frequency module and provides serial communication interface support.

By adopting the non-intrusive online monitoring technology, the state of the monitored motor can be reflected quickly and timely. This technology provides sufficient data support for further motor operation analysis, helps people realize energy management of the motor system, and achieves the goal of energy saving of small and medium-sized motors.

9.3.3 *Ship Electromechanical System*

(1) *Composition of the Marine electromechanical system*
The term "Marine electromechanical system" refers to the mechanical, electrical, and electrical installations that promote the navigation of a ship and maintain its performance, structural integrity, and life services for personnel. Figure 9.12 is the main component of the Marine electromechanical system, including Marine power plant, Marine pump, and valve parts, deck machinery, ship auxiliary machinery, auxiliary machinery and electrical, electrical control and other related equipment.
(2) *Non-intrusive load monitoring of ship fault*
In recent years, the U.S. Navy and some commercial ships have been reducing the number of personnel on board to better reduce operating costs in commercial shipping and munitions costs on military ships (Ljung and Lützhöft 2014). Therefore, to ensure the normal operation of the ship's electromechanical system has to rely on advanced scientific control technology (Logan 2007; Eruguz et al. 2017). These complex electromechanical systems often require a single sensor to measure the motor's speed and vibration (Schantz et al. 2016; Zachar et al. 2016; Dzapo et al. 2009). If non-intrusive monitoring of ship electromechanical systems can be realized, more vessel maintenance costs will be reduced.
On a ship, in addition to a separate electrical system, the diesel engines that power the ship also require auxiliary electrical equipment, such as pumps and heaters. The purpose of these electrical devices is to keep the engine running in standby mode (Lindahl et al. 2018).
The power supply stability of the auxiliary load is related to the operation efficiency of the Marine electromechanical system. On the USCGC SPENCER ship of the US Coast Guard, two electrical subpanels supply power to these and several other engine loads (like Jacket Water (JW) heater and Lube Oil (LO) heater) that are essential to the ship's operation. These loads work in conjunction

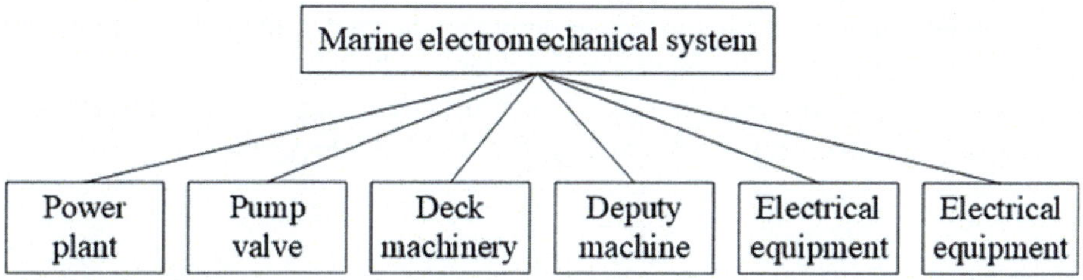

Fig. 9.12 The main components of the marine electromechanical system

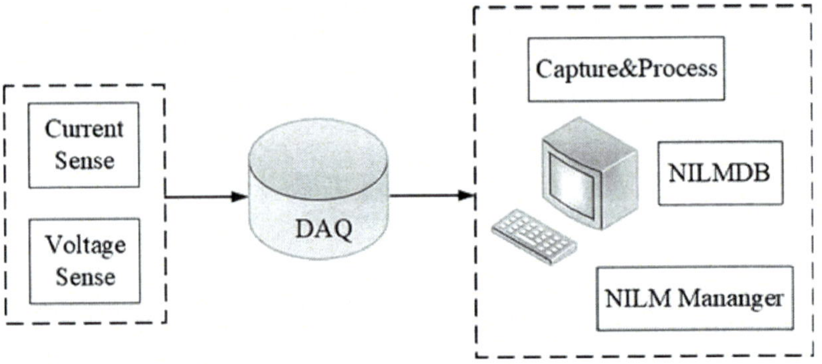

Fig. 9.13 The design concept of the NILM system

with other loads as part of the control system, such as the thermal insulation system for diesel machines. To track the operation of these cabin systems, NILM instruments were installed on two subpanels, and a state monitoring method was developed and tested (Lindahl et al. 2018).

Figure 9.13 shows the design concept of the NILM system. The three-phase voltage and current are first measured by NILM meters mounted on two electrical subpanels, and then the DAQ samples the current and voltage output of the meter. Finally, for the collected data, the data is transmitted to the host via Ethernet. In the host computer, the signals is processed into a power flow, which is then stored by the NILM Optimization database (Paris et al. 2014a) and supplied to the NILM Manager, which can create and run load recognition and status monitoring algorithms (Donnal et al. 2016).

In more detail, the data acquisition unit DAQ uses 16-bit resolution. And then performs high-frequency (8 kHz) sampling on the sensor output of each channel. With this high sampling rate and resolution, accurate measurements of power transients can be achieved, helping to identify the changing state of individual loads such as oil pumps and heaters. Meanwhile, in order to keep the transient information well and reduce the storage space of data in the host, the host needs to preprocess the original sensor data. Good results can be achieved by using the sinusoidal fitting method (Paris et al. 2014b), which extracts the spectrum envelope of each phase current in the period of each measured voltage line.

For the load decomposition algorithm, two neural network hidden layers are adopted for decomposing the load. The transient shape indexes are extracted from the active and reactive power flows and then input into the neural network. Based on the load information of each device, the NILM system computes the operation measurement related to the fault of the ship switch control system, and then implements the load equipment and fault detection article.

9.4 Environmental Pollution Monitoring Applications

9.4.1 Overview of Global Environmental Pollution and Environmental Monitoring

With the progress of the world, human civilization has reached a new height, but the energy-intensive economic development has brought great pressure to the ecological environment (Udemba 2020; Wang et al. 2018). According to relevant studies, the air quality in most cities in the world exceeds the standard (Samad and Vogt 2020; Sun et al. 2019). Meanwhile, the problem of water pollution is becoming increasingly serious (Kumar et al. 2019; Liu et al. 2018). According to the urban environmental monitoring network, people living in cities often face the problem of noise pollution (Yang et al. 2020; Chiarini et al. 2020). The problem of environmental pollution has become more and more serious, among which industrial production is the main culprit (Du and Li 2020; Han et al. 2019).

Environmental monitoring is defined as the use of biochemical methods to analyze the proportion of environmental factors and hazards in the environment, and according to the results to correctly evaluate the quality of the environment and the trend of change.

Relying on its advanced information system, the United States has built a complex environmental pollution monitoring system, which enables comprehensive, real-time, and accurate monitoring of the environmental pollution problems of the whole country (He and Wu 2011). The EU environmental monitoring system is mainly divided into three contents: data collection and information network for chemical pollution sources in the environment, ecological environment, and resident health information database, and coordination of information on the environment (CORINE) program. CORINE is a monitoring project aimed at collecting and coordinating environmental and natural resource data in the European Union countries (Fritz 2017). Environmental information systems in the Asia-Pacific region mainly include the United Nations Environmental Technology Transfer Center for Asia and The Pacific, Asia-pacific Environmental Protection Association, Japan International Environmental Technology Transfer Center, Singapore Environmental Technology Transfer Information Base, etc. (Martin 1993).

The current environmental monitoring system basically takes the form of air quality monitoring stations to monitor and predict air pollution status. However, the number and distribution of air quality monitoring stations are always limited due to the large space required and high cost. At the same time, due to the limitation of the monitoring range, air quality monitoring stations cannot monitor the concentration of air pollutants in every corner (Ma et al. 2019). In fact, the data provided by these stations do not reflect the real-time spatial distribution of air pollution in the region. Therefore, based on the idea of non-intrusive identification, it is of more practical significance to infer the air pollution status of areas without monitoring stations through the data collected by a small number of monitoring stations, and finally reflect the real-time spatial distribution of air pollution in this area.

9.4.2 Non-intrusive Spatial and Temporal Monitoring of Air Quality

As mentioned above, it is meaningful to realize air pollution monitoring in areas without monitoring stations. At present, satellite remote sensing and space interpolation techniques have been used to achieve this task. (Engel-Cox et al. 2004) proposed remote sensing technology to retrieve air quality and some scholars used satellite remote sensing to measure the concentration and distribution of PM2.5 (Gupta et al. 2006; Huang et al. 2019; Yang et al. 2019). However, satellite remote sensing technology also has some inevitable limitations (Martin 2008), so other scholars, inspired by the idea of non-intrusive, put forward the spatial interpolation/extrapolation method.

Similar to the research idea of non-intrusive smart device identification, spatial interpolation can use data from monitoring points to estimate the air pollution status of non-monitoring points. Common methods of spatial interpolation include distance weighting, spline , and reverse distance weighting (IDW), etc. (Lam 1983; Joseph et al. 2013). On the basis of spatial interpolation, some scholars also took into account the time series characteristics of pollution data and established a temporal-spatial prediction model for air pollution moniinterpolation, Kriging interpolationtoring based on machine learning and neural network, and realized the temporal-spatial distribution prediction of PM2.5 and ozone (Qi et al. 2019; Ma et al. 2019; Zhan et al. 2018).

The framework of non-intrusive air pollution monitoring based on the spatial interpolation method is shown in Fig. 9.14. The implementation steps can be summarized as follows:

Step 1: Monitoring station data acquisition and data preprocessing, including air pollutants and monitoring point locations (latitude and longitude).

Step 2: Time series modeling of historical data and temporal-spatial correlation analysis.

Step 3: Spatial interpolation/extrapolation of air pollution at non-monitoring points.

Through the above non-intrusive monitoring framework, only the basic air pollutant detector, GPS positioning, and spatial interpolation algorithms (such as Kriging algorithm, IDW algorithm, and Geo-LSTM, etc.) can realize the temporal-spatial distribution prediction of air pollution in the region.

Therefore, according to the above air pollution monitoring framework, it is not necessary to monitor every corner of the area to know the air pollution situation here, which is also the essence of non-intrusive identification technology.

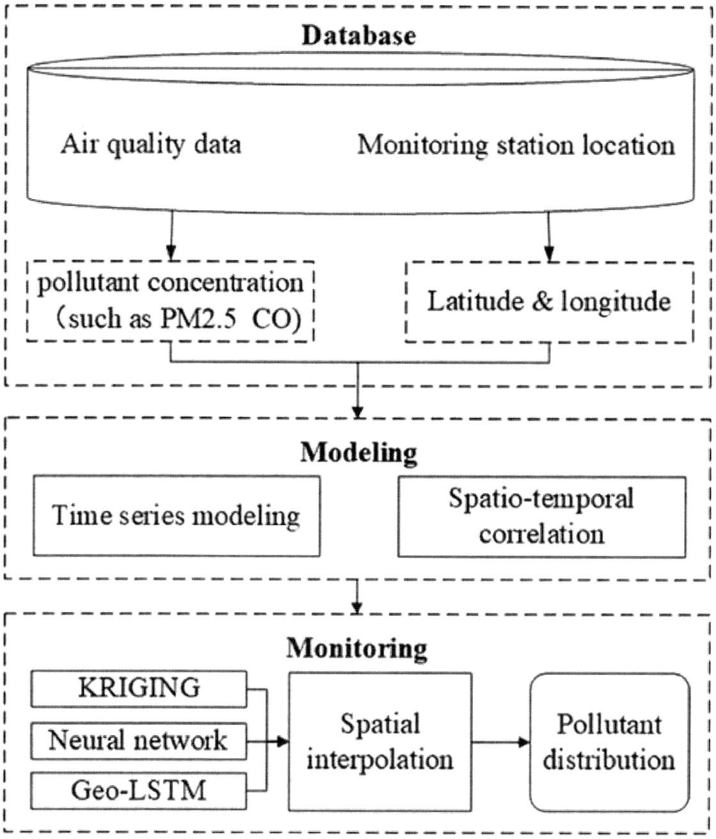

Fig. 9.14 Spatial and temporal monitoring framework of non-intrusive air pollution

9.4.3 Intelligent Identification of Environmental Pollution Sources in Industrial Parks

As is known to all, industrial parks have many types of industries, such as food manufacturing, textile, chemical manufacturing, storage, and so on. And the factory of each industry can produce a variety of pollutants, but the emission of pollutants from each factory has its characteristics. Chemical manufacturing, for example, emits more polluting gases than other industries, and textiles have a different composition of sewage than food manufacturing. Therefore, the collection and analysis of environmental pollutants can determine which factory or enterprise the pollutant comes from. Through the identification of pollution sources, and then timely implementation of their pollutant discharge management, to ensure environmental quality. Based on the technical idea of non-intrusive intelligent identification, a scheme design for environmental pollution monitoring and pollution source identification in industrial parks is presented below.

(1) *System structure design*

The system structure can be divided into two parts: environmental pollution monitoring terminal and environmental pollution monitoring management platform, among which environmental pollution monitoring terminal is composed of the multi-node data acquisition terminal. The environmental pollution monitoring and managing platform mainly include two parts: environmental pollution monitoring server and environmental monitoring and management center.

The environmental pollution monitoring terminal is mainly responsible for the real-time collection of environmental parameters, and then these environmental parameters are transmitted to the environmental pollution monitoring server through wireless network technology. The environmental pollution monitoring server mainly realizes the functions of data receiving, display, analysis, storage, and illustration, etc. The responsibility of the environmental monitoring and management center is to realize the unified monitoring and management of environmental pollution in industrial parks. Figure 9.15 is the structural design drawing of the environmental pollution monitoring and pollution source identification system.

(2) *System function design*

According to the functional demand analysis of the system structure design, the system functional modules can be divided into environmental pollution monitoring terminal and environmental pollution monitoring management platform. Among them, the environmental pollution monitoring terminal takes the computer processor as the core treatment unit, and realizes the real-time monitoring of the environmental noise intensity, harmful gas concentration, water quality, PM2.5 concentration, and other indicators in the industrial area through the external environmental pollutant measurement sensor.

The environmental pollution monitoring server is programmed with database technology. Its function is to receive, store, and display the data from the multi-node data acquisition terminal, and make an accurate judgment on the received data through the data analysis algorithm (namely NILM algorithm). When the

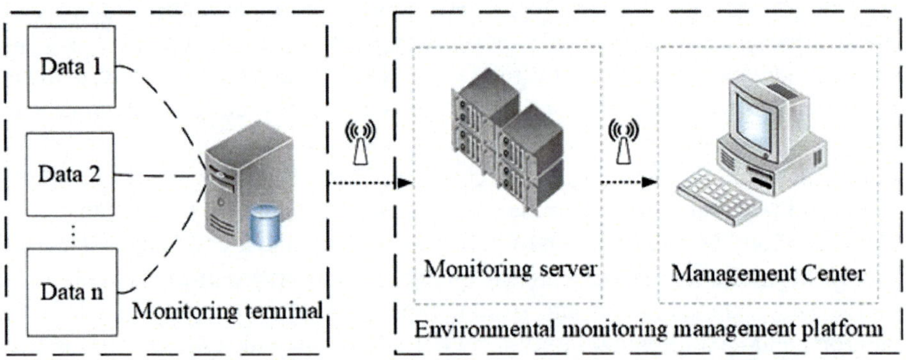

Fig. 9.15 Design drawing of environmental pollution monitoring and alarm system

data is abnormal (suspicious source location is found), an alarm will be sent to the environmental monitoring management Center on the server side.
The environmental monitoring management center receives the alarm signal sent by the environmental pollution monitoring server, compares the abnormal data with the historical data, and identifies the enterprises or factories that may pollute the environment. Then, according to the results of noninvasive spatial and temporal monitoring of air quality, the exact location of pollution sources is matched. Finally, the staff of the environmental management department is dispatched to patrol and check whether the environmental pollutants in the enterprise or factory exceed the standard, so as to realize the real-time control and monitoring of environmental pollution and improve the environmental quality. The specific system function design block diagram is shown in Fig. 9.16.
Therefore, with the help of the smart industrial park environmental pollution monitoring and alarm system, a number of environmental pollution factors and characteristic pollutants can be qualitatively and quantitatively analyzed in real time according to the set methods, and the environmental conditions of the industrial park can be controlled in real time to ensure that the environmental quality is up to the standard.
It should be noted that the application of non-intrusive technology in environmental monitoring is still rare, so the environmental pollution monitoring system proposed in this section is only a potential application of non-intrusive ideas.

9.5 Other Applications

Although the application of NILM is still concentrated in power and power-relaõted systems, NILM technology has been involved in various fields with the efforts of some scholars, which has opened up the way for the research and application of NILM technology in the future.

9.5.1 Substations and Distributed Energy Sources

Sustainable smart cities and smart communities have now become the common direction of the world, and distributed energy plays an important role in this regard. However, for a variety of reasons, monitoring the load with a distributed network is not a mature technology. Therefore, this is a more challenging task for the distribution network operator.

Moreover, installing sensors for each distributed energy source (DER) in a distributed energy network is impractical. Therefore, to monitor DER connected to regional substations, researchers have tried to adopt new techniques to detect DER (Wang et al. 2020). The purpose of this method is to decompose the regional comprehensive load into a typical load and a flexible load, which can be used to

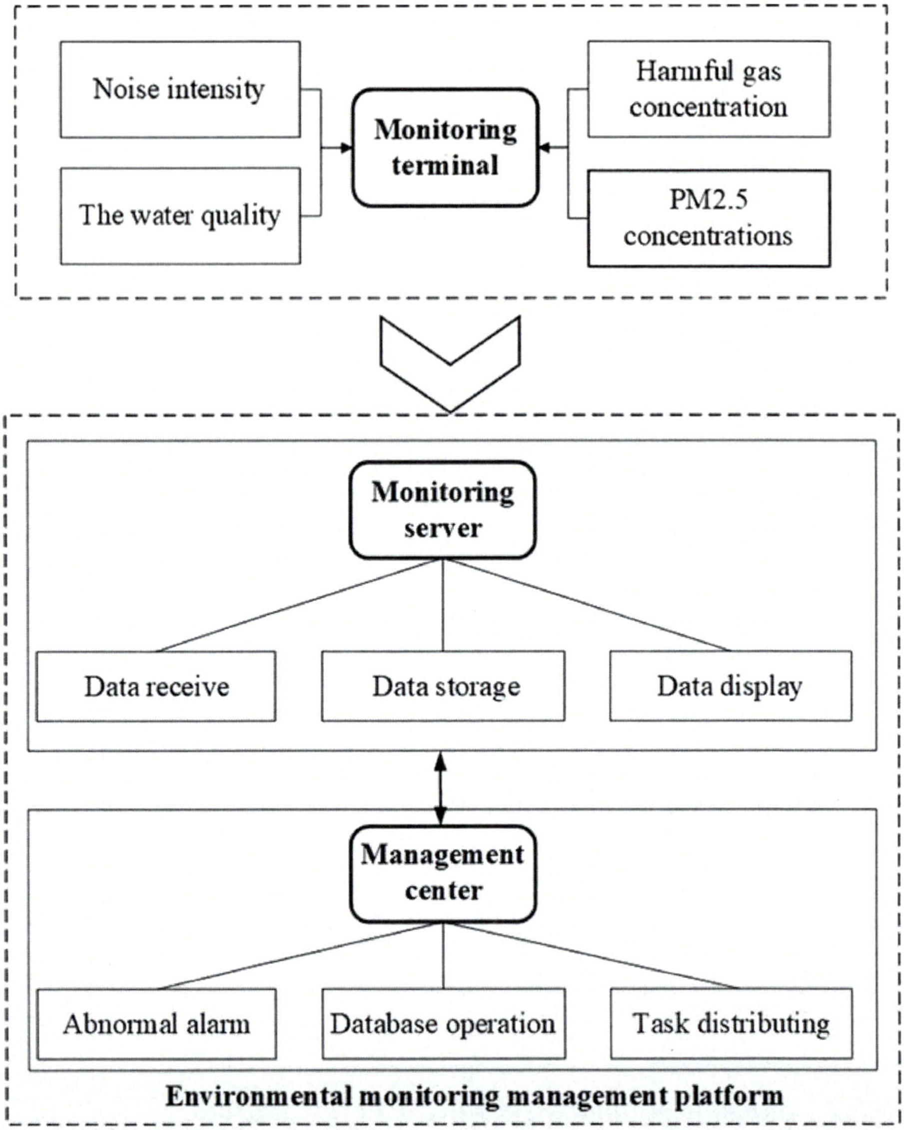

Fig. 9.16 Functional design block diagram of the environmental pollution monitoring system

estimate the number of electric vehicles. Similarly, to monitor the load's connection with local renewable energy sources, researchers also conducted NILM experiments. In the experiment, the decomposition of energy is realized by means of frequency-domain characteristics and total harmonic distortion (Aleksandrs.Suzdalenko 2013). But accurately decomposing the energy will be a challenge due to the presence of harmonics and noise.

Although non-intrusive identification has been applied in substations and distributed energy, there are still some challenges, such as how to reduce and minimize the error in traditional prediction, how to carry out effective regional decomposition and plenty of problems need to be solved (Wang et al. 2020). In addition, to improve the Electric Vehicle (EV) and Photo-Voltaic (PV) profiles, a cost-effective solution for monitoring the state of the distribution network may be provided by NILM.

9.5.2 Monitoring of Common Problems in Industrial Production

To measure the concentration of flowing liquid, Twu and Chen studied a compact displacement sensor based on the idea of non-intrusive identification (Twu and Chen 2017). The sensor adopts the phase query technology in the zero-difference interferometer to solve the real-time quality control problem of fluid in the production line. In addition, Jo et al. proposed a non-intrusive ultrasonic signal detection method, which can detect the gas-water interface in circular pipes (Jo et al. 2020). The experimental results show that the proposed non-intrusive acoustic method can be used for qualitative and quantitative analysis of the gas-liquid interface in a circular tube. Scaling is a major obstacle in pneumatic conveying systems. *Haugland* et al. studied the complex mechanism of scale formation using non-intrusive monitoring techniques and evaluated the feasibility of acoustic sensors for scale growth monitoring in pneumatic conveying systems (Haugland et al. 2019). In addition, to solve the problem of sand production in the process of oil and gas exploitation of sandstone reservoirs, based on the idea of non-intrusive, Gao et al. proposed the ultrasonic sensor sand production monitoring system, and developed a program to identify and filter acoustic noise (Gao et al. 2015).

9.5.3 NILM Health Services for the Elderly

As the population ages, the number of elderly people in the world population is increasing. While more elderly people pursue an independent life, which has subtly promoted the development of welfare undertakings and health services. As the existing health monitoring sensors are intrusive and limited by wireless sensor networks, their actual effect is not good. Based on non-intrusive load monitoring, Alcalá et al. proposed a new technique to monitor the activities of the elderly and provided a rating scale for normal behavior (Alcalá et al. 2017).

The daily activities of the elderly in their living environment are often accompanied by some potential hazards, so they may use some devices (such as electronic crutches) to facilitate their movement. In the absence of guardian care, pattern recognition of devices used in the residential environment of the elderly will help to identify their health conditions, such as sleep quality, memory problems (Ruano et al. 2019). In this case, NILM was able to monitor the activity of the elderly using normal patterns of everyday appliances. In the absence of a guardian, deviations from the behavior of the appliance can be used as a threshold to distinguish between normal and abnormal activities to identify and alert the elderly if they are at risk. In addition, more sensors can be combined to relate their health problems. Based on the fuzzy logic algorithm, Er and Tan proposed a non-intrusive fall monitoring method for the elderly, which can be used to detect accidental falls of the elderly (Er and Tan 2018).

9.5.4 Speech Quality Measurement

Voice quality assessment is an important part of service quality in a complex environment. Most of the traditional assessment methods are intrusive. In recent years, non-intrusive assessment is getting more and more attention in this field. In this study, to obtain clean speech, Zhou and Zhu proposed a non-intrusive online speech reconstruction algorithm based on the Bayesian non-negative matrix (Zhou and Zhu 2019). Based on the biologically-inspired computational model of the auditory system, Jassim and Zilany proposed a new non-intrusive speech quality measurement method, namely the neurogram speech quality measurement method (Jassim and Zilany 2019). Their proposed non-intrusive speech quality measurement method achieves better overall results than most existing methods in identifying some indicators (such as additional noise and channel noise). It's valuable to study the measurement of atonic speech. By adjusting the energy ratio of speech reverberation, Falk et al. achieved a non-intrusive measurement of speech and reverberation. The experimental results show that the modulation spectrum has a good understanding of atonal speech (Falk et al. 2010). In the communication network, in order to improve the quality of voice, the traditional method needs the network source voice signal. In order to simplify the processing of speech quality, Doh-Suk and Tarraf proposed a non-intrusive auditory assessment model on the premise of ensuring speech quality (Doh-Suk and Tarraf 2004). In the experiment, although the model no longer needs the source speech information, the experimental results show that the model is available.

Obviously, the application of non-intrusive identification technology for both industry and home, environment, and health goes far beyond these. In the trend of the era of artificial intelligence, more and better non-intrusive intelligent device identification technology will be used to serve human beings. People's quality of life, and health services are bound to improve with the prosperity of the UEIOT.

References

Alcalá, J. M., Ureña, J., Hernández, Á., Gualda, D. (2017). Assessing Human Activity in Elderly People Using Non-intrusive Load Monitoring. Sensors 17 (2). https://doi.org/10.3390/s17020351

Aleksandrs Suzdalenko (2013). Case Study on Using Non-intrusive Load Monitoring System with Renewable Energy Sources in Intelligent Grid Applications. In: 2013 international conference-workship compatibility and power electronics, pp. 115–119

Anderson-Cook, C. M., Klamann, R. M., & Morzinski, J. (2012). Modeling the Reliability of Complex Systems with Multiple Data Sources: A Case Study on Making Statistical Tools Accessible to Engineers. *Quality Engineering, 24*(2), 280–291. https://doi.org/10.1080/08982112.2012.641152.

Biansoongnern, S., & Plungklang, B. (2016). Non-intrusive Appliances Load Monitoring (NILM) for Energy Conservation in Household with Low Sampling Rate. *Procedia Computer Science, 86,* 172–175. https://doi.org/10.1016/j.procs.2016.05.049.

Bose, B. K. (2010). Global Warming: Energy, Environmental Pollution, and the Impact of Power Electronics. *IEEE Industrial Electronics Magazine, 4*(1), 6–17. https://doi.org/10.1109/MIE.2010.935860.

Briggs, D. (2003). Environmental pollution and the global burden of disease. *British Medical Bulletin, 68,* 1–24. https://doi.org/10.1093/bmb/ldg019.

Catelani M, Ciani L, Luongo V, Singuaroli R Evaluation of the Safe Failure Fraction for an electromechanical complex system: remarks about the standard IEC61508. In: 2010 IEEE Instrumentation & Measurement Technology Conference Proceedings, 3–6 May 2010 2010. pp 949-953. https://doi.org/10.1109/imtc.2010.5488034

Chiarini, B., D'Agostino, A., Marzano, E., & Regoli, A. (2020). The perception of air pollution and noise in urban environments: A subjective indicator across European countries. *Journal of Environmental Management, 263,* 110272. https://doi.org/10.1016/j.jenvman.2020.110272.

Colgan, J. D. (2014). Oil, Domestic Politics, and International Conflict. *Energy Research & Social Science, 1,* 198–205. https://doi.org/10.1016/j.erss.2014.03.005.

Delucchi, M. A., & Jacobson, M. Z. (2011). Providing all global energy with wind, water, and solar power, Part II: Reliability, system and transmission costs, and policies. *Energy Policy, 39*(3), 1170–1190. https://doi.org/10.1016/j.enpol.2010.11.045.

Doh-Suk K, Tarraf (2004) A Perceptual model for non-intrusive speech quality assessment. In: 2004 IEEE International Conference on Acoustics, Speech, and Signal Processing, 17–21 May 2004. pp iii-1060. https://doi.org/10.1109/icassp.2004.1326731

Donnal, J. S., Paris, J., & Leeb, S. B. (2016). Energy Applications for an Energy Box. *IEEE Internet of Things Journal, 3*(5), 787–795. https://doi.org/10.1109/JIOT.2016.2560123.

Du W, Li M (2020) Assessing the impact of environmental regulation on pollution abatement and collaborative emissions reduction: Micro-evidence from Chinese industrial enterprises. Environmental Impact Assessment Review 82. doi:10.1016/j.eiar.2020.106382

Dzapo, H., Stare, Z., & Bobanac, N. (2009). Digital Measuring System for Monitoring Motor Shaft Parameters on Ships. *IEEE Transactions on Instrumentation and Measurement, 58*(10), 3702–3712. https://doi.org/10.1109/TIM.2009.2019316.

Engel-Cox, J. A., Hoff, R. M., & Haymet, A. D. (2004). Recommendations on the use of satellite remote-sensing data for urban air quality. *Journal of the Air and Waste Management Association, 54*(11), 1360–1371. https://doi.org/10.1080/10473289.2004.10471005.

Er, P. V., & Tan, K. K. (2018). Non-intrusive fall detection monitoring for the elderly based on fuzzy logic. *Measurement, 124,* 91–102. https://doi.org/10.1016/j.measurement.2018.04.009.

Eruguz, A. S., Tan, T., & van Houtum, G.-J. (2017). A survey of maintenance and service logistics management: Classification and research agenda from a maritime sector perspective. *Computers & Operations Research, 85,* 184–205. https://doi.org/10.1016/j.cor.2017.03.003.

Esteves OM, Almeida F. (2014) The Eco-design of Complex Electromechanical Systems:Prioritizing and Balancing Performance Fields. Contributors and Solutions,Technology and Manufacturing Process Selection (Springer London):257–279. https://doi.org/10.1007/978-1-4471-5544-7

Falk, T. H., Zheng, C., & Chan, W. (2010). A Non-Intrusive Quality and Intelligibility Measure of Reverberant and Dereverberated Speech. *IEEE Transactions on Audio, Speech and Language Processing, 18*(7), 1766–1774. https://doi.org/10.1109/TASL.2010.2052247.

Fang, Y., & Zeng, Y. (2007). Balancing energy and environment: The effect and perspective of management instruments in China. *Energy, 32*(12), 2247–2261. https://doi.org/10.1016/j.energy.2007.07.016.

Fritz, S. (2017). Book Review. *International Journal of Applied Earth Observation and Geoinformation, 59,* 63–64. https://doi.org/10.1016/j.jag.2017.02.020.

Gao, G., Dang, R., Nouri, A., Jia, H., Li, L., Feng, X., et al. (2015). Sand rate model and data processing method for non-intrusive ultrasonic sand monitoring in flow pipeline. *Journal of Petroleum Science and Engineering, 134,* 30–39. https://doi.org/10.1016/j.petrol.2015.07.001.

Gupta, P., Christopher, S. A., Wang, J., Gehrig, R., Lee, Y., & Kumar, N. (2006). Satellite remote sensing of particulate matter and air quality assessment over global cities. *Atmospheric Environment, 40*(30), 5880–5892. https://doi.org/10.1016/j.atmosenv.2006.03.016.

Han, X., Sun, T., & Feng, Q. (2019). Study on environmental pollution loss measurement model of energy consumption emits and its application in industrial parks. *Science of the Total Environment, 668,* 1259–1266. https://doi.org/10.1016/j.scitotenv.2019.03.002.

Haugland, I. B., Chladek, J., & Halstensen, M. (2019). Monitoring of scaling in dilute phase pneumatic conveying systems using non-intrusive acoustic sensors – A feasibility study. *Advanced Powder Technology, 30*(8), 1634–1641. https://doi.org/10.1016/j.apt.2019.05.012.

Huang, C.-S., Lin, T.-H., Hung, H., Kuo, C.-P., Ho, C.-C., Guo, Y.-L., et al. (2019). Incorporating satellite-derived data with annual and monthly land use regression models for estimating spatial distribution of air pollution. *Environmental Modelling and Software, 114,* 181–187. https://doi.org/10.1016/j.envsoft.2019.01.010.

Jacobson, M. Z., & Delucchi, M. A. (2011). Providing all global energy with wind, water, and solar power, Part I: Technologies, energy resources, quantities and areas of infrastructure, and materials. *Energy Policy, 39*(3), 1154–1169. https://doi.org/10.1016/j.enpol.2010.11.040.

Jassim, W. A., & Zilany, M. S. (2019). NSQM: A non-intrusive assessment of speech quality using normalized energies of the neurogram. *Computer Speech & Language, 58,* 260–279. https://doi.org/10.1016/j.csl.2019.04.005.

Jo H, Song YJ, Jo D (2020) Non-intrusive detection of gas–water interface in circular pipes inclined at various angles. Annals of Nuclear Energy 139. doi:10.1016/j.anucene.2019.107267

Joseph, J., Sharif, H. O., Sunil, T., & Alamgir, H. (2013). Application of validation data for assessing spatial interpolation methods for 8-h ozone or other sparsely monitored constituents. *Environmental Pollution, 178,* 411–418. https://doi.org/10.1016/j.envpol.2013.03.035.

Kumar, V., Parihar, R. D., Sharma, A., Bakshi, P., Singh Sidhu, G. P., Bali, A. S., et al. (2019). Global evaluation of heavy metal content in surface water bodies: A meta-analysis using heavy metal pollution indices and multivariate statistical analyses. *Chemosphere, 236,* 124364. https://doi.org/10.1016/j.chemosphere.2019.124364.

Lam, N. S.-N. (1983). Spatial Interpolation Methods: A Review. *The American Cartographer, 10*(2), 129–150. https://doi.org/10.1559/152304083783914958.

Lindahl, P. A., Green, D. H., Bredariol, G., Aboulian, A., Donnal, J. S., & Leeb, S. B. (2018). Shipboard Fault Detection Through Nonintrusive Load Monitoring: A Case Study. *IEEE Sensors Journal, 18*(21), 8986–8995. https://doi.org/10.1109/jsen.2018.2869115.

Littlefield S, Mazzuchi T, Sarkani S (2012) Predicting reliability in design of complex systems with common-cause failures and time-varying failure rates. In: 2012 IEEE International Systems Conference SysCon 2012, 19–22 March 2012 2012. pp 1-5. https://doi.org/10.1109/syscon.2012.6189455

Liu, W., Yang, H., Liu, Y., Kummu, M., Hoekstra, A. Y., Liu, J., et al. (2018). Water resources conservation and nitrogen pollution reduction under global food trade and agricultural intensification. *Science of the Total Environment, 633,* 1591–1601. https://doi.org/10.1016/j.scitotenv.2018.03.306.

Ljung, M., & Lützhöft, M. (2014). Functions, performances and perceptions of work on ships. *WMU Journal of Maritime Affairs, 13*(2), 231–250. https://doi.org/10.1007/s13437-014-0057-x.

Logan, K. P. (2007). Intelligent Diagnostic Requirements of Future All-Electric Ship Integrated Power System. *IEEE Transactions on Industry Applications, 43*(1), 139–149. https://doi.org/10.1109/TIA.2006.886993.

Ma J, Ding Y, Cheng JCP, Jiang F, Wan Z (2019) A temporal-spatial interpolation and extrapolation method based on geographic Long Short-Term Memory neural network for PM2.5. Journal of Cleaner Production 237. https://doi.org/10.1016/j.jclepro.2019.117729

Martin, P. K. (1993). Energy and environmental management information systems. *Applied Energy, 44*(2), 175–183. https://doi.org/10.1016/0306-2619(93)90060-3.

Martin, R. V. (2008). Satellite remote sensing of surface air quality. *Atmospheric Environment, 42*(34), 7823–7843. https://doi.org/10.1016/j.atmosenv.2008.07.018.

Ohki M (2011) Application of Complex Series Dynamics to Electromechanical Coupling System. Japanese Journal of Applied Physics 50 (7):07HB05. https://doi.org/10.1143/jjap.50.07hb05

Omer, A. M. (2008). Energy, environment and sustainable development. *Renewable and Sustainable Energy Reviews, 12*(9), 2265–2300. https://doi.org/10.1016/j.rser.2007.05.001.

Panwar, N. L., Kaushik, S. C., & Kothari, S. (2011). Role of renewable energy sources in environmental protection: A review. *Renewable and Sustainable Energy Reviews, 15*(3), 1513–1524. https://doi.org/10.1016/j.rser.2010.11.037.

Paris, J., Donnal, J. S., & Leeb, S. B. (2014a). NilmDB: The Non-Intrusive Load Monitor Database. *IEEE Transactions on Smart Grid, 5*(5), 2459–2467. https://doi.org/10.1109/TSG.2014.2321582.

Paris, J., Donnal, J. S., Remscrim, Z., Leeb, S. B., & Shaw, S. R. (2014b). The Sinefit Spectral Envelope Preprocessor. *IEEE Sensors Journal, 14*(12), 4385–4394. https://doi.org/10.1109/JSEN.2014.2334618.

Pleßmann, G., Erdmann, M., Hlusiak, M., & Breyer, C. (2014). Global Energy Storage Demand for a 100% Renewable Electricity Supply. *Energy Procedia, 46,* 22–31. https://doi.org/10.1016/j.egypro.2014.01.154.

Qi, Y., Li, Q., Karimian, H., & Liu, D. (2019). A hybrid model for spatiotemporal forecasting of PM2.5 based on graph convolutional neural network and long short-term memory. *Science of the Total Environment, 664,* 1–10. https://doi.org/10.1016/j.scitotenv.2019.01.333.

Rashid, H., Singh, P., Stankovic, V., & Stankovic, L. (2019). Can non-intrusive load monitoring be used for identifying an appliance's anomalous behaviour? *Applied Energy, 238,* 796–805. https://doi.org/10.1016/j.apenergy.2019.01.061.

Resch, G., Held, A., Faber, T., Panzer, C., Toro, F., & Haas, R. (2008). Potentials and prospects for renewable energies at global scale. *Energy Policy, 36*(11), 4048–4056. https://doi.org/10.1016/j.enpol.2008.06.029.

Ruano A, Hernandez A, Ureña J, Ruano M, Garcia J (2019) NILM Techniques for Intelligent Home Energy Management and Ambient Assisted Living: A Review. Energies 12 (11). https://doi.org/10.3390/en12112203

Samad A, Vogt U (2020) Investigation of urban air quality by performing mobile measurements using a bicycle (MOBAIR). Urban Climate 33. doi:10.1016/j.uclim.2020.100650

Schantz, C., Gerhard, K., Donnal, J., Moon, J., Sievenpiper, B., Leeb, S., et al. (2016). Retrofittable Machine Condition and Structural Excitation Monitoring From the Terminal Box. *IEEE Sensors Journal, 16*(5), 1224–1232. https://doi.org/10.1109/JSEN.2015.2498626.

Schumann, J., Cate, K., & Lee, A. (2011). Analysis of Air Traffic Track Data with the AutoBayes Synthesis System. In M. Alpuente (Ed.), *Logic-Based Program Synthesis and Transformation, Berlin, Heidelberg, 2011* (pp. 21–36). Berlin Heidelberg: Springer.

Sun C, Zhang W, Fang X, Gao X, Xu M (2019) Urban public transport and air quality: Empirical study of China cities. Energy Policy 135. doi:10.1016/j.enpol.2019.110998

Twu, R.-C., & Chen, J.-Y. (2017). A compact displacement sensor for non-intrusive concentration measurements of flowing liquid. *Sensors and Actuators, A: Physical, 267,* 424–430. https://doi.org/10.1016/j.sna.2017.10.056.

Udemba, E. N. (2020). A sustainable study of economic growth and development amidst ecological footprint: New insight from Nigerian Perspective. *Science of the Total Environment, 732,* 139270. https://doi.org/10.1016/j.scitotenv.2020.139270.

Wang A (2010) Bond Graph Method for the Dynamic Similarity Analysis of Complex Electromechanical System. Chinese Journal of Mechanical Engineering - CHIN J MECH ENG 46. doi:10.3901/JME.2010.01.074

Wang, J., Wei, X., & Guo, Q. (2018). A three-dimensional evaluation model for regional carrying capacity of ecological environment to social economic development: Model development and a case study in China. *Ecological Indicators, 89,* 348–355. https://doi.org/10.1016/j.ecolind.2018.02.005.

Wang S, Li R, Evans A, Li F (2020) Regional nonintrusive load monitoring for low voltage substations and distributed energy resources. Applied Energy 260. doi:10.1016/j.apenergy.2019.114225

Yang, L., Xu, H., & Jin, Z. (2019). Estimating ground-level PM2.5 over a coastal region of China using satellite AOD and a combined model. *Journal of Cleaner Production, 227,* 472–482. https://doi.org/10.1016/j.jclepro.2019.04.231.

Yang W, He J, He C, Cai M (2020) Evaluation of urban traffic noise pollution based on noise maps. Transportation Research Part D: Transport and Environment 87. doi:10.1016/j.trd.2020.102516

Zachar, R., Lindahl, P., Donnal, J., Cotta, W., Schantz, C., & Leeb, S. B. (2016). Utilizing Spin-Down Transients for Vibration-Based Diagnostics of Resiliently Mounted Machines. *IEEE Transactions on Instrumentation and Measurement, 65*(7), 1641–1650. https://doi.org/10.1109/TIM.2016.2540944.

Zhan, Y., Luo, Y., Deng, X., Grieneisen, M. L., Zhang, M., & Di, B. (2018). Spatiotemporal prediction of daily ambient ozone levels across China using random forest for human exposure assessment. *Environmental Pollution, 233,* 464–473. https://doi.org/10.1016/j.envpol.2017.10.029.

Zhou, B., Li, W., Chan, K. W., Cao, Y., Kuang, Y., Liu, X., et al. (2016). Smart home energy management systems: Concept, configurations, and scheduling strategies. *Renewable and Sustainable Energy Reviews, 61,* 30–40. https://doi.org/10.1016/j.rser.2016.03.047.

Zhou, W., & Zhu, Z. (2019). A new online Bayesian NMF based quasi-clean speech reconstruction for non-intrusive voice quality evaluation. *Neurocomputing, 349,* 261–270. https://doi.org/10.1016/j.neucom.2019.03.051.

He, Z., & We, C. (2011). Research of Risk Assessment System on Tailings Pond Water Pollution. *Procedia Engineering, 26,* 1788–1797. https://doi.org/10.1016/j.proeng.2011.11.2368.